MEDICINE AND MODERNISM:
A BIOGRAPHY OF SIR HENRY HEAD

SCIENCE AND CULTURE IN THE NINETEENTH CENTURY

Series Editor: Bernard Lightman

TITLES IN THIS SERIES

1 Styles of Reasoning in the British Life Sciences: Shared Assumptions, 1820–1858
James Elwick

2 Recreating Newton: Newtonian Biography and the Making of Nineteenth-Century History of Science
Rebekah Higgitt

3 The Transit of Venus Enterprise in Victorian Britain
Jessica Ratcliff

4 Science and Eccentricity: Collecting, Writing and Performing Science for Early Nineteenth-Century Audiences
Victoria Carroll

5 Typhoid in Uppingham: Analysis of a Victorian Town and School in Crisis, 1875–1877
Nigel Richardson

FORTHCOMING TITLES

Domesticating Electricity: Expertise, Uncertainty and Gender, 1880–1914
Graeme Gooday

James Watt, Chemist: Understanding the Origins of the Steam Age
David Phillip Miller

www.pickeringchatto.com/scienceculture

MEDICINE AND MODERNISM:
A BIOGRAPHY OF SIR HENRY HEAD

BY

L. S. Jacyna

LONDON
PICKERING & CHATTO
2008

Published by Pickering & Chatto (Publishers) Limited
21 Bloomsbury Way, London WC1A 2TH

2252 Ridge Road, Brookfield, Vermont 05036-9704, USA

www.pickeringchatto.com

All rights reserved.
No part of this publication may be reproduced,
stored in a retrieval system, or transmitted in any form or by any means,
electronic, mechanical, photocopying, recording, or otherwise
without prior permission of the publisher.

© Pickering & Chatto (Publishers) Ltd 2008
© L. S. Jacyna 2008

BRITISH LIBRARY CATALOGUING IN PUBLICATION DATA
Jacyna, L. S.
Medicine and modernism : a biography of Henry Head. – (Science and culture
in nineteenth-century Britain)
1. Head, Henry, Sir, 1861–1940 2. Neurologists – England – Biography 3.
Poets, English – 19th century – Biography 4. Neurology – England – History
– 19th century 5. Neurology – England – History – 20th century 6. Science and
the arts – England – History – 19th century 7. Science and the arts – England
– History – 20th century
I. Title
616.8'0092

ISBN-13: 9781851969074

This publication is printed on acid-free paper that conforms to the American
National Standard for the Permanence of Paper for Printed Library Materials.

Typeset by Pickering & Chatto (Publishers) Limited
Printed in the United Kingdom by Athenaeum Press Limited

CONTENTS

Acknowledgements — vi
List of Figures and Tables — vii

Introduction — 1
Part 1
  1 The Making of a Neurologist — 11
  2 The Poles of Practice — 55
  3 'The Great Hard Road of Natural Science' — 99

Part 2
  4 Ruth and Henry — 153
  5 The Cultivation of Feeling — 195
  6 The Two Solitaries — 243

Notes — 293
Works Cited — 335
Index — 345

ACKNOWLEDGEMENTS

This study is heavily dependent on the use of archival sources. Thanks are therefore due to staff of the many libraries around the world who have been of invaluable assistance in the writing of the book. The archivists and librarians of the Wellcome Library for the History of Medicine, who bore the brunt of my demands, deserve special mention. I am also grateful to Anne and William Charlton for alerting me to the existence of Robert Nichols's letters to Ruth and Henry Head, and to Jamie Andrews of the British Library for giving me access to this uncatalogued correspondence. Mrs Ann Wheeler kindly searched the Charterhouse archives for evidence of Head's schooldays. Stephen Casper, Chris Lawrence, and Bernie Lightman read earlier versions of the text and offered many useful comments and criticisms .

LIST OF FIGURES AND TABLES

Frontispiece. Photograph of Henry Head   viii
Figure 1. Distribution of 'Head's Areas' on the surface of the body   116
Figure 2. Henry Head and W. H. R. Rivers   126
Figure 3. Lateral view of Head's forearm and hand   129
Figure 4. Photograph of Henry and Ruth Head   291

Table 1. Henry Head, Summary of Finances

Frontispiece. Photograph of Henry Head.

INTRODUCTION

This is the first book-length biography of the neurologist Henry Head. A number of biographical sketches of Head do exist;[1] these tend to have been written by fellow medical men as acts of homage and in an attempt to place their subject within the genealogy of their discipline. Studies of this kind can yield much valuable information as well as providing insights into how Head has been viewed by other neurologists over time. Given their orientation and priorities, such biographies cannot, however, do full justice to a life as rich and multifaceted as Head's. Clinical medicine and medical science were indeed central aspects of his identity. But his life merits the attention of those to whom neurology and the physiology of the nervous system may seem extremely abstruse topics.

Because of his association with such poets as Siegfried Sassoon and Robert Nichols, Head has also been mentioned in passing in a number of literary biographies.[2] But the full extent of his literary and artistic interests has not been described. Thanks to his collaboration and friendship with W. H. R. Rivers, Head and his wife Ruth have figured as minor characters in Pat Barker's *Regeneration* trilogy of novels.[3] Sadly the Heads were not depicted in the motion picture adaptation of these books.

I first developed an interest in Head while researching a monograph on the history of aphasia studies.[4] 'Aphasia' is the name given to a complex of language disorders arising from injury to the brain. I chose to conclude the study at 1926 because this was the year in which Head's massive work on *Aphasia and Kindred Disorders* was published. Two aspects of Head's writing on aphasia were in particular striking. The first was his account of the history of aphasia studies and of his own place within that narrative. Head maintained that the value of the great majority of the studies of the phenomena of aphasia undertaken in the previous fifty years had been vitiated by both basic conceptual errors and technical shortcomings. He excluded only a handful of his predecessors from these strictures: in particular, the English neurologist, John Hughlings Jackson was, in Head's view, a lonely genius who had sketched an alternative, more fruitful, approach to the subject.

Head's was clearly a highly polemical use of history; his account can be faulted on that as well as on other grounds. What was most arresting, however, was the view of historical discontinuity that it implied. In a period where a positivist account of the history of science as a gradual, progressive, accumulation of knowledge was prevalent, Head insisted that a radical departure from the most cherished assumptions and practices of the past was necessary.

Reading the case histories upon which Head based his understanding of aphasia one is indeed struck by how different they appear to those found in the monographs and research papers of his predecessors. The patients embodied in these narratives seem very distinct from those found in the aphasiology of the nineteenth century. They tend to be rounded characters that play an active part in the unveiling of the truth of their condition. The doctor, for his part, is in this company much more an involved *primus inter pares* rather than the detached representative of a disinterested medical gaze. The understanding of the disorder appears to occur through as a process of mutual negotiation and cognition. There is also in these case histories a novel interest in technique. Head held that the errors of the earlier generation of aphasiologists had to a large extent been the result of the inadequacy of the methods of examination and assessment that they had employed. To remedy these deficiencies, he elaborated an extensive series of tests and protocols designed to illuminate the true nature of the linguistic and other deficits from which his patients suffered.

Head's case histories thus seemed to confirm his assertion of the dawn of a new era in the understanding of the language disorders classed under the term 'aphasia'. Examination of Head's researches on the physiology of sensation and of his programmatic statements on the future of medicine revealed a similar insistence that his work was the harbinger of a new medical modernity. The present volume is an attempt to explore those themes more thoroughly.

My initial interest in Head as aphasiologist led me to examine the extensive archive of Head papers held at the Contemporary Medical Archives Centre at the Wellcome Library for the History of Medicine in London. There are disappointing lacunae in this archive. In particular, it contains little material relating to Head's clinical or scientific activities. With the exception of some records of his early research on herpes zoster kept at the Royal College of Physicians of London, most of Head's professional papers have vanished. What has survived, however, are documents that provide an often-startling insight into Head's personal life, his views of the world he inhabited, and of his own place within that cosmos.

The most important of these sources are the letters and other manuscripts charting the course of Head's relationship with Ruth Mayhew whom he was, after a lengthy courtship, eventually to marry. Their relationship was based in the first instance upon a shared interest in the arts and especially in literature. The correspondence thus provides insights into their tastes in books, music and the plastic arts. It is apparent, moreover, that this correspondence was also itself conceived as a work of art. Head and Mayhew also jointly composed a series of 'Rag Books' – collections of anecdotes, extracts from literary works, together with commentary – through which they further refined a shared sense of identity.

As a prominent member of his profession, Head left a mark on the records of such institutions as the Royal College of Physicians, the Royal Society, and the Royal Society of Medicine. The papers of acquaintances from the world of letters, such as Siegfried Sassoon, help to document his literary activities. The uncatalogued letters of the poet and playwright Robert Nichols have proven to be a particularly valuable resource. As well as sharing many of his literary interests, Nichols was for a time a patient; their correspondence provides an exceptional insight into the complexities of Head's clinical persona.

Taken together, these sources reveal Head as an aspirant poet and man of letters as well as a rising figure in the world of medicine. They show him to be interested in exploring artistic forms that challenged existing convention. But they also expose a sometimes-startling disparity between public and private personality.

Outwardly Head was a respectable member of middle-class society, advancing steadily through the ranks of his profession. Colleagues described him as a jovial, if sometimes bumptious, individual. Privately, however, he often expressed a deep sense of alienation from the society he inhabited. He found the values and mores of much of the medical profession repugnant; in particular, his commitment to the ideal of scientific medicine often seemed to put him at odds with colleagues. Head's ideal of a career that harmoniously combined clinical with scientific work was all too often subverted by the bland, dreary, and exhausting routines of practice and teaching to which he was obliged to conform. His patients – especially those he encountered in his hospital practice – tried his tolerance; sometimes, indeed, Head was unable to conceal the anger and disgust that they provoked. Moreover, the urban environment in which Head was obliged to spend so much of his time was also a source of spiritual gall. His letters often complain of the soul-destroying effect of the metropolitan life.

This negativity was, however, complemented and contested by a drive to transcend the deficiencies of the mundane world.

These archival sources thus provide an exceptional, perhaps unique, insight into the inner life, as well as the public presentations, of a man of science and of letters whose life spanned the last decades of the nineteenth century and the first four of the twentieth. They offer an opportunity to contribute to the history of consciousness. This biography is therefore an attempt to take the occasion of this rich resource to explore aspects of self-fashioning in the modern era. The notion that selfhood is not an essential entity but the outcome of negotiation, resistance, and accommodation between the individual and his or her cultural matrix has been essayed by historians.[5] Michel Foucault's later works have provided a critical impetus to this historiography.[6]

In some respects, Head's worldview is typical of the late-Victorian epoch. For example, in his publications he rehearsed familiar tropes of the hierarchical relations between men and women and between the civilized and the savage. In his private utterances he subscribed uncritically to the militarist and imperialist discourses of the era. Among Head's most bitter regrets was that he was in 1914 too old for active service. But other aspects of his persona seem to challenge and contradict such conventional postures.

In seeking to unravel the various strands of Head's own sense of self I have found the concept of 'modernism' a valuable resource. Dorothy Ross has provided an insightful analysis of the term and its cognates: in particular, she has distinguished between the notions of 'cognitive' and 'aesthetic' modernism. Aspects of Head's work – especially his rewriting of the physiology of sensation – place him as a cognitive modernist. His emphasis upon the contingent nature of the way that human beings experienced their world contributed to the questioning of the relationship of subject and object that Ross sees as definitive of the cognitive modernist.[7] Moreover, Head's deliberate fashioning of his own identity through the literary medium of the letter itself betrays a sense of the contrived nature of the self.

But he can also be viewed as an aesthetic modernist in his recoil from the vulgarity and constraint of bourgeois norms, and in his distaste for the metropolis. Head's insistence on historical discontinuity in science can also be seen as a rendition of the modernist reconfiguration of historical time to which Ross alludes.[8] Moreover, although Head's tastes in art were eclectic, he displayed a bias toward the avant-garde in music, drama, and the novel.

There was an obvious sense in which Head's character was stamped with traits that were explicitly anti-modern – although still consistent with a modernist outlook. 'Modernity' and 'modernization' are terms that historians use to refer to the interrelated series of economic, social, and political transformations that occurred in western societies during the period of the long nineteenth century. Urbanization, industrialization, and the spread of market capitalism were among the most salient features of these changes. New political ideologies and dominant classes emerged, at the expense of established elites and to the detriment of their value-systems. Such upheaval provoked a mixture of resistance and nostalgia among those who saw what they considered their rightful place thus usurped and their morality challenged.[9]

Henry Head was no social or political philosopher. With rare exceptions, he did not aspire to convert his standing as a man of science into the cultural capital that would allow him to make authoritative public statements on the great issues of his day. For most of his life, indeed, he showed a studied indifference to the political. But Head did articulate an almost instinctual account of his own place and loyalties within the social order. He identified himself with the 'educated' or 'professional' middle class. This was a cadre that based its claims to status and authority not upon wealth, whether derived from commerce or from the land, but upon its intellectual endowments. These endowments were in large part hereditary: in response to a questionnaire he received from Francis Galton, Head was proud to detail the eminent men of science (including Thomas Young and Joseph Lister) among his ancestors. His own sense of identity was thus to an extent contingent on this lineage.

Although he engaged in no sustained social analysis, it is apparent from many passing comments in his letters and other manuscripts that Head felt that the values of this professional class made up of individuals of exceptional talent was being undervalued and undermined by developments in modern society. Moreover, the general tendency of modern western societies was also the cause of concern and even anguish for him. The time he spent in central Europe and his exposure to trends in late-nineteenth-century Continental culture had a marked impact upon the way in which he viewed British society.

Head showed no animosity towards the landed aristocracy; indeed in many ways he sought to connect with them and to mimic their cultural practices. The 'manufacturing classes' he associated with the north of England, and deemed the fitting object of mild condescension and amusement. However, the moneylenders he encountered as a young man in central Europe excited real loathing. Head saw them as the most egregious agents of an encroaching capitalist system that was destructive of traditional societies and of the values those embodied. He appreciated the fragility of even respectable professorial prosperity in the face of the vagaries of the market. Head was in no doubt about the ethnic identity of the capitalist vanguard, referring to them simply as: 'the Jews'.

For Head the most obnoxious aspects of the metropolitan environment he was obliged to inhabit for most of his career was the urban proletariat. His work at the London Hospital, situated in one of the poorest parts of London, brought him into regular contact with this class. He made no secret of his disgust with their corporeality – lamenting having to endure the stench of the outpatients' waiting room – or of his contempt for their moral characteristics. Although he did not use the word himself, Head, like so many of his class, uncritically rehearsed many of the tenets of the discourse of degeneration.[10] He saw the poor of London as a symptom of social pathology. Because so many of the poor inhabitants of Whitechapel that he encountered were Jewish immigrants from Eastern Europe, this class hostility was commingled with a further strain of anti-semitism.

Head claimed that for as long as he could remember he had wished to pursue a career in medicine. Yet he did not feel entirely at home in the late-Victorian medical profession that he eventually joined. Head was of the persuasion that the intellectual foundations of clinical medicine lay in experimental science. This was an orientation he ascribed in part to the example of such 'kinsmen' as Joseph Lister and Marcus Beck, but also to his early indoctrination in German laboratory medicine in Halle and Prague. This understanding of the proper roots of medical practice was, however, far from uncontroversial in the London medical culture in which Head was obliged to operate.[11] At the London, Head's attempts to institutionalize the ideals of scientific medicine were too often frustrated by those of his colleagues who insisted that it was more important to appoint a 'practical man' to the staff than one who was versed in the ways of the laboratory. As a consulting physician, he also encountered attitudes among elite doctors in private practice that he found no less repugnant. The 'typical' Harley Street doctor seemed more concerned with appearances and fees than with the pursuit of scientific knowledge.

Head thus experienced the kind of alienation from his society that many other intellectuals of the period manifested. T. J. Jackson Lears has characterized this recoil from *fin de siècle* civilization as 'antimodernism'. Such a posture was associated with a repudiation of the supposedly artificial, over-refined aspects of modern life and a corresponding quest for primal, more 'authentic' experience, often involving extreme physical exertion.[12] It was also manifest in an interest in the past, especially in the medieval period. Such late nineteenth-century initiatives as the arts and crafts movement represented an attempt to recover something of the orientations and values of that lost world through a return to pre-industrial methods of manufacture.

Head manifested many of the characteristics of this antimodernist persona in his own self-fashioning. His revulsion for metropolitan life was matched by a passion for an idealized countryside. When his work commitments permitted, he would seek spiritual solace in some rural setting. Head took delight in the English landscape. But he also sought out wilder, more exotic settings in which he could touch the sublime. He took a keen interest in the people who inhabited these places, especially when there appeared to be a 'primitive', elemental, aspect to their existence. During his stay in the Bohemian forests Head even tried to live for a while the life of the peasant.

Head was also fascinated with medieval art – and in particular sculpture. Although an avowed secularist, he showed an erudition in matters of church architecture that amazed some of his auditors. Chartres Cathedral was among his favourite buildings. Through his friendship with C. R. Ashbee, Head became associated with the English Arts and Crafts movement.

Head saw bodily exertion as a necessary complement to intellectual application: as a young man he rowed and engaged in various other competitive sports. Even in middle age he would occasionally play tennis and other racquet sports. But perhaps his greatest enthusiasm was for cycling. He found that the vigorous effort that this activity entailed invoked states of consciousness that were themselves as worthy of contemplation as the delights of the scenery he traversed.

This contrast between 'external' sensory stimuli and 'inner' states of consciousness is, however, misleading. Head was always aware that the outer world only existed in so far as it impinged upon the mind. Each individual's impressions of that world were to a degree unique and determined by his or her psychological makeup. By meticulously documenting his perceptions and responses to nature, Head was thus seeking additional insights into the character of his own mind.

He thus sought solace from the egregious aspects of modernity by a withdrawal into an inner world. Within this secluded psychic sphere the highest value was the pursuit of beauty. Art in all its forms provided the most refined source of such stimuli. On his numerous trips to galleries, museums and churches, Head pursued the beautiful in as methodical and determined a fashion as he sought scientific truth.

Moreover, for him natural science in its most exalted forms, could produce effects that emulated those derived from natural or artistic beauty. Head found satisfaction in various aspects of the scientific way of life – even in those that might seem humble and mundane. Thus the manual facets of laboratory life – the preparation and manipulation of microscopic slides and laying the electrical wiring necessary for certain experiments – gave their own pleasure. Science too had its craft skills that were to be valued and relished. But the moment of scientific discovery, the instant when the mind gained a sudden and profound insight into nature, gave a higher kind of fulfilment: at such moments the scientist attained the same creative ecstasy as the artist.

For Head the path of natural science was, moreover, a momentous life choice. Science represented one of the principal means by which a 'refined mind' could confront and meet the challenge of existence. Head recognized that religion offered an alternative road; at an early age, however, he had made the decision that the truths of natural science were incompatible with the claims of revealed religion. He never had any doubt which of these competing claims should take precedence. His commitment to science was central to Head's sense of self. The degree of that commitment was perhaps most evident in a readiness, verging on the heroic, to suffer the self-mutilation involved in the extended experiment on nerve regeneration that he undertook in collaboration with W. H. R. Rivers.

Head was unabashedly elitist in his view of his place in the world. He maintained that few people possessed the highly refined sensibility with which he was equipped. When, however, he did encounter a kindred spirit he rejoiced.

Ruth Mayhew in particular he regarded as an equal in taste and erudition. The attraction was reciprocated. In the face of the constraints of bourgeois Victorian conventions, however, Ruth and Henry found it difficult to maintain the kind of intimacy to which they aspired. Because of Ruth's importance in Henry's life, this is necessarily for much of its length a joint biography. Indeed, by the final chapter Ruth has arguably become the protagonist. Drawing on the work of Lydia Ginzburg,[13] I have tried to show how, as their relationship developed, these two engaged in a process of mutual self-fashioning.

One challenge facing anyone writing a biography of Henry Head is the fact that in his sixties he developed Parkinson's Disease. This affliction cut short Head's medical and scientific career. The strenuous physical activities in which he had previously delighted were denied him. For the final fifteen years of his life he was indeed obliged to live the secluded life of an invalid. The nature of his affliction prevented him even from holding a pen. From the mid-1920s Head was therefore obliged to rely on others – and above all on his wife – to write letters in his name but eventually even the act of dictation became too tiring for him to sustain for any length of time. Head's own voice was in effect silenced by his disease: the historian is obliged to rely upon what others said about and for him.

In writing this book I have for the most part attempted to maintain the detached neutral voice of the academic historian. I have not concealed aspects of Head's persona that many will find distasteful. As the work progressed, however, it became ever more difficult not to develop an emotional engagement with my subject. In particular, it has proved impossible not to be moved by the courage and dignity with which Henry for many years faced the 'foul disease' that possessed his body while leaving his mind untouched. Ruth's fidelity to him in these years is also humbling.

The book is divided into two parts, the first of which is chiefly concerned with Head's youth and his professional activities. Chapter 1 covers Head's early years, his school days, and his time as a student at Halle, Cambridge and Prague. I have made an unfinished autobiography the starting-point for this account, supplementing the recollections Head set down in 1926 with contemporary letters and other sources. Chapters 2 and 3 deal respectively with the clinical and scientific aspects of Head's public persona. Given the nature of Head's scientific research, this is necessarily a somewhat arbitrary distinction, but one convenient for purposes of exposition. The second part is more concerned with the Head's private self, although no strict demarcation of the various aspects of his personality is possible. Chapter 4 documents the development of his relationship with Ruth and culminates in their marriage. Chapter 5 provides an analysis of Head's artistic interests with special attention to his aspirations to write poetry. The final chapter is more narrative in character. It begins with a discussion of the peak of Head's career as physician and scientist and deals with his social and recreational activities during this period. Later sections describe Henry's growing incapacity as Parkinsonism tightened its grip upon him. The book ends with an account of Ruth and Henry's last years as exiles first at Forston House and then at Hartley Court.

Medical biography is currently something of a derided form of writing. Somewhat unfairly, it is often associated with an outdated form of historiography that too often amounted to little more than ancestor worship and the celebration of the great man or – more rarely – woman.[14] I hope that the present monograph will do something to rehabilitate the genre. Among other things, biography provides a unique opportunity to explore the subjective side of the medical life – to gain some insight into what it was to be a doctor in late Victorian and Edwardian Britain. Because of his penchant for self-reflection, Head is of special value in this regard. As a contribution to the history of science Head's life, moreover, illuminates what might be termed the existential import of the scientific way of life at the dawn of the modern era. That is, the way in which a dedication to the exploration of the natural world could serve to provide meaning and direction in a Godless and indifferent cosmos.

1 THE MAKING OF A NEUROLOGIST

On 7 November 1926 Henry Head, MD, FRS, set out to write an autobiography. By this date Head was an eminent metropolitan physician who combined an appointment at the London Hospital with a private practice in Harley Street. His special interest was in the diseases of the nervous system, including such 'functional' disorders as hysteria and neurasthenia. He combined his clinical work with scientific investigations into the workings of the brain and nerves. In recognition of these researches Head had been nominated four times for the Nobel Prize in Medicine and Physiology. 1926 had seen the publication of Head's monumental two-volume study of *Aphasia and Kindred Disorders*, a work that was widely recognized as marking an epoch in the study of the subject. Head's devotion to science was matched by an enthusiasm for literature, music and the plastic arts. He was an authority on the poetry of Heinrich Heine and had published a volume of his own verse. For more than twenty years Head had been together with his wife Ruth née Mayhew, a relationship that by all accounts was exceptionally happy and enriching to both partners.

By November 1926 Head was also in the grip of Parkinson's disease. This affliction was to end his active career and force him into premature retirement. For the final fourteen years of his life, Head's exceptionally active mind was imprisoned in an ever more infirm and recalcitrant body. As a neurologist, Head knew better than any the nature of his condition and its likely prognosis. It seems plausible that a realization that his life had in a sense already come to an end was what led Henry to put down some recollections of his early life. These autobiographical notes are among the few typewritten items in the Head archive and had presumably been dictated to a secretary. By the time the document was composed, Head was unable to use a pen and relied on others to record his words.

Childhood

Henry Head was born on 4 August 1861 at 6 Park Road in Stoke Newington, London. His father was to recall: 'you came into the world at 20 minutes to midnight after a very long and anxious day'.[1] Head was in later life to ascribe some significance to the day of his birth. Because he was born on the Sabbath,

Head believed he had 'the good fortune to be what the Germans call a "Sonntagskind"...'.²

Harry, as he was known, was the eldest child of what was to become a large family. His bond with his mother was in his childhood and youth strong; as late as 1900, while on holiday in Egypt, she wrote: 'my first born, my first thoughts are for you now as always'.³ His relations with his father, although cordial, seem to have been more distant. Nonetheless, Harry was proud to report the judgment of one of Head senior's colleagues that 'the characteristics of my father's life were judgment and kindness'.⁴

In his 'Autobiography', Head noted that both his father and mother were of 'Quaker stock'. His father – also called Henry – was the son of Jeremiah Head, who had been Mayor of Ipswich. His mother Hester was a Beck and therefore connected by blood to the Lister family. A year after Henry's birth, his father left the Society of Friends to join the Anglican Communion. No reason is given for this conversion but Henry senior was building a prosperous career for himself as an insurance broker at Lloyds of London. Nonconformity in religion might have been an obstacle in the path of his professional progress. Many nonconformists converted to Anglicanism as they rose up the social scale.

Head was thus never a Quaker in the sense of observing the forms and practices of the Society of Friends. His parents' change of confession did not, however, impede continued social contact between the Head family and the close-knit north London Quaker community. Through these links Henry grew keenly aware of the scientific attainments of those who had come from a similar background to his. In particular, Marcus Beck (1843–93), his mother's cousin and an intimate of the Head household, was an ardent follower of the antiseptic revolution instigated by the surgeon, Joseph Lister (1827–1912). After working with Lister in Glasgow, Beck had helped introduce antiseptic surgery at University College Hospital in London. He also played a prominent role in making the bacteriological doctrines of Robert Koch accessible to the British medical community. 'Thus', Head recalled, 'I was brought up in an atmosphere of modern science and in an attitude of worship for the great man [Lister] who was connected with my own people'.⁵

This sense of belonging to a community with a proud scientific heritage⁶ formed an important aspect of Head's sense of self and did much to determine his own path: 'I cannot remember the time I did not wish to take Medicine as my career in life'.⁷ When in 1904 the eugenist Francis Galton circulated to Fellows of the Royal Society a questionnaire designed to identify the presence of 'hereditary genius' in their families, Head took care in his return to note his blood links not only with Lister, but also with Thomas Young (1773–1829), the author of the wave theory of light.⁸

Head may also have owed a less tangible debt to his Quaker heritage. Head lost his faith as a young man. As an adult he eschewed all religious belief and assumed an attitude of stern, unflinching scientific naturalism. One of the most striking aspects of his personal writing is the entirely secular nature of his outlook. Head did not even feel any need to argue against Christianity; for the most part, he simply ignored it.

Nonetheless, according to Gordon Holmes, although Head gave the impression 'of being a severe materialist, he was interested in certain forms of mysticism, probably due to the influence of the Quaker atmosphere in which he was brought up'. Holmes, who had collaborated with Head in scientific investigations, took a dim view of this aspect of Head's character. It had in fact, Holmes averred, detracted from Head's performance as a scientist: 'A rigidly scientific and objective outlook ... was in him combined with a vivid imagination which at times seemed to carry his ideas beyond the bounds of probability'.[9] The tone of disapprobation says much about Gordon Holmes's own conception of the qualities a scientist should possess.

Some of the most distinctive aspects of Head's scientific persona may indeed be traced to his Quaker background. Quakerism encouraged interest in the natural world. In particular, each individual was deemed to possess an 'inner light' that enabled him or her to discern a divine order in even the most mundane object. The Quaker stress upon direct personal observation of nature also encouraged an anti-authoritarian attitude to established hierarchies and orthodoxies in science.[10] At an early age, Head repudiated any notion of divine design in the world. Nonetheless, in his own researches he strove to attain a unique and original insight into the workings of the nervous system, one that depended in large part upon an ability to grasp the essential order that underlay a complex set of phenomena. His robust individualism and faith in his own insight often led him into direct conflict with the established scientific order. In his personal and aesthetic writings, Head, moreover, represented the natural world in lyrical and even rapturous terms as permeated with a transcendental beauty. The relationship between the laborious experimenter and the poetic Head was complex and he was himself equivocal on the relationship between art and science. The poet who enthused about German Romantic literature and who celebrated the glories of nature in both prose and verse was not, however, entirely banished when Henry Head took on the character of scientific observer and discoverer.

The 65 year-old Head recalled a variety of seemingly miscellaneous details about his childhood. His nurse was called Eliza. He had evidently retained contact with her because he noted that: 'at the time I write she is still alive'. At the time of Henry's birth, the Head family had inhabited a 'small semi-detached dwelling with a tiny garden behind it', and had made do with one servant. As the fortunes of Henry Head senior improved – and as more children arrived – the family moved to a series of more spacious houses. In 1865 the Heads arrived at a house in Albion Road with a good-sized garden and its own stables. From the age of six Henry and his brother Charlie 'had a couple of white ponies to ride, which was a great delight to us'. Often their father would join them on these rides.[11]

On 28 May 1866, at the age of four, Harry composed his first letter. He advised his mother: 'I have made some glue. The bottle is full now indoors ... Charlie send his love and a kiss. He is playing with a little horse. Little Hugh and baby are much better so are [Eliza] and Elizabeth'. Some fifty years later, Ruth Head found this letter and saw it as early evidence of Head's propensity to become a 'discoverer'. She assured her husband: 'you have not changed the weeest bit. How much I hope your bottle of glue is quite safe and beautifully sticky upstairs in your room at Middlecott'.[12] By this date, the glue had become a metaphor for the scientific manuscript Head was writing while staying at a friend's house.

In 1899 the adult Head had experienced a curious flashback to these early days. He told Ruth Mayhew that while wandering around a friend's house in Tunbridge Wells:

> I suddenly came on one [room] with bed, furniture and belongings that seem to awake memories of my childhood. I had that uncanny feeling of something burnt in on that memory in every small detail so long ago that one is in doubt whether it is not a dream or one of those freaks of recent memory. [?] the old nurse who still haunts the long empty nursery and who is dried up like a preserved pippen I found that in this room I had had the measles 30 years ago. Wandering round that garden I found a child's wheelbarrow dated 1866 and I remembered that this sturdy toy had so fascinated me that father had made me one like it – which has alas! long gone to dust.[13]

In 1870 the family moved to a larger house on Stamford Hill. They were by now sufficiently prosperous to employ the artist William Morris (1834–96) as a decorator. Among his many endeavours, Morris had set up a firm dedicated to introducing the fine arts to every detail of home decoration. Hester Head had taken the initiative in engaging Morris. Henry (who would have been eight or nine at the time) recalled that he had accompanied his mother on a visit to Morris's office in Red Lion Square; he even provided an account of the conversation that passed between them:

> Morris asked when she had put forward the her request, 'What's your husband?' She answered 'An underwriter'. To this Morris replied 'Oh something in the City[14] – I suppose you would call him a merchant. I've never decorated a merchant's house and I'll do yours for you'.

Decorating a merchant's house proved no easy task for Morris. The workmen engaged for the job failed to meet his exacting standards, and in the end Morris mixed and applied the paints himself. Henry remembered that Morris:

> Took infinite pains to the effects he desired and even painted the little panes of our toy cupboard. One day he arrived with a brass candelabra which he had bought on one of his journeys in Holland, saying, 'This will exactly suit your Library'.[15]

While many of these reminiscences appear random, or as the recounting of prized family anecdotes, Head endowed some of his childhood experiences with a peculiar significance. As with many Victorian families, 'children came to my parents thick and fast'. Before the arrival of the 'annual baby' Henry was despatched to his grandmother's house on Stamford Hill. His recollection of these stays is filled with nostalgia and affection. The grandmother wore traditional Quaker dress and an appropriate quiet reigned in her home – in stark contrast to the domestic hubbub with which Henry had usually to contend. At home he remembered 'creeping under the sofa to read the Arabian Nights in order to escape the annoyance of the riotous younger children'. At his grandmother's house no such extreme measures were needed: 'For the greater part of the day we sat in her morning-room where we also had our meals ... There were plenty of charming block puzzles to be put together, and spillikins were my great delight'.[16] This motif of a desire to escape from the clamour of the workaday world to place of calm where puzzles could be solved was to recur in Henry' later life.

These trips to Stamford Hill were not, however, moments of pure self-indulgence. Henry's grandmother was:

> extremely strict, insisting that whatever game was played the materials should be packed away always in an orderly manner. Moreover, she taught us that any game begun should be properly finished and I always remember her aphorism 'If it is worth beginning it is worth finishing'.

This lesson, the mature Henry Head maintained, had always stood him in good stead; indeed, it embodied the discipline and methodical habits essential to good scientific work. When years later he returned thanks 'for the Royal Medallists at the annual dinner of the Royal Society, I said that we Fellows of the Society were the sort of people who had been trained never to relinquish our tasks at the bidding of intrusive nurses and teachers'.[17]

Other of Henry's recollections of childhood also seemed to prefigure his future course. One of his earliest memories was of 'carefully preserving in a hidden drawer a scalpel and a piece of lint that I had succeeded in annexing from my cousin's [Marcus Beck] bag'. When he was eight an epidemic of scarlet fever swept through the Head household. After Henry had recovered, the family physician, Mr Brett, took him into his house for a few days. One day at breakfast Henry claimed that he startled his host 'by pouring a little tea into a tea-spoon and heating it over the spirit-lamp, carefully inspecting the result as I had seen him do so often during my illness to see if he could detect albumen in the urine'.[18]

Head's formal education began at the age of five. He and a small group of other local children took lessons from the proprietor's daughter at the back of a chemist's shop. Later he moved to a second day school in Stoke Newington before becoming a weekly boarder at Grove House in Tottenham, a Quaker school that catered for boys over the age of ten. There, Head recalled, his fellow pupils included a number who were to achieve distinction in later life.

It was at Grove House that Head was also to encounter the first of a number of teachers who were to nurture and guide his interest in science. A master named Ashford taught his charges the basics of physical science, including 'simple measurement and the use of weights; from this we went on to the various orders of lever and we were made to calculate the exact force that would have to be applied to overcome a given resistance'. The teaching was practical in orientation: 'When the correct answer was found, it was actually demonstrated on the lever in question'. At the same time Head was introduced to the basics of trigonometry and geometry. He recalled that Ashford 'had a method of teaching us Euclid's Geometry which made the solution of the corollaries and examples he set us, an object of passionate interest to us'.[19]

To this man, Head declared, 'I owe the fact that I was firmly grounded in the elements of Natural Science at an age when boys at an ordinary school were ignorant of the very existence of the subject'. Moreover, in his way of life Ashford exemplified the virtues of the fervent pursuit of science. Head remembered that the schoolmaster was 'an ardent conchologist and was always ready to demonstrate his collection of snail-shells of all kinds, together with other subjects of natural history'.[20]

At the age of thirteen, Henry and his brother, Charles Howard Head,[21] were sent to continue their education at Charterhouse. The two joined the school at the beginning of the Oration Quarter (Autumn Term) in 1875. Head noted that this represented a clear break with customary Quaker practice, which strove to find alternatives to the public schools. As with his conversion to Anglicanism, this departure was no doubt a token of Henry Head senior's determination to assert the new social status of his family at the expense of its nonconformist roots. Charterhouse had recently removed from London to new premises at Godalming. Head and his brother maintained contact with his parents by letter. This correspondence reveals that Henry was in most respects a typical public schoolboy.

He was subject to the various afflictions that were endemic in the school. Henry kept his mother advised of the state of a chronic discharge from his ear; at times this was viscous, at others 'nothing but flakey substance in the water ...'. He was also prone to occasional swelling of the feet. He also carefully catalogued the diseases of his schoolmates: scarlet fever, measles, bronchitis, as well as one boy suffering from 'congestion of the lungs and pneumonia'. Overall, Charterhouse in winter was a sickly place. Henry complained that: 'The amount of coughing in Chapel is appalling. I can hear none of the service in the Mornings'. The officiating clergyman was obliged to curtail his sermon, 'saying that he could not preach with so much coughing and that it would be too great a strain on the fellows stopping coughing if he preached long'.[22]

Henry took full advantage of the social and recreational facilities offered by the school. As well as games of whist and nap with his schoolmasters, after evening prayers there were regular debates where teachers and pupils practised their rhetorical skills. Motions included: 'Classical study is carried to too great an extent in modern education', a proposition with which Henry might have agreed although he maintained a studied neutrality in his account of the debate. At an earlier meeting, the motion 'was that no trust is to be placed in the present government'. Henry reported that: 'it was most interesting although I care nothing for politics'.[23] This lack of interest in the political was to prove to be a lasting trait.

The late nineteenth-century English public school sought to encourage manly virtues and the spirit of teamwork among pupils by means of an extensive programme of sports throughout the year.[24] Henry took full advantage of these opportunities. At Charterhouse he learned to play tennis, racquets, cricket and football. He also took part in events such as running the half-mile, jumping hurdles and the sack race. Head was to keep up these athletic pursuits at university and beyond. He was, while doing postgraduate work in Prague, reputed to have taught the Czechs to play soccer. Head was also eager to join the school fire brigade.

Henry cultivated his mind as well as his body in his leisure time at Charterhouse. He attended the plays and skits presented at the school, and showed an appreciation for the music played in the chapel. The teenage Henry Head was, moreover, already a voracious reader of novels. He found the school library stock inadequate and his mother was obliged to forward additional reading material along with the regular food hampers she despatched to Godalming. Among the books Henry enjoyed at Charterhouse were Jules Verne's fantasy *Journey to the Centre of the Earth* and Walter Scott's historical novel *The Talisman*.

Despite his apparent gregariousness, the studious Head recalled feeling during his schooldays a 'sense of loneliness liable to be produced by the fact that my own interests diverged so widely from those of other boys'.[25] At an early age he had gained a taste for 'fine literature, for my Mother would read aloud to us by the hour together both poetry and prose, thus awakening in me a real enthusiasm for the world of letters'.[26] This love of literature was matched by an early fascination with science.

The pedagogic regime in place at Charterhouse when Head first arrived in 1875 was hardly conducive to serious intellectual endeavour in either field. Several forms were taken in the same room encouraging an atmosphere of 'noise and disorder'. The pupils were supposedly seated in class order. As a new boy, Head was placed at the bottom of the class. When he successfully solved a problem that had defeated his classmates, 'I found myself at the top. But such was the want of discipline that, under the master's eye I was shoved down from boy to boy till I again found myself at the bottom, dismayed at the inexplicable injustice of such a want of system'. Head found this chaos especially disconcerting because he was accustomed to the quiet, orderly classes of his 'Quaker School'.[27] He also judged the curriculum offered at Charterhouse lamentably antiquated – and indeed inferior to what had been taught at his previous school.

Head omitted from his 'Autobiography' any reference to the more obnoxious aspects of school life. Nor did he alarm his mother by alerting her to the brutality he now encountered. But in a letter to Ruth Mayhew written in 1900 he alluded to the violence he met with and indeed participated in at Charterhouse. While running away from a bully, he found his way blocked by another boy: 'This boy I very nearly killed ...'.Years later, Head still recalled his 'feeling of maniacal fury' during this fight.[28]

Fortunately, shortly after Head entered Charterhouse, a new breed of schoolmaster began to effect a revolution within the school. In particular, 'a fundamental change occurred with the arrival in 1877 of a second Science Master, W. H. W. Poole, a Demy of Magdalen College, Oxford'. Poole, an energetic man, had a direct way of dealing with indiscipline; Head vividly remembered him: 'during one of his earliest lessons vault across the bench in front of him, cuff a boy over the head who was making trouble, and vault back again as if nothing had happened'. More particularly, Poole transformed science teaching at Charterhouse. Previously, the elements of physical science illustrated by simple experiments – the 'Stinks', as they were known – had been taught at the school. But these classes were 'universally despised, [and] very few boys learnt anything from them'. Poole introduced the teaching of biology to Charterhouse. Through his lessons, Head claimed that Poole instilled:

> into me the elements of the life of plants and animals. Gradually he led us to Physiology proper and in spite of the fact that most of the apparatus for teaching required to-day was absent and that he was compelled to make shift mainly with diagrams, he succeeded in laying a firm foundation for my subsequent work.[29]

Head also recalled Poole as a model on which he had sought to fashion his own character. He had evoked in his pupil 'a general outlook on life which became a happy blending of learning and practical achievement'.[30]

Poole seems to have recognized in Head a kindred spirit. He taught Henry the rudiments of botany. The two went out together to gather plants and other specimens. On occasion Head ventured out on such plant-gathering expeditions himself. He once discovered: 'a most beautiful wood full of bracken and flowers which I did not know of before; but I found, when I got out of it, a man who said that I was trespassing so I cannot go there again'.[31] Poole lent Head books on natural history and invited him to his house out of school hours where the two occupied themselves in dissecting earthworms and frogs. When a schoolmaster's fowl died, Henry took the opportunity to perform an autopsy; he concluded that the cause of death was probably 'an enlarged gizzard'. More particularly, Poole was a microscopist who taught his apprentice the fundamental skills of the craft such as the cutting and mounting of sections. He also initiated Henry in the mysteries of bell ringing, a pastime he was to pursue further at Cambridge. Evidently such special favour from a master was something Head thought it best to hide from his peers: 'To anyone who questioned me, I pretended I was going for a walk, and fortunately Poole's house lay at a little distance below a copse that covered one side of out hill'.[32]

Head also stressed his debt to his housemaster, Gerald S. Davies, who was 'interested in all forms of art and was particularly keen on establishing a School Museum, where we could meet for the discussion of scientific and artistic subjects'. Again Head evidently felt he had much more in common with this schoolmaster than with his fellow pupils. The other boys tended to have no interest in such intellectual pursuits and gave Davies's efforts a 'philistine' reception. Head, on the other hand, welcomed the invitations he received to breakfast with his housemaster and fondly recalled: 'the pleasure I felt in the rational conversation which it was the fashion of my fellow-guests to despise'.[33] At the time Head was rather less impressed by the quality of the food on offer at his housemaster's table complaining 'I don't think he himself knows what good meat is for the meat was awfully tough'.[34] Head's parents were on close terms with Davies and rewarded the attention that he paid to their son by, *inter alia*, gifts of apples.[35]

Head was active in the organization of the museum Davies had proposed. He spent at least one Saturday afternoon helping in the arrangement of the specimens and in carrying material from Davies's house. He also regularly attended the lectures given at the 'Science and Art Society'. These included 'a very interesting paper read by F F Daldy on primordial man they had some splendid flint instruments & Malay weapons the latter presented by one of the fellows' uncle'.[36] At an early age, Head was thus introduced to aspects of the theory of evolution, a doctrine that would figure large in his own later thought. The museum was also the site of a microscopical society at which Head was able to make use of newly acquired histological skills. Poole provided him with a variety of hydrozoa to mount on slides.

Despite his evident enthusiasm for science – he won a Natural Science Prize in each of his years at the school[37] – Head was not regarded as an outstanding scholar at Charterhouse. He later confessed to his mother: 'I never did a stroke of work till I had the prospect of superannuation before me – when I got into Charterhouse'.[38] In 1897 Head gave a rather more bitter account of his early travails at Charterhouse to Ruth Mayhew. He felt he had been victimized because it was considered 'a disgrace ... to learn natural science and history with avidity'. He was spared only on account of his 'moral character'. The experience was, Head remarked, 'a fine training in unhappiness'. It also led him to realize: 'On what threads one's life hangs – for what would have happened if I had been superannuated at 17 I scarcely like to think'.[39]

Although he had mixed feelings about his schooldays, Head was in later life a frequent visitor to Charterhouse. He found one visit to the school chapel in June 1899 particularly poignant. He went: 'into all the old haunts that I had not seen for 15 years read all the old names upon the wall and the brasses in the chapel. It is curious to notice that nearly all the brasses are to men of my own time who were either killed in battle or died of fever in some foreign part'. While ruminating on these sacrifices to the British Empire, Head was joined in the chapel by: '"the most successful soldier in the British army" Lieut-Colonel Baden Powell who was one of my contemporaries'. Unlike the less fortunate Carthusians commemorated in brass, Head noted that Baden-Powell was: 'splendidly set up and looks quite young except for a thin spot at the back of his head although he has been in every row for the last 15 years'. Baden-Powell had just returned from India because 'he thought there might be trouble in the Transvaal'[40] – a presentiment of a coming war that was to cost Head one of his closest friends.

In 1877 the headmaster seemed indeed, by Head's own account, to be finally on the point of expelling Head from the school because of what was seen as his disappointing progress, particularly in classics. He was dissuaded by Poole. Davies, in contrast, was unequivocal in his opinion of Henry's abilities. When Head left the school, Davies confided to his father that:

> As for Harry I shall miss him very much. I say 'shall' for I haven't in the least realized that he really has gone and shall not, I suppose, till next quarter. No boy ever could have passed through a school career more blamelessly and no boy has less to reproach himself for on the grounds of wasted opportunity. I can not give a better character than that.[41]

Head nonetheless found it something of a struggle to gain a place at Cambridge. He was fortunate that Charterhouse offered a prize in natural science intended to send the winner to university. In his final two years at the school, Head took extra science lessons from Poole in preparation for competing for this award. Even with this assistance, however, his success was far from assured. Head had to contend with a rival named Philip Waggett (1862–1939) whom he regarded as his intellectual superior. Waggett was later to gain distinction as a clergyman and theologian. Moreover, Head recalled that many years later the geneticist William Bateson 'told me that at that time Philip Waggett was the only man, not a professional scientist, who completely understood the Mendelian Doctrines'.[42]

Henry Head senior provided his son with a powerful incentive to succeed against this formidable adversary when he 'promised that if I could win the Scholarship I should go up to Cambridge and take up a medical career'. If Henry failed, his fate would be to join his father's burgeoning City firm. A career in medicine had always been Head's ambition. Thus encouraged – and with Poole's unflagging support – he triumphed in the contest. In a letter to his mother Henry revealed how close the competition had been: in the electricity paper both boys had sat, for instance, Head had scored 52% and Waggett 55%.[43]

Head's final two terms at Charterhouse were occupied with efforts to improve his chances of being awarded an exhibition at Trinity College, Cambridge. Even during the Christmas holidays, he sought further tuition from the London surgeon, D'Arcy Power who, at the time, was an assistant demonstrator at St Bartholomew's Hospital. Power was another of the significant mentors that Head dwelt upon in his 1926 autobiography. Head undertook various practical exercises under Power's supervision presumably to improve his anatomical and physiological knowledge. 'In this manner', he recalled, 'I succeeded in making considerable advances in what would now be called elementary biology'.[44]

At his first attempt in March 1880 Head, however, failed to win an exhibition at Trinity. Nonetheless, his marks were sufficiently high for him to be excused the entrance examination and gain admittance to the College. But instead of immediately taking up his place, the eighteen-year-old Head embarked on what he later described as 'one of the most important events in my life – my first stay in Germany'.[45]

Halle

This trip assumed such retrospective significance for Head as the starting-point of an admiration – sometimes verging on infatuation – with German culture that was to figure large in most of his adult life. Other Victorian men of science displayed the same trait. Head's near contemporary, Karl Pearson (1857–1936), for instance changed the spelling of his first name to make it more Germanic.[46] Like Pearson, however, Head was in 1914 forced to make a painful reappraisal of his lifetime esteem of the Germans. In Head's youth, however, Germany was not seen as an enemy. At a concert at Charterhouse, 'the band played and amongst the tunes was the 'German Hymn' [presumably the theme from Josef Hayden's "Emperor Quartet"]'. Prior to the performance there was a call for 'three cheers for the German Emperor which were given'.[47] Head's recollections of his first experience of Germany were coloured by hindsight. He wrote in 1926 of a prelapsarian, pre-industrial, country where even the functionaries of the state were essentially kind and decent. The observations that Head made in letters and the journal he kept in Halle give a less idealized view of German society in the 1880s.

Head set out on his journey at Marcus Beck's urging. Beck had presumably some inkling of Head's predilection for medical science. In the late nineteenth century, Germany was pre-eminent in this field. Beck therefore thought it expedient for Henry to obtain an early grounding in the German language, as well as some exposure to German science, before he proceeded to Cambridge. After making some enquiries, Beck advised that Head would benefit most from a stay at Halle. As one of the smaller German universities, this was not an obvious first choice: Berlin or Heidelberg might have appeared more eligible. But Beck reasoned that for this very reason Head was less likely to encounter other British students in Halle and would thus be forced to hear and speak German at all times. It was 'therefore decided that I should go to Halle on the Saale ...'.[48]

The young Head had first to get to Halle on the Saale. His father's contacts in the City were of use here. He was entrusted to the care of a sea captain named Jarman, often employed by Lloyd's of London, who was en route to Hamburg to deal with a wreck. On 13 April 1880, Henry had said farewell to his parents at Holborn Viaduct. After a train journey, the two took a steamer for Flushing in the Netherlands and then travelled on to Cologne by rail. Head's first sight of a foreign country was disappointing; he found Holland to be a 'dreadfully flat country with high banks and willows ...'.[49]

At Cologne Head was 'deposited' at a firm of shipping agents also known to his father. A clerk from this firm acted as a guide to the city. The young Head already demonstrated the interest in ecclesiastical architecture that was later to become a passion. He also already showed considerable confidence in his own aesthetic judgment. The tour began with the cathedral. Head noted:

> The tower is not yet finished but the inside is grand. The windows are lovely with the most perfect glass ... I had read Murray on the way so I knew a little about it. There is a very old wooden image of St. Christopher which is awfully ugly. There is also a most ghastly arrangement of stone figures representing the bearing of Christ to the Sepulchre; it is very old but extremely ugly.

While he admired aspects of Roman Catholic art, he found the folklore attached to it merely amusing. Upon leaving the cathedral he and his guide went to the

Church of St Ursula, who, Head noted wryly, 'according to the legend was killed with her 11,000 virgins by the Huns. They say they have the bones of the 11,000 but as half the bones are those of other animals than man it is not <u>very</u> probable'.[50] After visiting several other churches, Head spent the night at a hotel.

The following day he parted company with Jarman and set out on the final leg of the journey to Halle alone. Although Head had taken some German lessons before setting out, he could neither understand nor speak the language. As he set out, the clerk explained 'my helplessness to the Guard who displayed on my behalf the usual beneficent attitude of an official, common in Germany in those days'.[51] Head had come with a copy of 'Bradshaw' and noted that the train did not conform to the official timetable, passing by scheduled stops while halting at unexpected places. He took a keen interest in his fellow passengers. One of them offered Head 'some lovely scent like Eau de Cologne ... We all poured some on handkerchiefs and wiped our face; it was very refreshing for the day was boiling hot and the dust was dreadful'.[52]

The train arrived in Halle thirteen minutes early. Head found, however, that the address with which he had been supplied was unknown. Again, he recalled, a kindly public servant helped him. The local stationmaster marched Head off to the house of Carl Eduard Aue, a Reader in English at the university. Aue had taught at the Edinburgh High School for 10 years and was the author of a standard German grammar. He was to become Head's official tutor in the language during his time in Halle. Aue advised Head to spend the night at a local hotel.

The following morning Head eventually found the Niemeyer family with whom he was to lodge. He learned that the earlier confusion was due to a misspelling of their address: 'However I was none the worse and was very much amused'.[53] His host, Max Niemeyer – who, in appearance, reminded Head of his cousin Marcus Beck – was the university bookseller, and therefore well-acquainted with the faculty at Halle.

Head was assigned a set of rooms in the Niemeyer household, and dutifully sent details – including floor plans – to his mother. He assured her that: 'I do not at all mind the German cooking and I drink hardly anything but coffee & tea'. His accommodation possessed no bathroom. Head had, however, come to Germany equipped with an India rubber bathtub that elicited a great deal of curiosity and jollity from his hosts. They were 'very much amused at my bath & Mrs Niemeyer said "The Englishman always likes water"'.[54]

The Niemeyers were warm hosts. On a number of occasions Head joined Frau Niemeyer and her children on walks about the town. On one holiday, 'We went down by the Saale & fed some swans who have a nest there in the marshes. I heard the frogs croaking for the first time; they made a most dreadful noise like hundreds of ducks. We came home by the gardens and I looked at the bathing place on the way ... it looks delicious'.⁵⁵ Head and his companions often swam at this spot after lectures as a relief from the summer heat.

In 1926 Head recalled the Halle he had known as a boy as 'a typical corner of old Germany', untainted by the vices that were later to besmirch that nation. Each night Head listened as the watchman did his rounds calling out the hours. Some of the neighbouring population still wore a traditional dress consisting of 'waistcoats and coats with silver buttons and knee-breeches'. Certain of the local practices also seem to belong to a bygone age. Head remarked that: 'There are several peasants who go about with different things in carts drawn by great dogs and stop at little plank huts, put up on the high road, for the night turning their carts into kennels for their dogs'.⁵⁶ The physical fabric of the area, as well as its customs, also seemed, to a visitor who had grown up in Victorian London, to belong to an earlier era. There were no trams in the city. Moreover, 'the sanitary arrangements were of the most primitive character'.⁵⁷

The university, in particular, still preserved practices that even in 1880 seemed quaint and anachronistic. The student fraternities, or *Burschenschaften*, paraded through the streets of the town decked out in coloured caps and jackboots and wearing sabres. Nor were these weapons purely ceremonial. Duels, although nominally banned, were still fought each week at a restaurant just outside the town limits with tacit official approval. Indeed, local doctors attended to treat the wounded. One of the students Head met while matriculating confided: 'that he had had a fight about two weeks ago in which he had slit his adversaries cheek open'.⁵⁸ The German students whom he met early in his stay were anxious to know whether their English equivalents duelled with rapiers.

Head's curiosity was evidently piqued by talk of these duels. When the opportunity to witness some of these bouts occurred, he seized it eagerly even at the cost of (uncharacteristically) cutting a lecture. Head and his friends hurried to the *Wein Garten* restaurant some distance from the town where the bouts had already begun. He was struck by the grotesque appearance that their protective gear gave to the contenders:

> Over the shirt was a great pad like a butcher's apron, of leather from the breast downwards and his right arm was wrapped round enormously and every four inches was tied with leather so as to keep these enormous paddings on; he had also a tremendous glove with fingers the size of boxing glove fingers on the right hand put on first and firmly fixed on by the bandages. He had also a pair of spectacles made like a miniature pair of opera glasses put on with the big ends on the face and the little ones looking forwards. These were strapped on round the head. There was also a great deal of padding round the neck to protect it.

One shortsighted combatant sported a contraption that enabled corrective lenses to be fitted to his goggles: 'They had an optician in attendance with a pocketful of glasses ... [in the course of the duel] The short man had five or six pairs of glasses broken a splinter from one entering the spectacles and cutting his eyelid'.[59]

Head's account of the fighting shows little sign of a Quaker repugnance for violence. He was indeed appalled by the extent of the injuries inflicted – noting, for instance, that after one fight, a participant's face 'was perfectly horrible – one mass of clotted blood with a cut from his cheek through his lip and his light moustache caked with blood – his head was also cut open and blood streamed down over his face into his front guard'.[60] But the overall tone of his account is one of fascinated – almost ethnographic – observation of an alien ritual. Nor did he fail to remark on the different levels of fencing skill on display. As soon as he returned to his rooms, Head hastened to put his memories of the event down on paper.

On the whole, Head was impressed by neither the taste nor the mores of the student body at Halle. They were, he told his mother, 'rather peculiar; they walk about the streets thinking themselves very well dressed but their clothes are made mostly of bad stuff & are badly cut & many wear cotton gloves!' More objectionable in Head's eyes than their appearance was, however, the indolence and dissipation of the students – 'for they do nothing but drink the whole day & most of the evening'. His host Max Niemeyer assured him that 'some of them drink twenty five glasses [of beer] at a sitting....'. Nor did the native students show any interest in the athletic pastimes of their English counterparts. When Head went out for a walk with one German, he was constantly asked to slacken his pace; '& when I told him that in England we sometimes ran he was horrified'.[61] Head confessed his sense of alienation in this company to his mother: 'I feel very homesick when I am with these students as I feel so utterly different from them and I should hate to get into their ways....'[62]

He did make a number of friends during his stay in Halle, preferring other members of the 'English *Gesselschaft*', or other foreigners to the Germans. These included a fellow British student, George Lovell Gulland (1862–1941), who was to become a Professor of Medicine in the University of Edinburgh and William Bramwell Ransom (1861–1909), who was later a fellow undergraduate at Trinity College, Cambridge. Head was to be best man at Ransom's wedding. Head particularly remembered an American theological student called Bahnson who introduced him to a Moravian colony where a seminary for prospective missionaries was located. The atmosphere of this institution may have awakened in Head some memories of his Quaker upbringing; he recalled being 'greatly struck by the geniality and self-devotion of the young men I met, and much enjoyed the happy simplicity of the life in this Colony'.[63]

Head also made the acquaintance of a twenty-three-year-old Russian student named Woldemar Krivorotov whom he described as: 'a sallow fat & rather sickly looking man and I am sure he smokes too much'. What drew Head to this individual, despite these unattractive traits, was that he had already spent two years in Leipzig studying histology with the anatomist and embryologist, Wilhelm His (1863–1934). Head thus had much to learn from this Russian about the subject, and admired the 'beautiful sections he had made' while in Leipzig.[64]

Head recalled that he had been something of a harbinger of modernity to a corner of Europe that seemed in many respects still to belong to the pre-industrial age. He had brought with him 'a bicycle with a tall front wheel, 60 inches in diameter, the first to be seen in those parts'. So revolutionary was this new technology that whenever Head undertook the 20-mile rides to Leipzig, 'at every stopping point I was compelled to give a lecture on this unknown machine'. The bicycle struck terror among the local horses – even those bred for battle. On one occasion, Head 'routed a whole detachment of Uhlans as they were passing through the gates of one of the small neighbouring towns'.[65]

The Journal Head kept at the time reveals that the bicycle was in fact delivered about a month after he arrived in Halle. He housed it at the premises of Niemeyer's printer. Getting it there from his lodgings posed something of a logistic problem. Head was anxious to move his new conveyance when the local children were still at school:

> I asked Mrs Niemeyer what time the children came out from the school and she told me four; I hurried the bicycle downstairs with the help of the servant and by running & walking fast managed to get it to the house just as the children came out. I did not mind the people for they can only stare & make remarks but the children follow & I was afraid of going out of the town like the Pied piper of Hamlin. Of course the bicycle created a great deal of excitement as I went through the streets.[66]

It took Head a while before he learned to ride the bicycle, at the cost of several undignified and painful falls. Once mastered, however, this vehicle gave him the freedom to roam widely around the surrounding area; often, it seems, while singing extracts from Gilbert and Sullivan operettas. When Head's cousin Theodore Beck, who was on a tour of Germany, came to visit, the pair cycled to Leipzig where they surveyed the art galleries and attended a theatrical performance.[67]

Usually Head returned from these expeditions by rail. On one occasion, however, he missed his train and was obliged to cycle back to Halle by night. He penned an evocative description of the journey:

> I had the most ghostly appearance as though I was riding on a moon beam; for the man at [the printer's] had polished up the spokes beautifully and they shone in the moonlight though no other part could be seen except my light-coloured clothes & face. The dogs were wild and one either broke loose or was let loose on me but I went a bit faster and gradually left him howling and barking behind.[68]

In later life, Head was to return to Germany for cycling holidays, which were still to provide him with a peculiar exhilaration.

Nearly fifty years after this first trip abroad a miscellany of seemingly unrelated details appeared worthy of record. Head remembered a white porcelain stove that provided the sole heating in his sitting room. His breakfast, at 6.30 each morning, consisted of coffee and a bread roll. His other meals were taken at a *gasthaus* known as the 'Tulpe'. Head recalled the menu with odd precision: 'I obtained a dinner consisting of soup, the meat from which it was made with some sharp sauce, and sauerkraut or some similar vegetable, roast meat next with potatoes, and compote, the whole for 75pf: or 9d in our money'. In the afternoon Head often indulged himself with a cup of coffee or chocolate bought from 'a local confectioner's in the town celebrated for his cakes and other delicacies'. On Sundays he dined with his hosts, which gave him the opportunity to exercise his German.[69] His grasp of the language was soon sufficient for him to read fairy stories to the children of the family.

These activities were, however, incidental to the more serious business of Head's stay in one of the centres of learning in Germany. He enrolled – 'after the most elaborate and complicated ceremonies' – as a student at the University of Halle. These ceremonies were further complicated by Head's inability to understand many of the instructions he received leading to several days of frustration and fruitless wandering around the town. Head told his mother that the resultant anxiety brought on a recurrence of the problem that had plagued him since Charterhouse: 'My ear at first was nearly well but the last few days owing I think to the worry &c. it has come on worse....'[70]

The institution that Head attended was the result of the merger in 1817 of the old University of Halle and the University of Wittenberg. Both institutions had a history as schools of theology stretching back to the Reformation. By 1880, however, Halle, in common with other German universities, had also become a centre for advanced research and training in natural science and medicine. Thus although the university was 'the home of antique customs and ritual, not yet disturbed by the hurry and pressure of modern life, the teaching, especially on the medical side would hold its own to-day [1926] as an efficient programme for instruction in any Medical School.'[71] These fields were increasingly recognized to require extensive laboratory facilities. Head noted that during his time at the university, while some of these institutes were already in use, others were still in the course of construction.[72]

Halle was also well endowed with human talent. Among the university faculty at the time of his visit were: 'men who were not only actively engaged in original work, but were young enough to be interested in the didactic side of scientific teaching'. In the limited time available to him, Head decided to concentrate on histology and physiology teaching on offer at Halle. Despite his limited German he had little trouble following lectures and demonstrations, partly because he already had some grounding in both subjects, but also 'owing to the excellent demonstrations and diagrams' the professors employed. He had, moreover, come equipped with the English textbook of physiology written by his soon-to-be teacher, Michael Foster, 'and both my Professors would kindly indicate to me the passages which corresponded to the subject on which they were lecturing'.[73]

The Professor of Physiology was Julius Bernstein (1839–1917) whom Head sought out soon after his arrival at Halle. He described Bernstein to his mother as: 'very pleasant; he is quite young, not more than about thirty four or five but is half bald. I believe he is a Jew but I am not certain. His relation or whoever it was that came to speak to me had a <u>slightly</u> Jewish face'.[74] Jews were as a rule discriminated against in nineteenth-century German universities. A Jewish faculty member may simply have struck Head as a rarity. There is no hint of any overt anti-Semitism in Head's description of Bernstein. It is, however, noteworthy that he should think it important to remark upon his professor's likely ethnicity and that he revealed a belief that there was such a thing as a typical 'Jewish face'. Later in life Head was to display a keen interest in what he considered the distinctive features of the Jewish mind and character.

Bernstein was a member of the 'organic physics' movement in nineteenth-century Germany. This school sought to apply the methods of physical science to the elucidation of vital phenomena. Bernstein himself did fundamental research in the field of nerve and muscle physiology.[75] Bernstein taught an advanced course on biophysics at Halle; but Head attended the more elementary physiology lectures that he gave. He recalled Bernstein as 'an excellent lecturer, clear and conscientious' who 'illustrated his teaching by numerous demonstrations'. The syllabus began with the physiology of nerve and muscle – Bernstein's specialty – and moved on to the central nervous system and its relations to the other organ systems of the body.

Bernstein's physicalist orientation was evident in his approach to physiology. His lectures were often designed to show how organs operated according to purely mechanical principles. Bernstein used a range of ingenious devices to illustrate these claims. Thus, on one occasion,

> He ... made a tracing on smoked glass of the various contractions of muscle with different weights attached ... He began with nothing and went up to four hundred grammes which was the greatest the muscle could lift. He then measured the lines on the smoked glass which were made by a steel point attached to the muscle and showed that for one, two & three hundred grammes the contractions were the same; this showed that muscle contracted according to the work required to be done.[76]

The next day, Bernstein made use of a diagram to try to demonstrate the absolute force of a leg muscle. Head admitted, 'I could not quite understand it'.[77]

In addition to this standard teaching, Bernstein advertised a *'privatissime'* course for anyone willing to pay 'the enormous extra sum' of ten marks. Head, much to Bernstein's surprise, chose to sign up for this exclusive instruction. Henry Head senior was sufficiently prosperous to ensure that his son was well provided with funds during his stay in Germany. Head was, however, to have some difficulty persuading the local banks to accept the sophisticated financial instruments called 'circular notes' with which his father had supplied him. Confronted with Head's ten marks, Bernstein was obliged to confess that there was in fact no formal teaching attached to this most private course. He did, however, give Head free run of his laboratory.

In the event, this meant that Head took on the role of assistant to Bernstein by helping to prepare the experiments for the next day's teaching and himself repeating the experiments he had witnessed in class. Bernstein 'would come repeatedly from his private room to see how I was getting on and to criticize my work, occasionally questioning me with regard to the principles underlying the experiments I had performed'.[78] Within months of leaving school and before he had even commenced his undergraduate education at Cambridge, Head was thus receiving personal instruction in laboratory science from one of the foremost physiologists in Germany. Looking back on those days 'in the light of subsequent experience', Head was well-aware of his good fortune: 'I cannot be sufficiently grateful for the surprising kindness shown by a man of Bernstein's eminence to a young foreigner who was only a tyro in his subject'.[79]

It was while working in Bernstein's laboratory that Head underwent an important rite of passage for any experimental physiologist. Head had previously only seen experiments performed on frogs. On the morning of 26 May 1880, he witnessed his first vivisection of a mammal: 'We had a rabbit shown us to demonstrate that the note given out by the muscle on contraction caused by an interruption is the same as that given out by the machine...'. This was done by applying a stethoscope to the exposed muscle.[80]

That afternoon Head was required to try the experiment himself. Head's account reveals an equivocal response to this experience. He had kept rabbits as pets at Charterhouse and clearly had some affection for animals. Nonetheless, he did not allow such feelings to check his progress in the scientific way of life. Head sought consolation from the thought that: 'It is not a very cruel experiment (comparatively) but I tried to get my work over as soon as possible ...'. So successful was he at eliciting the required sound, however, that Head was at Bernstein's insistence obliged to repeat the procedure for the benefit of others present. Afterwards he was relieved to get on with 'other work about a piece of muscle & its natural currents which wants no vivisection'.[81]

His first attempt at these experiments on muscle preparations was not a success. But the next day 'everything succeeded beautifully....'. Before he left, Head felt the need to satisfy his curiosity about a small room 'with a darkened door opening into the "Auditorium" which I had often looked at'. He found it to be a 'regular menagerie', stocked with the frogs, rabbits, and guinea pigs Bernstein required for his lectures and demonstrations. In a hutch by itself Head found the rabbit on which he had experimented; he seems to have taken comfort in the fact that: 'It was none the worse except for its helpless leg and it eats enormously and is now growing very fat'.[82]

The following day Head found time to return to the menagerie. He helped the laboratory assistant to feed the inmates. Thereby he perhaps reverted to an earlier, nurturing, attitude to animals as well as further assuaging his conscience. When he turned around, 'there was the Professor watching me. He laughed a good deal but that was all'.[83] Head's evident relief that Bernstein did no more than laugh at his actions is an indication that such behaviour might be deemed inappropriate in a physiology student. In the first place, it transgressed the boundary between the proper role of the scientist and of laboratory assistant. The latter was crucial to the efficient running of the establishment, yet was assigned a menial status.[84] Moreover, by helping to care for these animals, Head might have been charged with having failed to achieve the attitude of cold objectivity to these experimental 'subjects' that was expected of the physiologist.

Whatever early qualms about vivisection Head might have felt were, however, soon dispelled. In his later career he was, indeed, to gain a reputation for his aptitude in devising ingenious new experimental techniques in order to elucidate problems in physiology. By the time Head took Theodore Beck to hear a physiology lecture later in the summer, Henry regarded his cousin as a layman whose sensibilities were different from those of the scientist. Head was relieved that on that occasion: 'we had only the effect of poisons – shown on two frogs ...'. However, even this procedure 'rather shocked him'.[85]

The histology teaching at Halle took the form of lectures, 'illustrated by excellent microscopical preparations and by diagrams drawn by the Professor's own hand'.[86] These lectures were supplemented by practical instruction in the use of the microscope and the preparation of specimens for observation. Histology was a combination of practical craft and theoretical science. We have seen that Head had already acquired the rudiments of microscopic competence while at school. These skills were in the early decades of the nineteenth century transmitted chiefly by means of personal contact from an experienced practitioner to the tyro.[87] At Charterhouse Poole had played the role of master to Head's apprentice.

By the latter part of the nineteenth century, informal systems of pedagogy of this kind had been supplemented if not supplanted by more structured institutional ways of producing histologists who would possess the requisite combination of practical abilities and theoretical knowledge. The German universities took a leading role in creating these systems of pedagogy.[88] At Halle practical classes were held three times a week and were attended by no more that eight students. An important function of such teaching was to ensure that histologists shared a common language: that they applied the same terms to the appearances they viewed through the microscope. This was in turn predicated on the possession of a mutual *visual* language, which was promoted by the use of specimen slides and 'diagrams drawn by the Professor's own hand'. In Head's case the process of ensuring that correct terminology was being applied to the structures under consideration was complicated by the fact that the teaching was taking place in a foreign language. He recalled that: 'At first I had some difficulty in convincing my teacher that I understood what was said until I adopted the plan of making drawings of the specimens given us, on which the various parts were named in German'.[89]

In his Halle journal Head jotted down his memories of one of these sessions attended by him and six other students, which shows this mode of pedagogy in practice. He noted in particular that:

> The first preparation was utterly different from what he said. Many did not find this out; I luckily had not looked at mine it was not ready or else I should have been one of them. After this we had to say what all the preparations were which he gave us and he questioned us about them. I hope he will do it again next time.[90]

It is not entirely clear whether the initial misnaming of a specimen was accidental or intended to test the competence of the students. The link between language and observation is, however, evident. Sadly, Head's hope that he would receive further instruction of this kind was disappointed. He learned three days later that the Professor of Histology had died the previous night.

Head's histological instruction at Halle thus served to equip him to engage in the scientific discourse about the microscopic structure of organisms as much as to enhance his ability to use the instrument and its associated technologies. The course was confined to the normal structure of the tissues; histopathology – the study of abnormal structure – formed a further branch of study. He also took away a more concrete resource from these classes. By the end of the semester, 'I had collected a valuable series of typical specimens which stood me in good stead in after life'.[91] A collection of slides of this kind thus served as an *aide-memoir* to the histologist in his subsequent career as an observer.

Attending the histology classes in Halle was, however, not without its hazards. One afternoon, as Head made his way to the laboratory, a whirlwind suddenly swept through the town:

> I managed to find my way into the Institute before the rain came; it simply poured and the wind increased. Ransom & I with the Professor & Dr Solgar were the only people there. We went with the Servant to shut the windows as we were going below

... the large plate glass window facing the wind had its middle blown in. This entirely imprisoned the professor in the microscope room where he was preparing for the afternoon lecture. We managed to close the window again but the servant in moving past after helping us close it had his hand cut open & two sinews cut through by a huge piece of glass which was blown in. He luckily had his hands over his head otherwise he would have had his head cut open. Ransom got a very little cut on the hand from the same cause and a small one on the cheek. So we had a pretty lively time of it.

Following this experience, Head was so shaken that: 'I could cut no sections so I was given a newly born child to prepare for histological purposes'.[92]

As well as these lectures and demonstrations in physiology and histology, Head attended daily German language lessons. For the most part, despite his professed aim of becoming a doctor, he gave little attention to clinical subjects during his time in Halle. Towards the end of his stay, however, he felt that his German was sufficiently advanced for him to attend the surgical lectures given by Richard Volkmann (1830–89) who was 'then at the height of his fame'.[93] Volkmann was indeed a surgeon of international repute. In particular, he was known as the leading German advocate of Listerian or antiseptic surgery.[94] He had in 1879 established a surgical 'Klinik' at Halle housed in one of the 'great new buildings' erected in the university, which drew students from around the world.

Head set out a detailed account of his experience of this Klinik in a letter intended, one suspects, as much for Marcus Beck as for his parents. He noted that Volkmann was 'a peculiar looking man with long red "weepers" such as a typical Englishman is supposed to have'. Volkmann's mode of lecturing was also remarkable: when expounding the ideas of others with which he disagreed, 'he states them and then answers them as if he were talking to the man himself'. As a convert to Listerian principles, Volkmann washed his hands carefully before conducting operations without interrupting the flow of his lecture. Head attended a lecture on lupus followed by some operations. He stayed behind when the other students departed and managed to attract Volkmann's attention. When he learned that Head was a relative of Lister,

> he immediately seized me by both hands and introduced me to his doctors who were amazed at his eagerness. He said, 'Why did you not tell me you were coming, I'd have had some operations for you to see then; not such things as these'. He was quite grieved that I had not seen any of his wonderful operations.[95]

Despite this experience, and his youthful admiration of Marcus Beck notwithstanding, Head never showed any inclination to take up surgery himself. Indeed the clear bias of his early studies was less toward the clinical and more orientated toward basic laboratory science – a field for which he had both an inclination and a talent. Indeed, as we shall see, at one time he seems to have seriously considered abandoning his boyhood ambition of becoming a doctor and devoting his career to biological research.

Head does not explain in his autobiography why this visit to Halle was in retrospect 'one of the most important events of my life'. But Germany, its science, language and culture were to become central to Head's sense of personal identity. Indeed, when in later life he travelled through Europe he was often mistaken for a German. One can infer that his first stay in Germany took on this significance because it gave him privileged access to a scientific way of life that had already exerted an allure upon him and which he was to pursue with great vigour in subsequent years.

Moreover, the months he spent living over a bookshop in Halle gave Head his first taste of the German literature with which he was later to become so enamoured. He kept his mother, to whom Head attributed his early love of letters, informed of his reading. At first he was obliged to confine himself to familiar English works. Thus on 21 April 1880 he advised her that: 'I bought a Aurora Leigh from Mr Niemeyer & am now beginning at the beginning again'.[96] Towards the end of his stay in Halle, however, Head began to feel sufficiently competent in his new language to begin to essay some of the classic German authors. He and a fellow student, for instance, read aloud Friedrich Schiller's *Turandot* over coffee. Head was not impressed by this particular work; in his view, it was 'very ridiculous to think of Schiller writing a sort of Pantomime as the play is'.[97]

More significantly, it was during these months that Head was first to become acquainted with the works of Goethe: once done with Schiller, he and his friend planned to make a start on *Faust*. Frau Niemeyer had also suggested that he improve his German by translating and learning some of Goethe's poetry from a book that had been forwarded by one of Head's Charterhouse teachers.[98] Head came to ascribe extraordinary importance to Goethe's writing as a key to understanding the world and one's place within it. So much so, that decades later he was to prescribe reading Goethe as a form of therapy to at least one of his neurasthenic patients.

This immersion in Continental literature, in Head's view, effected an irreversible transformation in his nature. On returning to England he commenced his Cambridge career 'steeped in "Peer [Gynt] and similar poems"'. Head's version of *Schwärmerei* did not, however, impress the 'cold quiet dons' who were obliged to read essays 'filled with execrable Teutonismus', and who 'thought I was mad'.[99]

It was also at the end of this visit to Germany that Head first manifested a lyrical reaction in the face of natural beauty. Once his course of study at Halle was complete he and his Moravian friend set out on a walking tour of the Swiss Alps. On 28 August 1880, Head wrote in his journal:

We woke up at ½ past 4 and at 5 saw a most lovely sunrise. I have never seen anything so fine. As the mist lay in all the valleys like snow and only the mountain tops were seen. First a little piece of the sun appeared and then it rose very quickly. The mist there looked lovely...[100]

The young Head was unable fully to express this sense of rapture. The passage ends in bathos: 'It was a very pretty walk'. In later life Head was to struggle with more success to find words for such feelings of ecstasy.

Cambridge

Head returned to England to take up his place at Trinity College Cambridge in October 1880. After Halle, his early months at Cambridge seemed in educational terms a retrograde step. He was expected to read Latin and Greek texts, study 'elementary Arithmetic', and – most galling – peruse William Paley's *View of the Evidences of Christianity*, which was still required reading for undergraduates. This was, Head recalled, 'a singularly disappointing beginning for us, who had already enjoyed the study of Natural Science elsewhere'.[101]

Something of the lassitude that afflicted Head in his early days at Charterhouse seems to have returned after he entered Trinity. He admitted to his mother that: 'I don't seem to be able to get on with my work at all; today & yesterday I have done nothing and very little on Saturday. If I don't manage to put on Steam I certainly shall not pass both parts of my little go this term'.[102]

In common with other undergraduates, Head engaged a coach – or 'grinder' – to help him pass his examinations in these preliminary subjects – or the 'Little Go' as it was known.[103] This gentleman thought it his duty not only to help his charges with their studies, but also 'to pay attention to our manners. His habitual greeting to his students was "Have you wiped your feet on the mat?"'[104]

Despite this grind, Head found time to pursue his real interests by attending lectures on chemistry as well as those given by his future mentor, Michael Foster (1836–1907), on physiology. At the chemistry lecture Head noted the presence of: 'Girton & Newnham[105] girls. There are a good many and the front rows are set aside for them'.[106] This was Head's first experience of coeducation. He was later in his capacity of demonstrator in physiology to teach women from Newnham, and gave his mother a jocular account of a reception at the College to which he was invited: 'One girl informed me she fell in love with all her lecturers ... I was sorry to find however she spoke somewhat slightingly of Demonstrators ....'[107]

Head combined his academic activities with active participation in the athletic activities of his College. He joined the Boat Club and spent his afternoons out on the Cam earning a place in the Freshman's Eight. His letters to his mother were full of reports of the races the Trinity boats undertook against other colleges. He also pursued his Charterhouse passion for campanology.

Head threw himself into student amateur dramatics. In later years he recalled in particular his part as Teiresias in a performance of *Oedipus Tyrannus*. The production was presided over by the eminent classicist, Richard Claverhouse Jebb (1841–1905), and other members of the cast included Montague Rhodes James (1862–1936), who later earned fame as an antiquarian and writer of ghost stories. The play evidently had a profound emotional impact on the cast: one 'burly athlete' in the chorus broke down in tears during the closing scene of the tragedy. A record of the production was among the mementoes Head cherished to the end of his life.[108]

At Charterhouse Head had also shown an interest in botany – one he shared with and may have derived from his mother. His account of a walk in the fens outside Cambridge that he, Ransom and D'Arcy Wentworth Thompson took in May 1881 shows that he was knowledgeable about the flora, as well as the bird life, of the area. He sent his mother a specimen of 'the bog violet (of which I enclose a flower) a very rare plant'. Head also continued to show the fascination for landscape he had manifested in his journey through Switzerland. He found the flat fen country in its own way as beautiful as the Alps. But he also revealed anxiety for how vulnerable and transient this landscape was. By the time Head viewed them the wetlands around Cambridge had been all but drained. He was accordingly thankful for the opportunity 'to visit the strange fen country before it is entirely swept away'.[109]

In his autobiography Head reminisced at length about the lively social life he enjoyed at Trinity. He and his fellow freshmen gathered in each other's rooms of an evening and tried various culinary experiments: 'Omelette á la Cour du Mâitre' was a particular favourite. Members of staff would also sometimes participate. These gatherings included men who were later to achieve eminence. In particular, Head remarked on two fellow Trinity men, working in different schools, who 'had a fundamental effect upon our method of thought'. The first of these was William Ritchie Sorley (1855–1935), who subsequently became Professor of Moral Philosophy at Cambridge. The other was Alfred North Whitehead (1861–1947), the mathematician and philosopher, with whom Head was to maintain a lifetime friendship. Head remembered that Whitehead's 'whimsical wit and delightful paradoxes added enormously to the gaiety of our little society and at the same time his profound knowledge and width of literary interest were a salutary corrective to our dogmatic specialism'.[110]

It is clear that these soirees possessed an intellectual as well as a culinary aspect. The collegial system at Cambridge encouraged interaction between undergraduates studying a variety of subjects. Head recalled in particular his friendship with Stanley Leathes who later became First Civil Service Commissioner and an editor of the *Cambridge Modern History*. Although Leathes 'was reading Classics and History we had much in common, particularly our interest in English and German literature'.[111] Communication between the two was further facilitated by the fact that the wall that separated their rooms was sufficiently flimsy for them to conduct bedtime conversations through it.

Whitehead later maintained that these informal conversations were of far greater educational value than the formal instruction offered by the University. At these gatherings, he recalled, 'we discussed everything – politics, religion, philosophy, literature – with a bias toward literature'. For Whitehead, looking back at these gatherings after an interval of more than fifty years, these conversations had:

> the appearance of a daily Platonic dialogue. Henry Head, D'Arcy Thompson, Jim Stephen, the Llewellen Davies brothers, Lowes Dickinson, Nat Wedd, Sorley, and many others...That was the way by which Cambridge educated her sons. It was a replica of the Platonic method.[112]

Whitehead added that this system formed part of the education of the professional middle class. This was a 'social grade' that played a dominant role in the governance of late nineteenth-century British society, 'influencing the aristocrats above them, and leading the masses below them, [it] is one of the reasons why the England of the nineteenth century exhibited its failures and successes'.[113] Insofar as he possessed any articulated political consciousness, Head identified strongly with this professional middle class. According to his wife, 'he believes firmly that all of us Upper Middle Class people (the Professionally Educated he calls us) are the Salt of the Earth, he regarded as the backbone of British society'.[114]

By Head's account, these Platonic symposia at Cambridge also possessed a critical, ludic, aspect. The young men who attended them may have been cadet members of the ruling social grade, but they saw no need to defer to conservative dogmas that they found archaic or ridiculous. In particular, the doctrines of natural theology, as exemplified by the works of William Paley, 'formed a source of amusing discussion at these symposia, and happy was the man who could prove by some authoritative arguments that Paley's arguments were ridiculous or false'.[115]

In his autobiographical sketch Head passed lightly over his abandonment of Christianity. This transition seems to have occurred while he was an undergraduate; at Charterhouse he still kept a prayer book and took communion. A reading in September 1902 of William James's *The Varieties of Religious Experience* (which he found 'a most disappointing book')[116] prompted Head to go into some detail about how he became a sceptic. He did not portray this process as a loss of his faith. Instead Head saw his change in worldview as a conversion experience with all the associated psychological states that involved. Conversion, he noted, could be either a sudden event or a slow process.

> My conversion to religious scepticism occurred gradually. At 19 I still prayed and clung to the emotional side of religious services. But neither prayer nor church brought rest. Deep down in my mind as yet unformulated lay the feeling that I was hugging the false and excluding the Truth. Then on going up to Cambridge I talked to other men and found that they accepted my personal doubts as commonplaces. Like a cloud religion fell away and it was as if the sun had suddenly begun to shine. How fine the world looked – I was like a colt turned into a field after being penned in a stable – I rolled in the sweet grass and jested without malice as a colt lashes out in playfulness.[117]

When he informed his parents of his change of heart, they arranged a meeting with a clergyman, presumably in the hope of dissuading their son from embracing infidelity. After a brief conversation, this spokesman for Christian belief admitted defeat. He asked Head for a glass of sherry: 'and from my return with the sherry on a tray spoke no further word with me on religion'. A subsequent discourse with leading members of the Broad Church movement, including Brooke Lambert (1834–1901), who had been 'the Demigods of my boyhood', also proved intensely disappointing. These liberal theologians conceded most of Head's arguments; they however refused to take them to what he regarded as their logical conclusion: 'From that evening I lost all faith in this school of thought for they seemed to me to be hugging to the last rags of a belief they had themselves annihilated fearing to walk naked in the sun'.[118]

Politics remained another lacuna in Head's range of interests. While at Halle he had conscientiously read the copies of *The Times* newspaper that his parents forwarded. He showed a keen interest in the fortunes of the British army in Afghanistan, claiming that: 'I am now ready to give anyone a lecture on Candahar [sic] fort!' However, he found the reports of parliamentary debates 'longwinded' and confessed he could form no opinion on the issues involved.[119]

Head's cousin, Theodore Beck—with whom he had cycled in Germany – had preceded him to Trinity. Beck evidently cut something of a figure in the University, going about dressed in flannels and sporting a 'straw hat crowned with hawthorn blossoms'. Indeed, Whitehead was of the opinion that: 'The man ... who had the most likeness to Shelley of anyone he had seen was Theodore Beck'.[120] Moreover, Beck 'had already made his mark as a speaker at the Union [i.e., the University debating society], and was deep in political and social movements of the more radical kind'. Head met many of Beck's associates in these movements, 'but I never found time to take part in these activities'.[121] He seemed, in particular, indifferent to the kind of social concerns that drew so many university men of his generation to seek to bridge the gap between the classes. Head was to spend much of his later career working as a doctor in the

East End of London – among the most socially deprived parts of the capital. The spirit of Toynbee Hall – the university settlement founded in Whitechapel in 1884 – however had no appeal for him.[122]

When it came to defending his own interests, however, Head was quite capable of playing the student militant. He noted that most of his fellow freshman had come straight from school and were somewhat naïve in money matters. He and a few of his friends, on the other hand, had experience of managing their own finances and expected to scrutinize their College bills – much to the astonishment of their tutors. Unlike most undergraduates, Head enjoyed a more than adequate income. In addition to his Charterhouse scholarship, he received £100 a year from his father. Henry Head senior had moreover furnished his son's rooms in college and continued to make occasional gifts; he seems also to have passed on something of his financial acumen to his son. The younger Head took exception to what he saw as a serious anomaly in the charges that Trinity levied upon him and his fellow natural science students. It was customary for the College to charge seven guineas a year in tuition fees from all its undergraduates. When Fellows of Trinity provided the teaching, this was reasonable. But in Head's day, teaching in natural science took place outside the College, and was charged separately. Head and his colleagues therefore 'rebelled and refused to pay tutorial fees for instruction we had not received'.[123] The College authorities eventually capitulated in the face of this defiance.

Once he had passed his preliminary examinations, Head was at last able to proceed to study for the Natural Science Tripos. He concentrated on the life sciences, taking courses in physiology, comparative anatomy, botany, and chemistry. He recalled in particular Foster's elementary lectures on physiology. Foster maintained that it was more difficult to teach the elements of a science than its more advanced aspects; he accordingly gave these introductory lectures himself, while leaving the higher teaching to his demonstrators. By 1880, Foster had established a system whereby second year undergraduates attended his elementary physiology lectures while also taking practical classes three mornings a week.[124]

Foster was also the author of a standard textbook of physiology, which Head studied in conjunction with the course; he had already made use of this work while studying in Halle.[125] In his autobiography Head lamented the changes that had been made to later editions of this work. In particular, in the edition he had used, 'the addition of the names of the men who had worked at each particular branch of the subject quoted was undoubtedly stimulating and useful, and served to remind the reader of the different opinions held by investigators of the same problem'.[126] This feature had later been discarded.

In 1926 Head could 'still see the tall, rather cadaverous figure with sunken cheeks and long beard, a lock of grey hair falling over his forehead standing sideways in front of the blackboard with his head drooping forwards and a piece of chalk rolled between the fingers of both hands'. Foster's 'curiously monotonous' style of delivery reminded Head of 'an inspired preacher'. Even the witticisms with which the professor sought to enliven his teaching were enunciated in the same solemn way, 'and without a change in his countenance'.[127] Not all undergraduates shared Head's appreciation of this understated style of pedagogy; Charles Sherrington (1857–1952) – later a distinguished physiologist – described Foster as an 'appalling lecturer'.[128]

Foster is remembered, however, not only as a teacher and textbook author, but also as the founder of a remarkable school of physiology at Cambridge. Two of his protégés – J. N. Langley and Sheridan Lea – acted as his demonstrators, and gave practical classes to complement the professor's lectures. Langley also gave an advanced course on histology, while Lea taught chemical physiology. Walter Gaskell (1847–1914), another notable member of the Cambridge school, lectured on circulation and respiration.

Gaskell in particular was to prove an inspirational and exemplary figure for Head. In his autobiography, Head undertook to deal in a later section with Gaskell's 'personal influence upon myself and my work ...'.[129] However, he abandoned this account of his life without dealing with these issues. Head's relationship with Gaskell must therefore be reconstructed from other sources. When Head first encountered him, 'Gaskell was just beginning his fundamental researches on the Heart, and from the first his geniality and enthusiasm captured our imagination'.[130] In the course of his childhood and youth, Head lionized a number of individuals – such as Beck, Poole, and Foster – whom he regarded as models and mentors. Gaskell was to retain this status in Head's later life, while the glamour of others in this pantheon faded.

In August 1903 Head and his collaborator William Halse Rivers Rivers (1864–1922) visited Foster at Great Shelford in Cambridgeshire. Both men had thought that the doyen of English physiology might want to hear about the experiments they were then performing on the sensory nerves of Head's arm (see Chapter 3): 'We gave several leads but the old man shied away each time and persistently talked of ordinary matters – I believe he was afraid to talk of things that would have interested him in the old days'.[131] Head and Rivers then went to see Gaskell who lived nearby. The three sat in the garden,

> and off went the talk – He heard everything we had to tell, brought it into line with general physiological principles and when I left at the end of a couple of hours we were the richer for a luminous idea – And all the time the iron grey man with the red face is treating you with the affection of a father.

The impression Gaskell left on Head was thus as much emotional as intellectual: 'The story we talked was the most abstract in the cold region of experimental psychology and yet to both of us his words left an affectionate memory'.[132]

Early in his career Head took from Gaskell an orientation that guided his early researches on the nervous system. A distinguishing characteristic of the Cambridge school of physiology was the evolutionary perspective it brought to its discussion of structure and function. Gaskell demonstrated this orientation more strongly than any of Foster's other protégés in his account of the myogenic origin of the heartbeat.[133] In a letter to Head written in November 1918, Sherrington acknowledged that Gaskell had exerted an influence on his own work by making him realize that: 'the [spinal] cord offered a better point of attack physiologically'. This bottom up approach was justified if it was assumed that the complex structures of the vertebrate brain had evolved from more primitive forms of organization and that the cord 'was originally and still must essentially be a chain of "ganglia" ...'. If this were true, then some trace of this originally metameric organization should be evident in the distribution of the afferent nerves in from the human spinal cord. According to Sherrington, Gaskell saw Head's early work on herpes zoster (see Chapter 3) as a vindication of this expectation: 'I remember how pleased he and I were about your observations and results. He had expected it would be something like that'.[134]

Although Head was to move away from laboratory science to concentrate on clinical medicine, he continued to view the phenomena of nervous disease with the eyes of a physiologist. In particular, throughout his career he sought to understand the workings of the human nervous system in ways that were informed by the theory of evolution. In this respect, he remained a true adherent of the Cambridge school of physiology and in particular loyal to the principles he had derived from Gaskell. Sherrington, like Head, emphasized that Gaskell's influence was as much personal as intellectual: 'He personified truth'.[135]

Another charismatic figure Head encountered in Cambridge was the embryologist, Francis Maitland Balfour (1851–82), one of Foster's most outstanding students. A professorship in animal morphology was created for Balfour in recognition of his achievements in the field of developmental biology, and Head recalled that: 'he ruled over a flourishing Laboratory filled with ardent disciples'. In 1882, Balfour had established a senior class, which Head attended. He recalled that on occasion he was also invited to Balfour's rooms in college. Balfour's career was cut short when he died during a holiday in the Alps in July 1882 while trying to climb Mont Blanc. Balfour therefore never delivered his assigned course as a professor. Head recalled how his senior students responded by organizing a series of private lectures, each of them contributing to the teaching. 'This remarkable effort' was, he maintained, 'evidence of the influence exerted by Balfour and our determination to atone in some way for his loss'.[136]

Balfour did not exert the same direct influence of Head's future scientific endeavours as can be ascribed to Gaskell. Head never showed any particular inclination towards morphology. But, like Gaskell, Balfour came to embody for Head an ideal of the scientific way of life. He was an outstanding example of a personal *virtue* that surpassed mere intellectual capacity. Forty-four years after Balfour's death, Head could still admire: 'the beauty of his character and his rare enthusiasm for the progress of knowledge'.[137] Both Balfour and Gaskell thus represented an epitome of the man of science – one that endowed the true scientist with an almost transcendent character, with a mission far removed from the sordid and petty concerns of mundane life. In his own career Head was to strive to approximate to this ideal.

By early 1882 Head was beginning to make the transition from being a student to being an apprentice member of the Cambridge school of physiology. He was appointed as a teaching assistant to Langley. On three afternoons a week he attended the laboratory to prepare for the following day's classes. Langley 'coached' him on the lines to take in his teaching. Head admitted that it was rare for someone to attain this kind of status before graduation: 'Our unusual good fortune was due to the fact that the class had reached the then remarkable number of 80 and there was a dearth of graduates familiar with the subject'.[138]

In the summer of that year Head took the first part of the Natural Sciences Tripos. He and the group of friends with whom he had pursued his studies in life science were all placed in the first class. Thanks to this success, he was awarded a major college scholarship. Head then set about study for the second part of the Natural Science Tripos, by now it seems, with a clear determination to pursue a career in physiology. On Michael Foster's 'insistent advice [I] gave up all thought of medicine, although I had already passed the first M.B'.[139]

A career in medicine had of course been Head's ambition for as long as he could remember. The fact that he was ready to yield to Foster's urgings testifies to the impact that such exemplars of laboratory science as Gaskell and Balfour had upon him. In the latter part of his time as an undergraduate, Head became ever more familiar with the leaders of the Cambridge school. He was given: 'free run of the Laboratory and was brought into intimate contact with the active research carried on both by Gaskell and Langley'. Sheridan Lea too 'treated us young men like brothers....'[140] This diversion from Head's original path was to prove only temporary. He eventually proceeded to University College London where he obtained the second M.B. and completed his training as a physician. For a while longer, however, Head was to pursue the classic course of an aspirant laboratory scientist and turn his steps once more to a German University.

In later years Head was a frequent visitor to Cambridge. Often he stayed at Whitehead's house at Grantchester where he experienced something of the former excitement of their undergraduate symposia: for there 'nobody yawns and suggests it is time for bed and bedgoing comes only at the first check in conversation'. Indeed, Head felt a curious kind of temporal dilation after these visits. When he looked back on his visits to Cambridge, he told Ruth Mayhew in 1899, 'I always imagine them twice or three times their real length – for so many people have been visited and so many subjects discussed and all these visits and discussion stand out so clearly and distanced that it seems impossible so much should have happened in so short a time'.[141]

Prague

As soon as he completed his final examinations, Head set out in the summer of 1884 on a tour of German universities with a view to determining which would best suit his requirements. A letter from Foster gave him ready access to these establishments. Head visited the physiology departments at various institutions before deciding on what was at first sight a somewhat eccentric choice. Rather than going to such obvious centres of medical science as Berlin, Bonn, or Leipzig, he elected to attend the laboratory of Ewald Hering (1834–1918) at the German University in Prague.[142] Head recalled that, after questioning him about his proposed line of research, Hering 'gave me an invitation to join his band of workers at the beginning of the next term, without payment of any kind, except for such animals as I might need in my work'.[143]

In his autobiography Head gives no clear explanation for this choice of laboratory. He claims that he was 'not satisfied with the opportunities' offered by the other universities he had considered. The facilities in Prague, on the other hand, were impressive. Every researcher was assigned his own room, which was fitted out to suit his requirements. Head was also struck by the quality of the design of the equipment used in Hering's laboratory: 'There was no unnecessary show of apparatus, but every appliance was exactly suited to the end in view'. He admired how old cigar boxes and blocks of metal were adapted to form 'beautifully executed mechanical contrivances, all made on the spot by the mechanician who occupied the basement'.[144]

This degree of craftsmanship evidently had more than a purely functional appeal to Head; he displayed a kind of connoisseurship in his appreciation of the quality of the instruments he saw in Prague. The 'mechanician' in question was an exceptional artisan named Rudolf Rothe. Rather than the industrialized style of scientific work that was typical of, for instance, Berlin physiology by the last quarter of the nineteenth century, Hering's institute thus appeared to preserve an older artisanal set of practices.[145] The fact that this 'arts and crafts' style of scientific work appealed to Head is an early sign of a modernist sensibility.

He seems, however, to have been chiefly drawn to Prague by the little he knew of the professor who presided over this establishment. Head's mentors in Cambridge were not personally acquainted with Hering. Moreover, 'the authorities there did not think highly of his theories on colour vision, and passed lightly over his fundamental researches on the respiratory reflexes'. Head, on the other hand, evidently found what he had learned of Hering's theories intriguing. He therefore: 'determined to close my journey with a visit to his Laboratory in Prague'.[146]

The research Head was to undertake in Prague focused on the respiratory reflexes. But he also took a keen interest in the revolutionary theory of vision that Hering had promulgated. He recalled that from time to time Hering would invite him into the 'Optical Department' of his laboratory:

> Once seated before the Spectroscopic colour-mixer or some other instrument, he would demonstrate some new advance in the views with which we were all familiar. This gave us a vivid realization of the progress of his theories of vision, and formed a superb lesson in patient and careful work.[147]

Hering had challenged Hermann Helmhotz's view that the phenomena of vision could be understood in a purely mechanical way. Instead he insisted on the need to understand the workings of the sensory organs in truly biological terms.[148] In particular, Hering insisted on the *active* role of the nervous system in the organization of man's experience of the outside world. This stay in Prague may be seen as the origin of Head's interest in the physiology of sensation. Indeed, it is likely to have been the source of some of his fundamental orientations in the study of the higher functions of the nervous system. Head in his 1926 reminiscences was proud to declare himself Hering's 'most devoted disciple'.[149] A token of this allegiance was the fact that Hering was more than thirty years later, along with Gaskell and John Hughlings Jackson, one of the dedicatees of Head's monograph on *Aphasia and Kindred Disorders*.

Hering thus became for Head another embodiment of scientific virtue; indeed he described him as a 'physiologist of genius'.[150] As with most of his heroes, Head provided a description of Hering's appearance:

> He was powerfully built but somewhat below middle height. His hair was brushed straight back from his brow over his beautifully shaped head, and he wore a black beard that was rapidly turning grey. His voice, a low baritone, was always perfectly modulated, and he spoke an unusually pure German.

Like Gaskell, Hering's character was marked by 'his transparent honesty'. When one of his protégés made a statement 'not based on fact', Hering's demeanour suddenly became ominous: 'At such moments his voice deepened, and his whole expression became that of a severe judge passing sentence on a crime'.[151] As was customary in representations of the natural philosopher, Head listed some of the great man's eccentricities and quirks.[152] Hering was orderly in his habits, 'but when deep in thought would become forgetful of all around him'. He made his own coffee in the morning. He liked to join in games of skittles with his wife and colleagues as an occasional relief from intellectual labour.[153] Although overtly austere and grave, the professor was also capable of displaying an impish sense of humour.

Head returned to Prague in September 1884 to take up his internship at Hering's laboratory. The city then formed part of the Austro-Hungarian Empire and was plagued by the ethnic conflicts that beset the Habsburg domains. Prague was dominated by a German-speaking minority. In the course of the second half of the nineteenth century, however, the demands of the Czech majority for equal cultural expression became ever more vociferous. One result of this nationalist movement was that in February 1882 the Charles-Ferdinand University in Prague was divided into two independent institutions: in one, German was the language of instruction, while in the other all teaching was conducted in Czech.[154]

Head could not fail to notice that the city in which he now took up residence 'was torn by internecine quarrels between its German and Czechish inhabitants'. So fierce was this hostility that each party had its own shops, theatres, and places of recreation. When members of the two nationalities did come into contact, 'outbursts of patriotic violence not infrequently occurred'.[155] Head seems to have been merely bemused by these hostilities and to have formed no opinion on the issues that divided Czechs and Germans. Theatres and concert halls were, however, also segregated – a fact that did cause him some annoyance. As a guest of the German university he was expected to avoid Czech establishments and events. This was irksome to an admirer of Czech music: 'and on the occasion when I heard a Dvorak Opera I was compelled to do so by stealth, pretending I was an Englishman who had just arrived in the city'.[156]

In terms of his research, however, he recognized that he was fortunate to be on the German side of the academic divide. The German university possessed superior facilities. Moreover, it was able to attract staff not only from Austria, but also from Germany and Switzerland. As a result, 'during my stay, our University was the home of a brilliant set of Professors and assistants...'[157]

When Head arrived in to Prague to commence his researches Hering's return from vacation had been delayed by illness. Head made use of this time to read up relevant physiological literature. He also took the opportunity to explore the city. He found it:

> a very quaint place; for as you walk along between the stuccoed houses the town appears much like any other German town. But it is built on several hills and it is when on one of these that you see how picturesque Prag really is. The houses are of the most varying heights & capped with the most different roofs. There are innumerable churches and though the walls round the old town have gone the huge watch towers

& gateways still remain. In addition there are many boulevards & great open squares planted with trees & laid out with flowers...

He was especially struck by the sight of huge barges, some 300–400 feet long, carrying timber down the river. Even before he set out Head had given thought to the waterways that flowed through Prague and made arrangements for his personal boat to be sent from Cambridge.[158]

Head had found lodging with Frau Linke, the widow of a professor of Latin at the university, and a friend of the Hering family. Frau Linke served Head as a guide through the fraught social relations of the city and in general treated him 'like a favoured child'. She liked to play the piano while her guest listened. In the evenings they often sat together while Head read to her. Head's recollections of his host in Prague are remarkably affectionate; he was 'certain that the uniform health and happiness I enjoyed during my two years in Prag were due largely to the unfailing goodness of this admirable woman'.[159]

The titbits of information about Frau Linke that Head passed to his mother at the time were mostly of a humorous nature. He had decorated the wall of his sitting room with family photographs including 'that splendid one of Father and the Frau Professor has fallen in love with it – "What a delightful face! ... He hardly looks like an Englishman"'.[160] The account of the Frau Professor that Head composed in 1926 is, in contrast, remarkable for its recognition of the pathos of her condition. She was evidently an intelligent and gifted woman who possessed 'a natural feeling for good German literature ...' She had, however, been kept from cultivating those interests by her late husband's 'prejudices against promiscuous reading for women'. Even though the husband was no more, Frau Linke still felt constrained by this inhibition: she possessed many books, 'which she had never read, in consequence of the embargo laid on them by the dead hand'.[161] Hence the service Head performed by reading the works of Ibsen and Turgenev to her.

Such sensitivity to the repressed and restricted role assigned to women in late nineteenth-century society was to surface again in a letter written later in Head's stay in Bohemia. While in Prague Head became acquainted with Wilhelm Weiss (1835–1891), the professor of surgery at the university. Weiss invited Head to stay with him, his wife, and his daughter at his villa in the Böhnewald near Eisenstein. Head took the opportunity of this vacation to take long walks among the trees. Head noted that for the most part he went on these hikes alone: 'for the Professor is so dreadfully nervous that he will not leave the women alone in the Villa – the Frau Professor is too delicate to go out when the weather is not settled – and "es schickt sich nicht" [it is improper] for the Fraülein to go alone with me'. Head was in fact glad to be able to roam without the impediment of Weiss and his wife. But as he returned to the villa 'glowing' at the end of a day wandering the forest he became increasingly haunted by the sight of: 'Fraulein Weiss in an opera cloak shivering in the house and the Professor pacing up and down on the small platform before the villa. I must say I pity her ...'. He was aware that she loved the city and the company of people her own age. Because of her parent's preferences, she was instead forced to spend two and a half months a virtual prisoner at the villa – 'without speaking to a living soul except her parents'. The girl had no taste for country life. Even if she had wished to go abroad, 'she cannot stir from the platform on wh. the villa is built owing to having no one to go with ...'.[162] These observations led Head to try to imagine what passed through Fraulein Weiss's mind as she sat day after day on the platform staring out at the trees. He surmised that she brooded on the dismal fortunes that likely awaited her:

> Every day she sees money laid out on the land wh. unless the professor can hang on at least 20 years will be absolutely lost and which of course entails an ever increasing economy when the family is in Prag – that takes the form of refusal of invitations and absence of all gaieties during the winter – She has to hear from this friend in Marienbad, that in Johannesbad – of this festivity and of that picnic and must herself remain absolutely isolated on the Spitzberg – It is a comedy which has a distinctly tragic side for if the Professor dies before his land rises in value the family are beggars – and the invalid mother and the daughter, ornamental but useless, will have to earn their bread in some way or other.[163]

There is a novelistic character to the passage – which has more than a hint of Flaubert (whose novels fascinated Head) about it. But it also reveals a keen awareness of how fragile was the security and comfort of the bourgeois family in general and of its female members in particular.

During the same vacation on the Spitzburg, Head recorded some remarkable observations on the customs and social conditions of the local peasantry. He had already shown an ethnographic eye during his time in Halle. But these later remarks lack the detachment of the account of the student *Mensur*; they show instead an indignation and sorrow at the passing of an ancient way of life.

Head had the opportunity to study peasant life at close quarters. Because the Weiss villa was so small, he was housed with a peasant called Wudy. Head took a close interest in the culture of his hosts. Indeed, he became something of a participant observer of peasant life – joining the woodsmen in cutting trees in the forest and rising at 5a.m. to milk the cows. He told his mother: 'I could write for hours of the peculiarities & customs of the peasants which I got to know pretty closely'. He noted, for instance, that they retained belief in the existence of magical creatures; several of the peasants claimed to have seen 'Wassen Nixen' in the nearby Schwarzen See. As a scientist, Head felt obliged to check these claims. After making observations of his own in the region of the lake, he concluded that these nixies were in fact beavers. Head was also intrigued by what he saw as the paradoxical notions of property rights these peasants displayed. Theft from a house was regarded as reprehensible; stealing wood from another's holding was however perfectly acceptable. But what especially engaged Head's attention was the realization that the peasant way of life he was observing 'is so rapidly dying off'. In the past, society had been organized around a system where *Koenigbauern*, or 'King Peasants', had occupied large areas of woodland subletting plots to common peasants 'who received no wages, but were simply housed and fed'. This system had for some time been undermined by the encroachments of the market economy. The common peasants began to require wages from the King Peasant who, in order to meet these demands, was obliged to cut down trees and raise money by selling the timber. The society and the physical environment were thus simultaneously under attack: 'For just as there are now scarcely any trees over 200 years old in the woods so the peasants who are a similar relic of a bygone age are fast disappearing'.[164] What had finally, in Head's view, completed the ruin of the *Koenigbauern* had been 'the advent of the Railway and the Beer house'. The peasants were forced into debt and 'many have got into the hands of the Jews ....'.[165]

Head did not have an overly sentimental view of the way of life that was now passing away. He recognized the abject poverty of the society he was describing. He noted that even though most of the King Peasants had the noble 'von' prefix to their names, 'they lived in a way a dog would scarcely tolerate ...'. Head was nonetheless clearly moved by the plight of this class. Head prefaced this account of peasant life by advising his mother that: 'This is not intended to be shown. I have simply written a small portion of what has been running in my head & what I have already written down for my own use – It will perhaps not interest even you much less others'. By writing down these observations Head was thus seeking to gain a clearer insight into why and how what he had witnessed had so affected him. Intellectually, he chose to view the fate of the peasants in Darwinian terms. The peasants 'were perfectly adapted to their surroundings 2 or 3 hundred years ago but owing to the suddenness with wh. a new life has been forced upon them they have been unable to adapt themselves and are fast being swept away'. What had destroyed this ecology was the inexorable penetration of even the most remote corners of Europe by capitalism: 'The very fact of their being obliged to pay for the labour they previously obtained for the price of the food & lodging of the labourer was a terrible blow but to have their previous "slaves" competing with them utterly ruined [the King Peasants]'.[166]

Head recounted a scene that epitomized the tragic end of a class that had been overtaken by history:

> As we stood outside the villa one evening flames were noticed in the direction of one of the Höfe – we thought the house was on fire and pointed it out to the Shepherd – He at once fell on his knees in prayer and then told us it was the death fire of the

farmer – His bed & clothes had been taken out of the house & set on fire and he had been laid on the painted board at once after the death.[167]

Head had already shown the intimations of what might be called a nostalgic conservative consciousness during his time in Cambridge. The East Anglian fens were, like the Böhmewald, being destroyed because of the depredations of modernity. In both cases an age-old way of life was disappearing along with the environment that had sustained it. Head showed no developed political understanding of these processes but he did manifest an almost instinctual revulsion for the corrosive effects of capitalism upon long-established societies and landscapes. His remarks may, however, reveal that he had in the course of his stay in Prague been exposed to some of the anti-capitalist, anti-liberal, and anti-Semitic discourses that proliferated in the Austro-Hungarian lands during the 1880s.[168] As we shall see, Head's attitudes in later life to Jews were complex.

The remarks on the impact of economic 'progress' on traditional societies that Head felt moved to write after his stay in the Böhmewald were an early sign of a sense of alienation from the modern world that was to become more marked in later years. As Ross has pointed out, this was a typical feature of what she describes as an 'aesthetic modernist' sensibility. The complement of this retreat from what was ugly and repugnant in contemporary society was a growing preoccupation with the aesthetic delights of the 'inner life'.[169]

With Hering's belated return to Prague, Head was able to commence his investigations in earnest. The Physiological Institute of the German university was housed in the *Wenzelsbad*, an old structure that had once served as a bathing establishment. The university had taken over the premises when it was deemed necessary to provide separate premises for physiological teaching and research. Some of the original baths survived and proved useful to the new occupants as a vivarium for frogs and other amphibians kept for experimentation. The rooms on the first floor were equipped with the latest instruments for various kinds of physiological research. The basement housed the 'mechanician' and his assistants who manufactured these devices. Head noted that this was an enterprise of considerable size because Rothe 'not only made all the instruments we required, but did a considerable business with other Institutions'.[170] The Institute contained a lecture room so positioned that the students never passed through the laboratory, which thus remained dedicated to advanced research. The first floor was also where Hering's apartment was located. He lived here with his family 'who made much social use of the garden, in the corner of which lay a fully-sized skittle alley'. In this garden the Frau Professor and her friends would gather on sunny afternoons for coffee and Head and his colleagues would often 'be invited to tarry a few moments to take part in the symposium under the shade of the trees'.[171] The Institute was thus a mixture of a physiology 'factory',[172] with its own workshop and highly developed division of scientific labour, and a *gemütlich* bourgeois dwelling.

Hering imposed strict rules about how the workers within this system were to organize and conduct their research.[173] The physiologists in his laboratory were allowed a considerable degree of autonomy. Indeed, the Professor insisted that each of those engaged in experiments requiring the use of electricity should set up and use their own equipment and forbade them from touching that belonging to another. This meant that at the outset Head was obliged to expend considerable time laying the wires for his apparatus; he came, however to see the wisdom of this injunction: 'For if a mechanical fault appeared, each investigator could be certain that it lay somewhere within his own circuit'.[174]

This material autonomy was complemented by a strict system of central surveillance and control. Each month Head and his fellows were required to submit a written report on their work for Hering's scrutiny. This would be followed by a private interview in the Professor's office. Often Hering would also require the researcher to demonstrate practically some point for him. He was particularly insistent that experiments should be carefully planned in advance. Head remembered that: 'This careful daily preparation formed a most valuable training in orderly habits'.[175]

These 'orderly habits' formed part of the assimilation of a young scientist into the experimental way of life. Head had already been exposed to this culture through his experiences in Halle and Cambridge. The rigorous standards that Hering imposed on those working in his laboratory formed a culmination of that training. The end-product of such research was a publication – more particularly, publication in the form of a research paper, a genre that by the late nineteenth century had acquired most of its modern characteristics. Hering also supervised this aspect of the production of scientific knowledge at his laboratory. When he came to inscribe his own results in research paper form, Head told his mother, 'The Professor wrote out a small portion of the first section as he wished to have it written up to his Standard ...'.[176] While he clearly took the literary aspect of laboratory life very seriously, Hering was not oblivious to the comic aspect of some of the conventions of scientific prose. On one occasion, Hering and his son sent Head: 'a most amusing poetical effusion ... with an explanatory prose introduction, a parody on the physiological writings of the professor – He is always extremely accurate and cautious and every sentence contained "possibly" "it is not improbable that – " "we must guard ourselves from – " "Let us suppose" &c'.[177]

The Professor also set Head the subject for his research, which was an extension of an aspect of Hering's own work. Early in his career Hering had isolated what came to be known as the 'Hering-Breuer' reflex. This was a feedback mechanism mediated by the vagus nerve that regulated the inflation and deflation of the lungs during respiration.[178] It was, however, hard to demonstrate this phenomenon conclusively because of the difficulty of registering these effects: 'the change of tension within the lungs produced so much disturbance in the chest-wall and diaphragm that every method of registering the movements of the respiratory apparatus was thrown out of gear'.[179] Hering set his new disciple the task of solving this problem.

Head accordingly spent his first month in Hering's laboratory acquainting himself with the literature on the subject and testing existing methods of measuring respiratory movements. Then by his own account almost by chance it occurred to Head that he might create a more sensitive registering device by transferring the movements of two 'stout muscular strips' attached to the ensiform cartilage to a lever. Head used rabbits as his experimental subjects and had by this time entirely overcome his earlier squeamishness about engaging in vivisection. When he proposed his new technique to Hering, the Professor was, however, dubious; he had, Head recalled, 'an almost morbid horror of what I might call "chance experimentation"'. Evidently Hering viewed Head's idea as an example of such random testing. But he did not forbid Head from proceeding. The professor's disapprobation was instead signified by his electing to go out for an uncustomary morning walk while Head made his trial.[180]

The experiment proved to be a success. While Head was busy recording his first set of results, Hering returned to the laboratory. His 'eye caught the moving lever; hat and stick were thrown into the corner, and he sat on my stool with his eyes glued upon the smoked paper, whilst I showed him the effect of the various reflexes one by one'. Afterwards, Hering begged Head to curtail his lunch break so that the two could repeat the experiment in the afternoon. Replicating the experiment proved difficult. Eventually, however, Head's 'patience was rewarded, for on the sixth day I obtained a perfect result and from that time onwards never failed to make the new method do my bidding'.[181]

Head's account of his early success conveys something of the excitement of discovery. His pride in his achievement is apparent, as is his admiration for Hering's eagerness – his early reservations notwithstanding – to share in the experience. But within a few days Head was to encounter a less edifying aspect of scientific life. Through an intermediary he received a message from Philpp Knoll, the Professor of Pathology at the University, 'saying that my work must stop'. Knoll claimed that he had already been working on the same subject and accumulated a large body of data that he had been so far been unable to publish: 'because he was so much engaged with politics and with the duties of his professorship'.[182] If Head proceeded with the publication of his own results, Knoll maintained that he would be obliged to 'anticipate' them: that is, to assert that he had precedence in anything that Head claimed to have discovered.

Head was 'extremely angry at the threat'; he was not, however, intimidated. He had read Knoll's papers 'and felt no fear on this score'. He based his claim to priority chiefly upon the technique he had employed to explore the respiratory reflexes. Using the same intermediary, Head informed Knoll that: 'I had discovered a new method capable of giving me an insight into the regulation of respiration, which he could not obtain by any of the older means at his disposal…'.[183] The pathology professor's bluff was called. It is tempting to speculate that there may have been some political tension between Hering and Knoll. Head's *chef* had played a prominent part in the division of the University in 1881 and was strongly aligned with the German nationalist movement in Prague. Head went on to publish the results of his experiments in 1889 after his return to England.[184]

Hering evidently saw it as part of his role of mentor to try to guide Head through minefields of academic politics. Head recalled that his 'master, Hering' had in 1885 sent him to a conference in Strasbourg in order to meet 'a certain Professor Rosenthal' – the author of the theory of respiration that Head now proposed to challenge. Hering advised this move on the grounds that: 'if he has drunk a glass of beer with you he will not say you lie when your paper appears'. It was thus that the young 'beginner sought out the holder of the torch'.[185]

While in Prague, Head also undertook some experiments at Hering's suggestion on the positive and negative electrical variation produced in a nerve by stimulation using a new type of rheotome. The results of these investigations were published in 1887.[186] While his work on the respiratory reflexes had given Head sense of mastery over the phenomena, these researches produced a humbling effect. He confessed to his mother:

> My work has not turned out nearly so pretty as I hoped and all my theories have disappeared – However as I did not create Nature I can't compel her to conform to my ideas as to how she should act and must do my best to describe what takes place. I have discovered nothing but simply arrange old facts and show how they are connected – whereas at Easter I thought I had hold of a very pretty explanation.[187]

This show of modesty is especially interesting in view of the fact that Head's later work was often criticized precisely upon the grounds that he sought too hard to compel nature to conform to his ideas.

In 1900 Head gave Ruth Mayhew a rather more lyrical account of his mood as he came to the end of his first piece of original scientific research. He recalled that once his paper was complete he made the pilgrimage to Bayreuth in order to attend a performance of Richard Wagner's *Der Ring des Nibelungen*. He remembered these as: 'good days…with life before me and my first work finished but as yet unpublished'. At this 'time of joy', Head 'felt like a woman about to be a mother that the eyes of many friends & rivals are fixed upon her fruition &

who knows from the movements of her child that she will bring forth a lusty infant'.[188]

Towards the end of his time in Prague there are signs that Head was beginning to tire of the place. 'The quiet and constant life in the laboratory is most congenial to work but otherwise very demoralizing – Since I have become the only person of my acquaince [sic] in Prague I have lived the life of a Nazarite'. He had, he confessed to his mother, given up shaving and was 'bearded like a pard'. Head's linguistic practices had moreover strayed far from the plain speech befitting a Quaker; he was 'full of strange oaths'. To allay maternal alarm, he promised: 'Before you see me I hope to be shaved & in my right mind. His landlady, 'Frau Prof. Linke, says it is "männlicher" – but she always finds an excuse for any laziness or insanity on my part'.[189]

Head left Prague in 1886 to complete his medical training in England. However, he retained fond memories of the place. In April 1892, when returning from a vacation in Italy, he took the opportunity to pay a brief visit to his former mentor. In his journal Head recalled the 'grand discussion' he had with the Professor, as well as a meeting with Hering's protégé, the psychologist Franz Hillebrand.[190]

This appears to have been Head's final visit to Prague. But in June 1897 an event occurred that was to shake the foundations of the memory palace into which he had transformed the *Wenzelsbad*. While working in Prague, Head had befriended the Professor's son, Heinrich Ewald Hering (1866–1948). He remembered him as: 'a charming bright boy' with whom he had 'boated, skated, played football & discussed the latest Ibsen'. The young Heinrich was thus one of the fond memories Head cherished of those days. This image was, however, shattered when: 'an insignificant looking bearded man in a grey suit and one of those straw hats that is made up to look like a fort[?] turned up in my rooms'. The 1897 version of Heinrich disgusted Head not only by his appearance. Heinrich had followed his father by pursuing a career in medical science. He had come to London in the hope of promoting that career, and evidently set about it in a single-minded way that incensed his host. Head complained: 'He can talk no English and talks nothing but Shop in German'. Even when Head tried to turn the conversation towards scientific topics, Heinrich showed little interest. He had drawn up 'a list of the heads of the Profession in London and wants to call on the busiest and give them his views on locomotor ataxy'.[191]

The letter is important as giving an early hint of the contempt in which Head held medical men of a certain kind: namely, those who were so blinkered in their pursuit of narrow objectives that they were oblivious to the full possibilities of life. But his rage against Heinrich Hering was chiefly occasioned by the fact that he corrupted and betrayed memories from which Head drew much satisfaction. He complained that Heinrich: 'tells me nothing about his people or the old surroundings ...'. Instead he only wanted to talk about 'the time of the trains of the metropolitan'. It made Head 'very sad that what was once to me a flower should turn out such a weed ...'. Moreover, the discrepancy between memory and reality left him with 'an uneasy feeling that it may have been all a Klingsor's Garden'.[192] In other words, Head was now afraid that the entire edifice of his recollections of Prague, which evidently formed a valued aspect of his inner life, might be founded on illusion.

Head concluded his attempt at autobiography with a final accolade to Ewald Hering: 'I became', he insisted, 'his most devoted disciple, reading everything he wrote, and ready at any time to listen to his pithy conversation and wise aphorisms'.[193] This final instalment was written on 28 December 1926. It may be that Head thought this an appropriate point to conclude his recollections of his former life. We know, however, that the extant text is unfinished: Head intended to add sections on the Natural Sciences Club in Cambridge, in which he had been active that were never written, and on Gaskell's contribution to his scientific outlook. These were, however, never written. A likely explanation is that because of the debilitating effect of his Parkinsonism, Head found it impossible to pursue the project further. The remainder of Head's biography must thus be written without reference to a narrative of his own making.

2 THE POLES OF PRACTICE

The time Head spent in Prague appeared to confirm that he had taken Michael Foster's advice to abandon medicine in favour of a career in laboratory science. Upon his return to Britain, however, he seemed to reverse this decision and completed the stages necessary to qualify as a doctor. After taking courses in anatomy and physiology at Cambridge, he proceeded to University College London for clinical training. He found time to participate in the social life of the hospital, serving on the Executive Committee that organized a fancy dress ball and as steward at a festival dinner to raise funds for the institution.[1]

Having completed his studies, Head became a Member of the Royal College of Surgeons and Licentiate of the Royal College of Physicians in 1890. He received his MB and MD from Cambridge two years later. He served as a house physician in the obstetric and ophthalmology departments at University College Hospital before acting as clinical clerk to Thomas Buzzard at the National Hospital for Nervous Disease between December 1891 and October 1893.

Head's 'Autobiography' does not deal with this stage of his career and he seems to have left no other explanation for this apparent change of mind. One possible reason is pragmatic: in Victorian Britain there were few job opportunities in physiology. Although Henry Head senior was an affluent man, he had a large family and his children were obliged to make their own way in life. Medical practice offered his eldest son the promise of a steady income.

There may, however, have been other, more intellectual, considerations at work. Although ostensibly concerned during his time in Prague with the physiology of respiration, Head had through his interactions with Hering begun to take note of more psychological questions. In particular the mechanisms of sensation had begun to engage his interest. Investigations into these subjects could only be conducted on human subjects. Moreover, certain issues had to be addressed through the study of individuals whose sensations were impaired by disease or injury. In Gordon Holmes's words, 'Throughout the rest of his active career Head's work was mainly on problems capable of investigation by the observation of the symptoms of clinical disease, and to the solution of these problems he brought the physiological methods he had learned and used at Cambridge and

Prague.'² In other words, the volte-face was more apparent that real. He remained committed to addressing fundamental physiological issues, but required for this purpose materials available only to a clinician.

Although his clinical posts were always in general medicine, Head became a 'neurologist' – a term that had somewhat different connotations at the turn of the twentieth century than it does today. The 'Neurological Society', for instance, consisted both of doctors with a special interest in the diseases of the nervous system and laymen concerned with psychological and epistemological issues.[3] Head spelt out his own understanding of the nature and attraction of neurology in the Page-May Lectures on 'The Afferent Nervous System' he delivered at the Institute of Physiology at University College London in 1912. There he explained that: 'for the study of sensation ... man is necessary. For we must be able to hear from the percipient the consequences which follow stimuli – Disease is nature's experiment on man and in many cases the results are as precise & delicate as any laboratory experiment'.[4] But he went on to insist that to fully understand these questions, the investigator must be clinician as well as physiologist. In the middle of the nineteenth century, 'the physiologist and the physician were the same person'. More recently, however, due to the growth of knowledge, 'these two activities became specialized in separate individuals to the irreparable detriment of the two branches of Science. For the interest of the Physiologist became confined to those problems which could be settled in the Laboratory and the physician became frankly utilitarian'. Neurology, however, was an exception to this general trend; this branch of medicine alone: 'preserves the old alliance between the Laboratory and the Wards'. Head thus identified himself as 'a clinician but cases of disease will be considered simply as the means by which certain laws are demonstrated & certain physiological activities are laid bare'.[5]

This commitment to a clinical practice informed by physiological principles seems to have made Head something of an oddity when trying to establish himself in London. He applied unsuccessfully for a post at the Hospital for Nervous Diseases at Queen's Square – an institution that might have seemed the most obvious base for someone with his interests. When reflecting on the reasons for this failure, the eminent neurologist W. R. Gowers reflected:

> As far as I could discern it was simply that, whenever I spoke to Head about any subject (or rather he spoke to me) I always felt that I knew *nothing* about it & was only too glad to escape. I had little contact with him. I think I remember hearing, formerly, that those who did had an analogous feeling about practical matters. Every one recognizes his great ability.[6]

From this one can infer that Head appeared as a somewhat alien figure in the world of London medicine – even among other 'neurologists'. His idioms, formed in the laboratories of Cambridge and Prague, were not immediately intelligible to the established members of the community he now sought to join.[6]

Head, in contrast, saw his brief time at Queen Square as significant in shaping his outlook as a clinician. Indeed, retrospectively he maintained that it was this experience that convinced him of the relevance of the outlook of the laboratory scientist to clinical neurology. Head claimed that working at the Hospital for Nervous Diseases had taught him the necessity of approaching disease from a functional rather than from a structural point of view. In 1924 he: 'recalled vividly his own early experiences at Queen Square, where he had to revise many things he had previously learned'. Among his early encounters at Queen's Square was with his future collaborator, W. H. R. Rivers who was at the time (1891) house physician at the National Hospital. On one occasion Rivers enquired about what case Head had seen that day. 'Just a straightforward hysteria', was Head's reply; '"Then", said Rivers, "I am sure it is not"'. A chastened Head later realized that the case in question was, in fact, one of disseminated sclerosis. However, it required 'many such shocks ... before he was sufficiently humbled to look at the work in the wards from the old point of view of the physiological laboratory'.[7]

Holmes noted that the hospital posts that Head picked after he qualified could be seen as further preparation for the programme of physiological research in a clinical setting that he contemplated. These 'posts were obviously chosen for the purpose of gaining experience in those branches of medicine which he later made subjects of investigation on physiological lines, in fact throughout his active career the practice of medicine served largely to provide him with material for research'.[8] His stint as House Physician at the Victoria Park Hospital for Diseases of the Chest, for instance, provided Head with material for the study of the psychic changes that attended visceral disease he later published. While the time spent as a Clinical Assistant at the Rainhill Mental Hospital in Lancashire in 1895 served to broaden his knowledge of psychiatric disorders.[9]

It was while he was at Rainhill, however, that Head engaged in a curious diversion from the disciplined and planned progress that Holmes discerned. He became acquainted with Francis Henry Everard Feilding (1867–1936), a barrister and member of the Society for Psychical Research. Feilding tried to inveigle Head into his fascination with the paranormal. For instance he invited Head to join him in a 'ghost hunt' at a supposedly haunted orphanage in Pantasaph in Flintshire, claiming that: '2 or 3 days ago, ... the sister last in charge of the children at the house, heard the phantom coach drive up every night!'[10]

Head's response to this invitation is unknown. Feilding did, however, persuade him to take an interest in another of his preoccupations: the investigation of the alleged power of Roman Catholic shrines, such as Holywell and Lourdes, to produce supposedly miraculous cures. Feilding made elaborate arrangements – including procuring a letter of introduction from Herbert, Cardinal Vaughan[11] – for Head to take part in the examination both of pilgrims on their way to Lourdes and those who had purportedly been cured there. Head spent part of the summer of 1895 in France engaged in such investigations.[12] Head's curiosity in the topic occasioned surprize among the Catholic doctors whom Feilding approached on his behalf. As one bluntly remarked: 'The fact of his being a protestant will perhaps be an objection. Why should the question interest a protestant?'[13] Unfortunately Head left no explanation of why he was willing to expend considerable time and effort in this pursuit. After this brief excursus, he returned to his secular concerns.

Head's idealized notion of the clinic as an extension of the laboratory was not always realized among the exigencies of making his way in the medical marketplace. For practical reasons he was obliged to engage in a range of forms of medical practice, some of which he found tedious or even repugnant. The present chapter deals with Head's activities as a clinician. His attempts to turn these labours to scientific use will be addressed in the one that follows.

The Nerve Doctor

After qualifying Head spent the whole of his active career in London. His life as a clinician was organized around two principal poles: a private practice at Montagu Square and Harley Street in the West End and a hospital post in the London Hospital sited in Whitechapel. These topographic extremes, which Head traversed daily by underground train for many years, lay upon a steep social gradient.

In the West End Head practised as a consultant physician. This role required the acquisition of suitably impressive accommodation in which to meet patients. At the outset, his mother presented him with: 'the beautiful pencil sketch of Lord Lister ..., as I am sure you will like to have it at once for your consulting room'.[14] As a consulting physician, Head relied on general practitioners to refer patients to him whom Head would see for a fee. Ruth Mayhew in 1903 reported how the Oxford antiquarian Paul Oppé 'has been consulting more London doctors. The last said to him, "The man you ought to consult is Head".'[15] Oppé's problem appears to have been dyspepsia.

Head established, however, a reputation as a specialist in 'nervous' disorders. These could be organic complaints such as disseminated sclerosis. He was also consulted for more minor afflictions such as neuritis. Head was, for instance, often called upon to treat this condition in the right arm of lawn tennis players – experience that he put to use to diagnose and prescribe by letter for Charles Ashbee (1863–1942) who had been afflicted in the left arm.[16] 'Neurotics' were also directed to a neurologist of Head's kind.[17] Indeed, according to Macdonald

Critchley, Head's private practice comprised 'perhaps an unusually high percentage of neurotics'.[18] In these 'functional' cases there was no evident lesion of the nervous system; the symptoms of such patients were therefore often classed as 'neurotic' or 'hysterical'.

For example, in 1902 Head noted that: 'A young woman who had spent many years in training herself for a profession in which she had reached a lucrative & honourable position consulted me for a slight ailment'. In the course of the consultation it became clear to him that the ostensible reason for her visit was a mere pretext; upon questioning, the patient confessed:

> When at work in London surrounded by pupils and friends in the midst of all she valued and enjoyed most highly in life she was haunted by a feeling of homelessness so intense that on several occasions she had nearly committed suicide. Her mother lived in a small country town, old & feeble and entirely out of sympathy with any of her daughter's pursuits. Yet in spite of its intellectual dullness this, her only home, was the only place where she was entirely free from suicidal impulses.

On Head's advice, 'she gave up her profession and devoted herself entirely to the care of her mother's house – She lost most of her old interests but no longer feels the old feeling of homelessness and desire for suicide'.[19]

The distinction between 'organic' and 'functional' cases was, however, not always clear. In 1921 a girl was referred to Head 'as a possible Hysteric with pains in her arms and shoulder that no doctor could explain'. Because she was 'a perfectly strong, hefty, athletic girl', Head doubted this diagnosis. Instead, he proceeded:

> to test all the Headischen Zones and finds out she has a superfluous rib which presses on nerves and naturally causes pain. X rays prove he is right and a Surgeon whips out the offending rib. Miss came to see him last week perfectly well with such a neat tidy scar it would not offend the eyes of the most exacting Jew husband.[20]

This case shows how Head's research interests might serve in a clinical context. By the turn of the twentieth century, Head had indeed gained, by virtue of his research, a reputation as a neurologist with special competence in disorders of sensation. Thus he noted in 1902: 'Lately it has occasionally become necessary in private practice to make use of the results of my new work on the structure & function of sensory nerves'.[21]

This status was to bring him patients from as far afield as the United States. A cousin of the American pathologist and bacteriologist Simon Flexner (1863–1946) consulted Head in July 1919 – apparently on Flexner's recommendation – because of burning sensations in his lower back. Head performed an extensive battery of tests on this patient, including investigating his threshold to feelings of heat. Head finally diagnosed the condition as a form of 'paraesthesia'. He was able to assure Flexner that: 'None of the few cases in my records have had organic signs and although I cannot say I have been able to follow them up I have not heard that any of them has developed organic disease'. Head suspected that the case was, at least in part, 'functional' in character. He suggested: 'If possible I should like to see him take up some work again for he is terribly introspective. I should prefer some business that was not too responsible but was at the same time interesting'.[22]

Access to a consultant such as Head was confined to those who could afford his fees. The patient was, to use Nicholas Jewson's terminology, a 'patron' with expectations and demands that the doctor was obliged to heed.[23] H. H. Bashford in his, *The Corner of Harley Street: Being some Familiar Correspondence of Peter Harding. M.D* (1911) maintained that 'a practice consisting almost entirely of prosperous and middle-class patients' was for a young doctor 'perhaps the most difficult to manage'. This class of patient 'possesses an extremely accurate appreciation of the cash value of services rendered, and its consideration for a gentleman is by no means going to interfere with this when he comes before them as a salesman of physic and incidentally of advice'.[24]

Such patients would not hesitate to drop a practitioner who failed to give satisfaction. Thus after a number of consultations with Head, the artist Jacques Raverat, who was suffering from disseminated sclerosis, abandoned the arrangement because: 'he wd. only tell me to rest and I am anyhow spending 18 hours out of 24 in bed'.[25]

Moreover, even when the doctor gave satisfaction and the patient was wealthy, payment for consultation was not guaranteed. The supposedly 'hysterical' patient described above, who had been cured by surgery, returned to Head for a follow-up examination. Upon submitting a bill to her titled family, he received a note from their secretary saying, 'The visit ... was a friendly one only'. Ruth Head could not forbear from exclaiming: 'Can you conceive such meanness?'[26] Even as eminent a person as the German ambassador to London could renege. Prince Karl Max Lichnowsky 'bolted off when war broke out owing H.H. large sums of money for long series of treatments!' Head took this abscondence philosophically, remarking that: 'it was well worth treating L. gratis to have the fun of his conversation & stories!'[27]

In stark contrast to private practice, the clientele of the London Hospital were drawn from among the poorest inhabitants of the city. As a voluntary hospital, funded by charitable donations, it was intended to meet the needs of those who were unable to pay for their medical care, or who could only make partial payment. In particular, the hospital had a long-standing association with the local Jewish community whose numbers were swelled at the end of the nineteenth century by an influx of east European immigrants. These charitable hospital patients might be supposed to have been more tractable than those encountered in private practice. Head was, however, to discover that they presented challenges of their own. In addition to his clinical duties at the hospital, Head also taught students from the medical school and became active in routine administration and institutional politics. He was, moreover, active in a number of medical societies. In the interstices between these commitments, he endeavoured to pursue research.

In some of his letters, Head described a typical day oscillating between these extremes of practice. Thus 16 March 1900 comprised: 'A lecture at nine, a long day at the Lab followed by a series of troublesome interviews aimed at circumventing a woman who considers it her moral duty to cut her husband's throat'. His evening was then interrupted by an Irish doctor who insisted on engaging Head in lengthy conversation about the behaviour of his son.[28]

A day's practice could involve travel within and around London, with occasional excursions further afield. In this journeying, Head was in later years aided by modern modes of transportation. Thus he recounted to his wife how on 8 September 1907:

> At 12.30 came an urgent call to Earl's Court to see a poor wretched woman dying of morphia. Then to Clapham to see an alcoholic woman whose husband is teetotaler and trusted her entirely.
>
> Thence by rail cross country to Penge to see a little fool who believe she has a genius for the stage and has developed hysteria because she did not win the medal at her dramatic school. I had just time to get home by eight – snatch a hasty dinner and away in a taximeter cab to see a Jew who had gone mad and was filling the street from his bedroom with bursts of insane laughter. I went from one door to a mile beyond that hospital in 2 minutes arriving a few minutes before two instead the best part of an hour late. These cabs are wonderful and not really expensive.[29]

The morning of 14 March 1915 was spent at the home of a private patient at Tring where Head was offered breakfast. The meal was, however, interrupted by the patient's wife who insisted on talking about her husband: 'I don't think it was the desire of getting the last moment out of her doctor; it was pure kindness and sweetness – but she could not keep off the man who fills her life'. Upon his return to London he was met by his manservant at the railway station, and:

> So to the Hospital where the work was prodigious and will become more so later. Then I rushed back to see two patients had a quiet dinner and off to the Neurological Club. It was an amusing evening – not much work but a great deal of talk and wagging of wise heads over the times and the slackness of our colleagues.[30]

On occasion Head was also called upon to serve as an expert witness. His evidence was, for instance, required in a case where the testamentary capacity of an individual was in question.[31] In June 1923 he also testified when a patient of his filed for divorce on the grounds of the cruelty and adultery of her husband. The woman had consulted Head two years previously. Head told the court that: 'She was extremely agitated and very nervous, and was in a general state of irritable weakness'. The judge seems to have taken this evidence into account when making a decision. Before granting a decree *nisi*, he remarked: 'I cannot doubt that a calculated course of treatment such as is shown here is bound to have had an adverse effect on the health and nerves of a person such as the petitioner'.[32]

Such experiences occasioned reflections about the incommensurability of legal and medical discourse. On 6 March 1901 Head recorded,

> I have been in Court all day with a very pleasant Judge (Bucknill) and have learnt a great deal. As usual our Profession came out extremely badly. Medical men have so little power of expressing themselves that they are quickly flamed by expert barristers. Then again vague words like 'paralysis' which means nothing to the scientist represents everything to the lay mind that does not recognize that they are vague and ancient terms.[33]

In the course of his career Head would have seen thousands of patients. He kept careful notes of at least the most scientifically 'interesting' of these; his office was lined with files containing these records. Head bequeathed this archive to his protégé George Riddoch (1888–1947). Following Riddoch's premature death the files appear, however, to have been lost. The historian is therefore left with only occasional insights drawn from miscellaneous sources into the patients Head treated. These tend to give the patient's perspective on the encounter.

Head's literary interests made him especially visible to those who moved within artistic circles. On 31 August 1928, for instance, Virginia Woolf, recorded:

> Morgan [Forster] was here for the week end ... One night we got drunk, & talked of sodomy, & sapphism, with emotion – so much so that next day he said he had been drunk ... Morgan said that Dr. Head can convert the sodomites. 'Would you like to be converted?' Leonard asked. 'No' said Morgan, quite definitely.[34]

Even if E.M. Forster was not tempted to make use of Head's expertise, other members of the Bloomsbury Group did seek his help. For instance Head was one of the doctors Virginia Woolf consulted in 1913 when she suffered a nervous breakdown.[35] Roger Fry turned to Head to treat his wife Helen when she became mentally ill.[36] Head was also the 'neurologist' entrusted with the care of Janet Ashbee when in 1908 she suffered a breakdown.[37] She evidently appreciated his help through this crisis, writing twelve years later: 'nearly all the sense & balance I have, I have learnt from you--& so I am always your faithful ex-patient'.[38] Head's attempts to aid Roger Fry's wife, Helen, were less successful. After all attempts at treatment had failed, Head was forced to concede: 'Unfortunately the disease has beaten us'.[39] Helen Fry spent the rest of her life in an asylum.

Head gained at least one well-known patient through his wife's literary connections. Ruth Head edited a compilation of the writings of Henry James, which led to some correspondence with the novelist.[40] In November 1912, when James was ill following a stroke, Ruth recalled that: 'H.H. went down to spend a weekend professionally with Henry James at Rye ...'[41] On the eve of the visit, James thanked Ruth for 'the beauty and generosity of Mr Head's so benevolent proposal', and offered to accommodate Head during the visit. James seems to have been struck by the fact that his erstwhile editor should prove to be married to an eminent neurologist. He remarked:

> what is wonderful and interesting indeed, what is, as you say, miraculous, is the concatenation of circumstances that, beginning with your beautiful letter to me of several years since, will have gone off in a day or two to the grand climax of Mr. Head's arrival. We must do all justice to the Miracle.[42]

Head also figures – albeit anonymously – in the *Journal of a Disappointed Man* written by the naturalist Bruce Frederick Cummings (1889–1919), also known as 'Barbellion'. On 30 April 1913 Cummings recorded that he went to see: 'a well-known nerve specialist – Dr. H –. He could find no symptoms of a definite disease, tho' he asked me suspiciously if I had ever been with women'. Cummings' account of the examination he endured from Head has a farcical tone:

> H – chased me round his consulting room with a drum-stick, tapping my nerves and cunningly working my reflexes. Then he tickled the soles of my feet and pricked me with a pin – all of which I took like a man. He wears a soft black hat, looks like a Quaker, and reads the *Verhandlungen d. Gesselschaft d. Nervenartzen*.

At the end of this ordeal, Head's sole prescription was that Cummings should take: 'two months' complete rest in the country'.[43]

Like Raverat, Cummings was diagnosed as suffering from disseminated sclerosis. Another account of the same consultation, by Barbellion's brother, Arthur J. Cummings, gives an altogether different impression of the proceedings. After several doctors had seen and 'tinkered' with Barbellion, Arthur finally imposed upon him to consult 'a first-class nerve specialist. This man quickly discovered the secret of his complex and never-ending symptoms'. Head apparently told Arthur that his brother was: 'a doomed man, in the grip of a horrible and obscure disease of which I had never heard'. Although Head told Arthur the name of this disease – and suggested that it had a bacterial cause – he advised against revealing the truth to Bruce. By such a revelation, 'we shall assuredly kick him down the hill far more quickly than he will travel if we keep him hopeful by treating the symptoms from time to time as they arise'.[44] According to Arthur Cummings,

Head also gave Bruce's fiancée a stark account of his prognosis to enable her to decide whether or not to proceed with the marriage.[45]

Such cases might be seen as falling naturally within the remit of a 'nerve doctor'. It is apparent, however, that both in hospital and in private practice Head dealt with a much wider range of conditions. On occasion, he ventured into the field of obstetrics. When C. R. Ashbee's daughter Mary Elizabeth suspected she was pregnant she went to Head for advice.[46] Head was a friend of the family and had evidently known her from an early age. After the child was born, Head wrote to Mary Elizabeth in uncharacteristically effusive terms: 'No event has given me, for a very long time, such complete pleasure as this coming into this wicked world. Thee canst not know with what anxious attention I have studied the signs which anticipated the arrival'.[47]

In his letters to Ruth Mayhew, Head made a determined effort to withhold information about his working life, feeling that such details might sully their correspondence. Occasionally, however, he felt moved to share a particularly poignant experience. On 23 February 1900 he told Mayhew:

> I am attending a poor fellow who has been an engineer in America & is now lying in a top room in the East End with Locomotor Ataxy. I went to see him on one of those bitterly cold days of last week and found him in bed in his little room overlooking the ramshackle tiled roofs and irregular chimneypots, no fire in the grate curled up like an injured animal – There was not a book in the room and it had an uncomfortable late in the morning look instead of looking half sitting room half bedroom as a cheerful sick place should – It has brought that room up five flights in the Avenue Matignon home to me with a vividness that is painful in its intensity.[48]

Significantly, what seems to have led Head to recall this case in particular was the strength of the impression that it made upon him – together with the literary association that it invoked.

In a similar way, Head related his futile fight to save a sick London Hospital colleague in terms of vivid images that he took away from the event: 'An attic where the huge cross-beams of the roof ran from ceiling to floor, with a skylight & a tiny gable window. Here four days & four nights we fought death for a man after my own heart'. The patient had realized that his case was hopeless and had made his mother promise to allow an autopsy: 'It was horrible'. Head made clear that he did not find this scientific detachment horrible in itself; it was only 'horrible in the presence of those quiet godfearing commonplace people'. He gave Mayhew an evocative account of his departure from distressing scene:

> As I turned into the Mile End road the East was rosy pink – The west was hung with grey that softened the outlines of roofs & streets like a veil. Lamps glowed white, orange & red in the empty streets – points of colour and not of light in grey dawn. The plane trees shivered and I wrapped my cloak close round me.[49]

Again Head was anxious to express in words the full range of sensory and emotional impressions he had derived from what had clearly been a deeply moving experience.

In such passages Head articulated an idealistic, if tragic, vision of the nature of medical practice, as well as a heroic image of the medical calling. In July 1912 he quoted with approval 'a beautiful speech' delivered at the London Hospital by Lord Roseberry. Roseberry had spoken of: 'the duty of the physician to lead a perpetual 'forlorn hope' and to continue fighting the Angel of Death up to the End'. Head held that this was 'wonderfully true and explains why some men never seem to help even in obvious cases where help can be given. They are not ready to fight to a finish'.[50]

After his death, Arnold Mayhew tried to analyse why Head was a physician who *was* determined to fight to the finish:

> There must be hundreds that owe him health or happiness – or the chance of a start on the right lines. But his reason for helping them was not 'thou shalt love thy neighbour as thyself'. He helped them because he hated waste and wanted to see every one 'functioning normally'.[51]

Mayhew was a clergyman who lamented that Head had divorced himself from ethics based upon any form of religious belief. Instead Head appeared to be guided by a naturalistic ethical code that viewed human beings as corporeal entities that could function more or less normally. The duty of the physician was to maximize the efficient operation of every body that came under his care. A neurologist like Head was often also required to have care for his patient's mind – as Mayhew knew at first hand. He recalled:

> Just after I left Oxford I nearly broke down through overwork in unorganized and inappropriate surroundings. He saved me from that and got his brother to take me on a tour of India. But as he carefully explained he did it because I was wasted where I was – and he hated seeing energy misplaced.[52]

However, Head drew no sharp distinction between these two branches of medicine, seeing the psychic and physiological as aspects of a biological whole, governed by analogous principles. Health and happiness, in Head's worldview, were therefore inseparable facets of an efficiently functioning organism. This conviction grew towards the end of Head's active career. In an address to the London Hospital Medical Society delivered in March 1922, he complained about the growing division of labour in medicine between those who purported to treat mental and somatic disorders:

> the treatment of the functional neuroses has become a special branch of medical practice carried out by men who see comparatively little of organic disease. At the same time the general physician is scarcely familiar with the psychical aspect of medicine;

he and his colleague, the surgeon, rarely consider how large a part the mind plays even in the symptoms of gross structural disease.[53]

These reflections on the interface between functional and organic disorder led Head to perhaps his most explicit statement of the ethic that should govern clinical practice. He exhorted his fellow physicians to remember:

> Nature's moral code, under which we work, is cruel and unrelenting. There is no forgiveness of sins; but, in the medical man, this knowledge should be tempered towards the patient by clinical curiosity and human sympathy. In conclusion, I would say to all who have to deal with these morbid conditions, be as honest in thought as you would be naturally in deed. Act without fear and never lose courage; finally call nothing common or unclean.[54]

The Financial Aspects Of Practice

Medicine possessed, however, a severely pragmatic side. Head's private practice was a business from which for most of his career he derived the bulk of his income. For a successful West End consulting physician, these rewards could be considerable. But Head expressed disdain for doctors who became preoccupied with the riches that a fashionable practice could yield. He wrote in 1898 of an encounter with 'a colleague & successful practitioner who insisted on sharing my journey'. This unwelcome companion 'sniffed at my country clothes and said the neighbourhood of my Laboratory was a nice one "for it was just sufficiently far away from London to get a good fee"'. This obsession with 'the financial aspects of practice' led Head to the fervent plea: 'God forbid that I should ever be successful if success means such thoughts'.[55]

Head professed to be more concerned with the scientific interest of a case than with the pecuniary potential of a patient. Thus in 1902 in one instance, 'where the patient afforded an excellent example of my new law [of the structure and function of the sensory nerves] the fee seemed almost an insult. I had much rather that he had given me simple thanks like my hospital patients'.[56] On a visit to London, the American surgeon, Harvey Cushing (1869–1939), appeared to confirm Head's indifference to fees. Cushing noted that his wife was: 'much amused at Head's lively answering of the telephone: he being totally uninterested in the guinea'.[57]

Head could not, however, be entirely indifferent to money matters. Indeed, his family background had endowed him with some expertise in the world of finance. Ruth Head remarked in 1916 that her husband was prone to 'abound' about the intricacies of investment: 'of course it is in his blood and [he] loves to be very clever about War Loans and Exchequer Bonds and such like'.[58]

While his own needs might be modest, Head was prevented for several years from marrying Ruth Mayhew because he could not afford to support her (see Chapter 4). In January 1904, shortly before the marriage finally took place, Head drew up a meticulous record of his finances, which showed how fine the calculation had been (see Table 1). His private practice yielded £600; his hospital work provided a further £100. This income was supplemented by money – presumably a dividend – derived from his father's firm. £600 per annum was not an impressive income for a London consultant. Head blamed this relative poverty chiefly upon the time he devoted to scientific work that was not merely unremunerative, but also the source of additional expense, costing Head around £66 per annum.

TABLE 1

PP/HEA/D4/17, CMAC

Estimated Income for 1904

Professional	600.0.0
Hospital	50.0.0
Lectures to class	50.0.0
H.H. & Co. Ltd.	360.0.0

	1060.0.0

With a possible bonus of £50.0.0 in addition from H.H. & Co. Ltd. Maximum total income with bonus £50   £ 1110.0.0.

Estimated Expenditure for 1904

Fixed professional Expenditure of H.H.	311.7.0
Estimated personal cost for two persons.	350.0.0
Estimated cost of 6 Clarence Terrace	655.15.0

	£1317.2.0.

Estimated total cost if living at Flat

Fixed professional	311.7.0
Estimated personal cost for 2 persons	350.0.0
Cost of Flat	350.0.0

	£1011.7.0

Expenses

Professional

Rent of 143 Harley Street	£175
Telephone	5.0.0.

	180.0.0

Research

Wages of boy at 8/.	21.0.0.
Keep of Animals	25.0.0.
Incidental Expenses Such as Illustration	15.0.0.
Laboratory material	5.0.0.

	66.0.0.

Fixed Personal Expenses

Scientific Societies & Savile Club	11.7.0.
Insurance	44.0.0.
Season Ticket	10.0.0.

	65.7.0.
	311.7.0.

Estimated Cost of 6 Clarence Terrace.

Fixed charges

Income tax, Rates, Water rate,	115.5.0.
Ground Rent.+ Sinking fund policy	say 55.0.0.

Lighting & heating

    Electric Light   30.0.0.
    Coal   21.0.0.
    Gas   9.0.0
        60.0.0.

Wages.

    Cook   25.0.0.
    Housemaid   18.0.0.
    Under housemaid   12.0.0.
        55.0.0.

Washing.

    Personal & house   40.0.0.
    Servants   10.0.0.
        50.0.0.

House Expenses

Painting, Repairs, &c.   120.0.0.

Food.

    Baker   15.0.0.
    Milk Butter & Eggs   20.0.0.
    Butcher   71.0.0.
    Green grocer   32.0.0.
    Grocer   65.0.0.
    Fish &c.   52.0.0.
        255.0.0.

After their marriage Ruth Head was able to contribute to the running of the practice as an unpaid receptionist. On 2 December 1905 she announced:

> Well! I have secured two patients for my little man. No.1 an old patient, Mr Gregory who begs for 10.30 on Monday. That has been promised conditionally. You may settle otherwise on your return, which the gods hasten propitiously! No. 2. A new patient, a Mr Claplin, sent by Dr Stride, to whom I have höchst eigenhändig given 12.30 on Tuesday, because your new Miss Savile is coming at 11.30. Was I right? Praise me.[59]

With the outbreak of the First World War, Ruth appears to have taken an ever-greater role in the running of the practice. Moreover, she took over from her increasingly harassed husband the task of worrying about the household finances. In October 1915 she wrote to Janet Ashbee: 'I have been writing lots of cheques and <u>all</u> the patients are gratis!'[60] Two months later she lamented: 'People like me know what it is to have had to live on nothing a year and keep tidy'.[61]

These concerns became more acute when in 1916 Head decided (apparently without consulting his wife) to withdraw from private practice in order to concentrate on his war work. Writing to Harvey Cushing, Ruth conceded the pressure her husband had been under:

> Since September he has never had one half-hour to himself and it has been horrible. Now the great sacrifice is made he is as jolly as a school boy home for the holidays and he promises himself a gorgeous year of solid work for his poor and his wounded.[62]

Head was supposed to be paid a salary for some of the war work he undertook for the government; but 'he has not had a sou yet and is too proud to ask – so silly of him'.[63]

Ruth's pleasure in this improvement in her husband's mood was however tempered by fear of its financial consequences. She told Janet Ashbee:

> You cannot think how hateful it is to sit down and write letter after letter to would-be patients saying Dr Head regrets – etc. Then too it is quite maddening to hear Harry night after night at the Telephone quietly turning off patients to the other neurological doctors who will fatten on his leavings and never even know how they get their rush of new patients.[64]

Matters were made worse by the fact that Head seemed oblivious to these pecuniary pressures. Delighting in his newfound freedom, he gaily invited people to lunch and dinner leaving Ruth to find the money for such entertainment. Even in these straitened times the Heads did not go hungry. Their manservant, 'the admirable Pratt takes all the burden of shopping off my mind and Harry and I feed like Princes'. They were, however, in 1918 obliged to curtail their holiday plans because of their 'empty exchequer'.[65]

Matters improved somewhat after the War. For a while after the Armistice, Head was 'blissful, [he] writes all the morning, goes to Committees and Hospitals in the afternoons and writes all the evening'. What made this idyllic lifestyle possible were the regular dividends he received from his father's insurance firm, Henry Head and Co. In 1904 these had amounted to £360 a year. During the War this flow of income had increased. Ruth worried at what would happen when the level of dividends returned to a more normal level. She earnestly hoped: 'he will ... begin to practice again with fewer people and with higher fees!'[66]

Soon Head was indeed again 'actively earning our daily bread'. When the Heads felt the need to move into larger premises, they were obliged to find 'a suitable abode in the old neighbourhood, which was a necessity for the practice'.[67] They therefore simply moved into a larger house in Montagu Square. To help in the running of his practice, Head now employed a 'snub-nosed but invaluable Secretary'.[68]

These remained, however, what Ruth called 'dreadful days of Peace and Penury'.[69] Coal was scarce and prices high. She was, moreover, outraged at the post-war rate of income tax her husband was obliged to pay: 'first instalment £997!! Pretty isn't it? And me with just 15/ balance at the Bank and lots of debts – too odious'.[70] Head's reluctance to accept the Directorship of a proposed new research Unit at the London Hospital (see below) derived at least in part from a disinclination to accept the financial loss such an appointment would entail.

A Steadily Paralyzing Coil

We have seen that Head embarked on his career with an ideal conception of medical practice: one in which clinical experience would inform basic physiological research and vice versa. By the mid-1890s, he was to find it increasingly difficult to sustain this model in the face of the reality of the daily demands made upon him. The hospital side of Head's practice was to prove particularly trying for him.

Through his long association with the London Hospital, Head became part of the collective memory of that institution.[71] Russell Brain collected some of these reminiscences after Head's death. Thus Henry Howard Bashford (1880–1961) recalled:

> I was a young dresser of Henry Head's and remember him as a rather tubby teutonic-looking figure with a beard, wearing a very tight frock coat and, as I rather think, a turned-down collar of a non-conformist minister type. Every now and then, when he got excited, his voice became falsetto, as when he would demonstrate something 'typ-ic-al, typ-ic-al'. This became a sort of ragging slogan with us, but he, himself was never ragged, we all instinctively recognized him as a great man.[72]

Bashford also noted that Head 'was a man who attracted legends to himself'.[73]

Donald Hunter, a Physician to the London Hospital, compiled a larger collection of such anecdotes, often of a humorous nature. In one of these, Hubert Maitland Turnbull remembered:

> I had the good fortune when first going to the hospital to meet daily in the mornings on the steam-engined underground railway Dr Henry Head. He told me to buy Gee's little book on percussion, and kindly taught me throughout our journeys about physical signs, much to the annoyance of our fellow travelers; indeed in his characteristic keenness he spoke so loudly that as we walked to the hospital from St. Mary's station (now closed) people on the other side of the wide Whitechapel road would turn to look at us.

Despite such eccentricity, Turnbull maintained, Head 'was a great man with a boundless and infectious enthusiasm, and a great and inspiring teacher'.[74]

Given the acrimony that often obtains between medical men, there are remarkably few negative representations of Head. Another London Hospital colleague, the bacteriologist William Bulloch, was among the few who seem to have actively disliked Head. Bulloch took a pleasure in deflating what he saw as Head's excessive self-regard. Donald Hunter reported:

> Returning from F. J. Smith's funeral Henry Head in superb conceit said 'Remarkable man, F. J., he always liked me!' But Bulloch says he had heard F.J. say of Head 'I hate all of him, from his black trousers to his great belly to his top hat, blast him'.[75]

The overall picture that emerges from this body of anecdotes is, however, of an engaging, enthusiastic – if slightly bumptious – figure who was generally viewed with affection and whose intellectual endeavours were respected, if not always comprehended. This is reflected in a caricature that appeared in the London Hospital magazine that, as is the wont of medical students, identified and made gentle fun of one of Head's verbal traits: his fondness of the word 'typical'. Ruth Mayhew's brother encountered a student from the London who epitomized this view of Head: 'We think all the world of him ... he is always there and always keen. Of course we feel he is often chasing butterflies, but then we know with work like ours that must be: it makes us believe in him'.[76]

One particular story, in various versions, figures especially prominently in the mythology surrounding Head. It demonstrates both his reputation for devotion to science and the somewhat comic character ascribed to him:

> A Jewish patient had an obscure disease of the spinal cord and Henry was very keen on getting a p[ost].m[ortem]. He watched the case until death but the people wouldn't agree to a p.m. even with bribery. He conspired with Sherren to go to the mortuary and remove a small portion of the spinal cord. He got hold of a porter who was an alcoholic and took down a lantern. They dared not turn on the gas for fear of being discovered. Sherren got to work and suddenly heard a crash. The porter had fallen down in an epileptic fit. Henry glanced at him, said it was typical epilepsy and pushed something into his mouth. Ultimately, Sherren had to go and as Henry gathered up his bits and pieces he found suddenly that the porter was advancing towards him balancing a p.m. knife as if to strike. The whole plot was revealed as Henry dodged round the two tables in a figure of eight shouting for help. 'Typical, typical. The post-epileptic automatic state'. (Another version has the actors in this story locked in the p.m. room for fear of being discovered.)[77]

Some contemporary private accounts by hospital staff of Head's persona, however, depict a somewhat opaque and enigmatic figure. When in December 1904 Geoffrey Slade consulted Head about the advisability of applying for a post at the London, he received an apparently encouraging response. But Slade felt moved to add: 'H.H. has always been rather inscrutable to me: and I can't be at all sure what he really thinks'.⁷⁸

Head's own writings show him taking a pride and active interest in aspects of his work at the London, especially that related to the training of medical students. In January 1903 he wrote to Mayhew: 'You know that our [London Hospital] Medical Society in its present form is my child'. He recounted a session of the Society at which he had demonstrated cases of General Paralysis of the Insane, which had cost him considerable effort, but from which he derived great satisfaction: 'All day we worked like the stage managers of a play on the first night and we had a wonderful meeting. The room was packed with 110 students (in the old days I have presided over a meeting of 6 people)'. He returned home around 11p.m., 'worn out, horribly hungry but very happy. It was fine to see the enthusiasm of the men as this most difficult subject unrolled itself before their eyes. It would reconcile me to any sacrifice of time and energy to think that in any way this body of enthusiastic men were my spiritual children'.⁷⁹

But even when teaching these 'spiritual children', Head sometimes found his mind wandering. In the 'Rag Book' of miscellaneous extracts and reminiscences he compiled with Mayhew (see Chapter Four), Head recalled a dissociative episode that occurred during an auscultation class he gave at the Hospital:

> I am now taking a class in the elements of clinical medicine. I teach them to examine the chest and to use the Stethoscope. This morning attempting to explain to them the characteristics of an abnormal sound heard in a diseased chest I likened it to the wind in the trees.
>
> At once I was transported to Buckingham [Head's holiday retreat in Sussex] and it was night. Huddled together I saw the outline of the stables and our houses – Black against the cold blue sky rose the bare limbs of the elms and the impenetrable soft blackness of the great [?] trees – I felt the cold night upon my face and drew a deeper breath. Over the silence of all living things I heard the soft high-pitched sighing of a lazy wind through the evergreen trees.
>
> Then one of the class asked the meaning of this sound and I woke.⁸⁰

Elsewhere Head wrote to tell Ruth Mayhew how: 'when Saturday evening comes round and the strain of the four hours teaching is over – whether I have taught well or badly – whether I have made shameful mistakes or narrowly escaped hideous pitfalls is all for one for the remainder of the day is to be devoted to you'.⁸¹

This was an early hint of Head's growing resentment at the pressure of routine work that was required of him at the Hospital. His letters indeed reveal an emerging sense of alienation from the forms of practice to which he was obliged to devote so much of his time. This discontent derived in part from the urban environment in which he was obliged to work. Occasionally, he might find relief from this sense of oppression by focusing on such aspects of the natural world that were still manifest in the city:

> I have longed to sing, the last two days, on my early journey to the Hospital; for overhead the sky was coldly blue with stationary flocculent clouds – The crocuses are out in the scanty remains of our hospital garden and the lilacs in the square have thrust out long light green groups of leaflets from their black stems – No words came to me but a constant underlying lilt destroyed all comprehension and all fascination of Dr Weir Mitchell's 'Injuries of Nerves', the constant companion of my journeys of late.[82]

But for most of the time, the squalor of the East End was an inescapable weight upon his consciousness.

He became in particular fixated upon the Jewish inhabitants of Whitechapel. The Jewish population of the area had been swelled during the 1880s by an influx of Russian Jews fleeing from tsarist persecution.[83] Head's protégé, Redcliffe Nathan Salaman (1874–1955), noted that: 'Stepney and Whitechapel were rapidly converted into solid Jewish Ghettos whilst dominating and yet serving them stood the great London Hospital'. Salaman estimated that around 50% of the patients attending the London were drawn from this population, 'most of whom could hardly speak English, whilst many by their dress and manners declared their foreign origin and alien habits of life'.[84]

The prominence of Jews in the patient population could lead to difficulties when religious sensibilities clashed with the practices of clinicians. In 1904 Salaman received a request from the office of the Chief Rabbi that Jewish 'watchers' should attend autopsies, 'as complaints had been made ... that, during the examination organs etc. had been removed from the body and were not buried – a circumstance that gave great umbrage to the poor of the East End'. Unless such concerns were addressed, calls would grow for the establishment of a 'specifically Jewish hospital' in the area.[85] In the anecdote cited above Head had apparently sought to circumvent such resistance by conducting a clandestine autopsy on a Jewish patient without the consent of the family.

While in Halle, Head had already shown an interest in Jewish physiognomy. He had, moreover, during his stay in Bohemia ascribed to Jewish moneylenders a role in the downfall of the *Koenigbauern* (see Chapter 1). Despite these signs of prejudice, he expressed a high regard for some Jews – notably Salaman. Salaman was, like Head, a Cambridge man who had been inspired through his acquaintance with Walter Gaskell. Head clearly viewed him as a kindred spirit. He exulted in having 'a Colleague who being a Jew has all that astonishing love of beautiful things that is so common in his race ...'[86] Salaman, for his part, considered Head: 'a most intimate and treasured friend of myself and family',[87] as well as an invaluable professional mentor. He named his first son 'Myer Head Salaman'. Head was moved by the gesture, writing: 'my proudest joy would be to think that you and your wife held me worthy to be associated with one so precious to you as your first born'.[88] Head was also godfather to another of Salaman's children.

But the East End Jews that Head encountered in and about the hospital evoked different responses. Sometimes these embodied an element of compassion and respect, and even identification. One Friday evening when riding home,

> from the Hospital on the top of an omnibus my eye was caught by a wretched first floor front room in Whitechapel – A clean cloth was spread upon the table and the candles in the branched candlesticks were lighted. The candlelight misty from the dampness & grime of the windowpane fell upon an old Jew sitting at the head of the table – His long beard was grey and a skull cap covered the bald crown of his head. There seemed something regal in this old man seated in silence in the candlelight and my heart went out to Moses Lump on the one night when a Gentile can envy his repose.[89]

The fact that he could associate this figure with a character drawn by his favourite German poet, Heinrich Heine, no doubt contributed to Head's sympathetic reaction.

On other occasions, however, Head turned a cold, quasi-ethnographic, gaze on the Jews he encountered. Thus he asked: 'Why is it that a Jew baby is so much more characteristically jewish [sic] than a boy or girl of the same race?' The question was occasioned by 'a tiny jewess of about 4 years old' whom he frequently passed on his way to the Hospital. Head found the child's mode of locomotion particularly fascinating: she 'waddles along as only Jewish women waddle'.[90]

Head showed a similar detachment when observing what others might have found a harrowing episode he observed while working at the London. He noted in his 'Rag Book':

> My laboratory looks out upon a back street entirely inhabited by Jews. Yesterday work was interrupted by a noise resembling the howling of an animal caught in a trap. Shortly a middle aged Jewess appeared, supported by a younger woman on either side, her arms flung about their necks, walking slowly. Suddenly she stopped arched her back and throwing up her chin gave vent to a long drawn howl. A few steps further and this was repeated. Behind at a respectful distance followed a crowd of Hebrew women of all ages, wringing their hands, weeping & sobbing in irregular accompaniment to the rhythmic lamentation of the mother of the dead child. For a child has been run over, taken to the Hospital & pronounced to be dead. And this was the return of its mother to her home.[91]

A similar incident occurred in June 1915 after a German bomb landed near the London Hospital. Ruth Head recounted how: 'A Jew mother so outdid Rachel mourning for her children that she roused every single person in the Hospital, including a poor wounded soldier of Harry's who has such pain in his leg, his moments of sleep are worth their weight in gold'. In this case, however, when the repose of one of his soldier patients was disturbed, Head no longer showed his previous calm detachment: 'Harry was so angry'.[92]

Head's ire and contempt was especially excited when Jews presented themselves as patients at the London. In November 1900 he complained to Mayhew:

> The world in general is entirely snobbish – if you succeed in what you do it will fawn on you – if you are unfortunate, no matter what your merit, it will trample on you ... The world in general is singularly like my East End Jew patients – Some little while ago a Doctor in the East End sent a Jew & his son to see me at the Hospital. I took a great deal of trouble and had much bother with them but felt amply repaid by my interest in the disease. Shortly after a young neurologist, a friend of mine, showed a case of this disease at our society and I found that it was the one I had taken so much trouble over. As usual the father had had his boy under two general practitioners, two consultants (gratis) and a hospital. I rated the whole family in my best Yiddish and thought I should never hear of them again. But one of the children fell ill – The Father sent for none of the people who had treated the boy & expended time & trouble on him for nothing but insisted on having me down in full state for a Consultation – I had insulted him and I was the only one he believed in sufficiently to make him part with golden guineas.[93]

Head came, in effect, to scapegoat Jews for the growing sense of angst with the realities of his professional life. In September 1902 he lamented the loss of his idyllic Claybury laboratory (see Chapter 3). In consequence, he was forced to conduct his research as well as much of his clinical work at the Hospital:

> Now my laboratory looks out on a street and windows must be closed to keep out the dirt. It smells of hot air and gas ... After a long day the moon would shine behind the young poplars and whiten the newly mown grass behind the Station [at Claybury] – Now it is the first hours of the Jewish Sabbath and the streets are alive with a parasitic race like a sleeping dog with fleas – [94]

A further sign of Head's growing disillusionment with his working life came in 1899 when he wrote to Mayhew:

> the week preceding this little holiday had been rather a bad one – I had had that horrid but salutary sense that time was flying that comes to me occasionally. First comes the feeling of how much I have set out to do which if only half done is worthless. Then a consciousness of the increasing demand on my working hours made by money getting and teaching with a sense as if one was being gradually wound up in a steadily paralysing coil.[95]

The happy synergy between clinical work and research that Head had imagined was thus replaced by a growing flood of routine cases and teaching responsibilities that threatened to overwhelm all creative activity. In October 1900 he complained that: 'During the last two months I have had 80 beds at the Hospital, my private practice and the paper to bring out and throughout this time I have not made a single observation'.[96] He even began to question the value of his scientific vocation. Head confided to Mayhew that, in moments of exhaustion, 'I sometimes wonder if it is worth all this giving up. Ever since I was 20 I have perpetually given up harmless healthful & beautiful pursuits at the nod of my mistress'.[97]

In his 'Rag Book' Head quoted a letter from John Ruskin to Henry Wentworth Acland (1815–1900), describing the routine of the Oxford physician and professor of medicine:

> you never were able so much as to put a piece of meat in your mouth without writing a note at the side of your plate – you were everlastingly going somewhere and going somewhere else on the way to it – and doing something on the way to somewhere else, and something else at the same time that you did the something – and then another thing by the bye – and two or other things besides – and then wherever you went, there were always five or six people lying in wait at corners and catching hold of you and asking questions, and leading you aside into private conferences and making engagements to come at a quarter to six – and send two other people at a quarter-past – and three or four more to hear what had been said to them, at five-and-twenty minutes past – and to have an answer to a note at half-past – and so on'.

Head attached the weary comment that this letter: 'gives a vivid description of the life of a busy consulting physician'.[98]

In this context he sometimes perversely came to identify with the very Jews that he on other occasions despised. 'Like Heine's Moses Lump', he wrote in April 1900, 'I go about all the week in wind & rain with my pack on my back poking about amongst the filthy rags of humanity to earn a few shillings ...'. His weekend meetings with Mayhew provided precious relief from this drudgery; but 'the stain of the ghetto cannot be wiped off in an hour ...'.[99]

It was an exception to the rule when Head was able to enjoy 'a leisurely morning seeing an interesting patient & able to take my time'.[100] He was sometimes able to take some solace during 'the rush in which I usually live began' by turning his thoughts to the literary associations of the places to which work took him. Thus while Head 'was off on a carrion crow expedition into the wilds of East Ham', he came across a street named Plashet Grove –

> which seemed to carry one back into the seventeenth century. For was it not in the Plashet Grove at Moor Park that Swift was pelted with apples by the child Esther Johnson? Then came the foul business in a little room and a race to the Hospital to be in time for my own patients – However I travelled in the company of Swift ...[101]

Such relief in the company of a favourite author was, however, only temporary. The same day he made his first visit to inpatients at the London; Head realized that this was a routine into which he was likely locked for the next twenty years.

Among the bustle of the hospital Head had a sense of 'intense loneliness' coupled with a sense of the futility of all his effort – ' A feeling that I had toiled and given up everything and gained nothing; that I knew nothing & had planned greater things than I could ever carry out'. The noble aspirations with which he had begun his career in medicine were overwhelmed as, 'the dirt & disease of this vile town is beginning to close round me ...'[102] In January 1901 he lamented:

> Life seems to be flying by in but clutches of recurrent engagements and duties and I often long for the old days when I was a free lance. This afternoon alone, between my various engagements, I have written 20 letters and notes. The post is indeed a curse except when it brings a letter from you. When I get home tonight, I have still more letters to write for tomorrow. I shall be at the Hospital until 11pm.[103]

In the face of the demands of routine work, every day became 'the same attempt to pluck something from the wreck of time'.[104]

Head looked back wistfully on the days when: 'A long period of hard investigation & happy thinking was closed by a short period of unpleasant toil with the prospect of a pleasant holiday to follow'. Now scientific investigation, 'writing & production have to be done in the moments snatched from public & private practice'. He felt life slipping away: 'As year by year the meshes of the net grow closer thinking becomes less & less frequent – action more & more so until at the end of the week I can now often count on half a hand the hours in which I have thought about other thing than what to do or what to say'. His greatest fear was that because of the pressure upon him, he was 'obliged to let pregnant opportunities slip for fear of neglecting my daily duties. I am like a cattle drover who dare not leave the high road to follow up some trail that leads to gold for fear lest his drove should stray'.[105]

He began to cast envious eyes on those who were able to pursue their intellectual aspirations under more favourable circumstances. Head had become acquainted with the philosopher Bertrand Russell (1872–1970) through their mutual friend, Alfred North Whitehead. In December 1900 Head remarked wistfully:

> Russell and his wife lead a very wonderful life ... They have given away all their income above £600 a year and live in a house at Hazelmere. Here he thinks & writes but in order that he may not become too great a solitary he & his wife take a house in Cambridge for the greater part of the Lent Term. This he looks upon simply as a time of rest & amusement. Sometimes he lectures if a lecturer in his subject be absent, sometimes he takes no part in teaching – Then a little spring journey & back again to the country and work for the rest of the year.[106]

This seemed to Head an ideal form of existence.

He came more generally to establish a polarity between a Cambridge that embodied the values of intellectual freedom and a London where drudgery was the rule. In October 1903, Head took a holiday in his old university town. On the final day of his vacation, he attended some lectures on metaphysics delivered by John Ellis MacTaggart:

> As I sat there ... listening to his plea for the study of metaphysics because although it was no guide to conduct and led to the acquisition of no practical learning yet it brought comfort to those who desired knowledge of reality in the universe I could not help contrasting my life up here with the vortex into which I shall be plunged next week. Here things are well done because there is time to do them and because only a moderate amount is demanded. In London everything is badly done because there is no time to do them in and no human being could get through what is demanded of even the weakest.[107]

During the researches on nerve regeneration that Head undertook with W.H.R. Rivers, Head made regular weekend trips to Cambridge, which he configured as escapes from this vortex to a place where there was time to do things well (see Chapter 3).

The outbreak of war in 1914 shattered many of Head's most cherished beliefs. It also added many new responsibilities to his workload. He became immersed in the treatment of wounded and traumatized soldiers. He also served on various government committees. Paradoxically, however, the War also provided the occasion for him to extricate himself from the coil he had felt tightening about him for so long. He resigned his hospital post and for a time abandoned private practice. As we have seen, the latter decision involved considerable financial sacrifice. But even Ruth Head, who bore the brunt of these losses, acknowledged the positive impact this move had on her husband's mood:

> H. is happy too – no Hospital, no bother with tiresome obstructionists – no tiring journeys out East; only a fair number of paying patients and long hours of work at his own things diversified by Committee Meetings at which his opinion is listened to with flattering consideration – and sometimes followed![108]

Sadly, this period of contentment was all too soon to be shattered by disease.

Details Of Administration

While he clearly saw most aspects of his routine life at the London as uninspiring, Head showed surprising enthusiasm when it came to performing some of the most mundane duties that came his way. In February 1901 he told Mayhew, 'I find that although I have no business training or aptitude I get enormously interested in details of administration'. He had no curiosity about 'high politics;' however, 'departmental questions ... excite my keenest interest'. For instance, when assigned the task of reforming the diets fed to patients at the London, he: 'accumulated a large amount of information on the subject that no one seems to have troubled to collect'. In order to present this report to the Hospital Council, Head was prepared even to postpone his usual weekend trip to see Ruth Mayhew because, 'If I am not there the whole movement will go to pieces'. In this commitment to practical, if unglamorous, matters Head felt he was following in the footsteps of Leonardo da Vinci, who had recommended: '"Do not engage in studies that lead to nothing". He must have been tempted to engage in all sorts of political questions but preferred to direct canals and to build locks'.[109] Head evidently saw no inconsistency between this fervour for severely utilitarian pursuits and his reverence for the contemplative life of the idealized Cambridge don.

Head was not, however, content to restrict himself to the minutiae of hospital administration. He also saw himself as endowed with a mission to ensure that what he saw as progressive causes should triumph in the internal politics of the London. This sense of mission was indeed central to his professional identity. In particular, he set out to ensure that the hospital should realize its full potential to become a site at which medical knowledge could be advanced, as well as a place where patients were treated. This ambition came to a head in 1900–02 when he became involved in the discussions about the creation of a new Pathological Institute at the Hospital.

Head described this new initiative as 'my pet pathological Department'.[110] He envisaged the new Institute as a place where medical men relieved, at least partially, of the pressures of routine work, would undertake fundamental research into the nature of disease. In his 'Rag Book', Head had opined that: 'Medical education in England suffers from the fact that the great hospitals are manned by practitioners of medicine who sometimes teach, instead of by professors of that science who occasionally practice'.[111] This comment applied *a fortiori* to medical research. Head's aim was to make the Pathological Institute an exception to this lamentable state of affairs; he planned 'to make that place a centre of light in the darkness of Egypt'.[112]

In July 1900 the Hospital Medical Council supported in principle the establishment of the new Institute and appointed the bacteriologist, William Bulloch as the first Director.[113] This was, however, only a short-term measure; Bulloch was not expected to hold the post for more that three years. Head, however, attached far more importance as to who should occupy the post of Assistant Director. In a memorandum Head submitted on the organization of the Pathological Institute, he pointed out that the 'Assistant Director now to be appointed may in a few months and at most in three years become the head of the whole pathological anatomical and clinical research work of the Hospital'. It was: 'therefore essential that he should be a pathological anatomist of wide training and experience'.[114] Head wanted this post to be held by a junior man imbued with the ethos of modern medical science: for him Salaman was the obvious candidate. Head recognized, however, that certain of his colleagues had other plans for the Institute: 'Some are making a dead set at its efficiency and I shall be fighting with my back against the wall and shall have to try to hit hard without making permanent Enemies and without alienating those who habitually sit on the fence'.[115]

These conflicts culminated in a meeting of the medical school Council held on 31 May 1901. The key issue appears to have been whether the Assistant Director would be subject to the Hospital Pathologists and therefore required to devote much of his time to routine post-mortem work. The minutes of the meeting are bland; they note only that the issue of the Pathological Institute 'was discussed first in a very wide manner ...' Ultimately, the point that the Assistant Director should not – except in special circumstances – be expected to perform routine autopsies was conceded.[116] In private, Head revealed that this had been a 'hard-won battle'.[117] The Pathologists at the London continued to see the new post as detrimental to their own interests.

In July 1901 the new Pathological Institute was opened. Head helped in the shaping of the speeches that were delivered on the occasion. Sydney Holland, the Chairman of the London, was determined that these should appear in *The Times* newspaper to gain maximum publicity for the Hospital. Head was pleased that the event was indeed fully reported and that 'the leading article the Times caught up most of the points I wanted emphasized'.[118] The article stressed that the advance of modern medical science made it essential for a hospital such as the London to have elaborate laboratory facilities staffed by suitably trained experts. Head's influence can be detected in such passages as: 'Men capable of making such [microscopic and bacteriological] examinations could not be created in a day, and it became necessary to have a centre where pathology was studied for its own sake and not for the purposes of immediate practical application, still less for mere examinational purposes'.[119] Head thought it especially important that bacteriology and pathology should be combined in the new establishment; thereby: 'We shall form a compact body of opinion located in the upper floor of the institute which I hope will be able to work reforms'.[120]

The question of who would occupy the key post of Assistant Director however remained to be decided. On 31 January 1902, Head told Mayhew that: 'today I have got into hot water again about the appointment at our Laboratory – All the people who were against my views last summer have stirred the matter up again and are violently attacking the whole scheme ...'. He had hoped to spend the day doing research. Instead, while examining an 'excellent case', he found himself 'run to earth and for an hour compelled to defend myself with my back against the wall ... Ten years hence no-one will understand what I was fighting for or what I had to oppose'.[121]

At a Council meeting on 3 February 1902, the merits of the candidates for the post of Assistant Director were debated. Finally, Salaman was approved unanimously. The Council, moreover, further resolved: 'that the House Committee be asked to allow Mr Salaman to spend the next six months in improving his knowledge of Pathology in such manner as shall meet with the approval of Dr Bulloch. His duties to commence in the Pathological Institute on October 1st., 1902'.[122] The Hospital House Committee accepted these recommendations on 10 February 1902.[123] The following day Head wrote to congratulate his protégé – and to coach him on how best to use his time before he took up his post. Salaman was to: 'learn as much about Pathological Anatomy as you possibly can. It would be a mistake to do any extended research'.[124]

After this success, Head confessed: 'I had hoped that all our troubles were at an end but the defeated party has rounded & proposed that a most undesirable person be appointed during [Salaman's] absence – Moreover I am in hotter water than ever. They have called a special meeting to tell me what they think of me'.[125] The Pathologists in particular continued to resent the independence that the Assistant Director would enjoy. Head briefed Salaman on the position in a letter that revealed just how acrimonious the conflicts between the medical staff at the London had become. He revealed that he had been subjected to 'a severe reprimand for having told the truth about the present system of P[ost].M[ortem].s and have learnt that the slightest breath from either of us may wreck everything and throw the whole reform back years'. Salaman was therefore enjoined that the 'definitive secrets' about these political manoeuvrings that Head had confided were strictly confidential and 'must be kept even from your wife'.[126] Head had previously warned Salaman to be 'very careful in any letter you may write in answer to congratulations &c. to make no allusion to the present Pathologists or to the (obvious) fact that we have spoken together of the condition of Pathology at L.H'.[127]

Head nonetheless had got his way in the matter of the Pathological Institute and was to do some of his own future research there. His triumph over the Assistant Directorship was, however, to prove short-lived. Soon after he took up the post, Salaman fell ill with pulmonary tuberculosis and was obliged to leave professional life altogether. In an effort to ensure that Salaman received the best available treatment for his condition, Head consulted with his associate George W. Ross about the advisability of inoculation therapy as a means to 'eradicate' the disease.[128]

Although Salaman was to live until 1955, he never returned to medical science – a deficit that Head felt keenly. 'It is all very well to say there are heaps of good fish in the sea', he complained, 'but there is one fish whose loss we shall ever lament. Indeed there is not a day when I do not think of you and your work'.[129] There were few medical men with whom Head felt so close a spiritual as well as intellectual rapport.

Salaman's replacement at the Institute was another of Head's protégés, Hubert Maitland Turnbull – the man whom he had tutored in physical signs while commuting to the Hospital. However, in later years a certain coolness entered into their relationship. Turnbull alleged that Head was sometimes overly anxious to receive confirmation of his preconceived views from his juniors. After he became Director of the Institute of Pathology at the London Hospital, Turnbull recalled: 'I did a great deal of hard work for him, but if I did not find the condition he expected I got no thanks; in fact I so seldom found what he wanted that I believe I lost his friendship'.[130]

In the years following Salaman's retirement, Head kept him apprized of developments at the Institute. He continued to lament the obstacles placed in the way of its work by the 'known opposition' within the London to this establishment. Head depicted the medical staff as strongly polarized; the Pathologists were all arrayed against him and Percy Kidd was one of his few allies. Head gloomily predicted that: 'we shall have all the old difficulties over again'. He confessed to Salaman: 'I feel terribly alone amongst these scientific philistines'.[131]

In 1919 Head became embroiled in his last controversy at the London. A plan was floated in response to the recommendations of the Haldane Commission to establish academic medical and surgical units at the Hospital to be led by full-time professors.[132] This was part of a wider strategy with the avowed aim of modernizing British medical education[133] – a programme with which Head had identified. Head was nominated to be the first director of the medical unit. He had by now resigned his post at the London. It is clear, however, that there were few alternative candidates; as Walter Morley Fletcher (1873–1933), the Secretary of the Medical Research Committee put it: 'There are barely three men in the country fitted to take the post, and all but Head are claimed elsewhere'.[134] Head's longtime associate at the Hospital, Sydney Holland, now Lord Knutsford, was also anxious to appoint Head to the new position. Indeed at one point in the negotiations surrounding the appointment, Knutsford told Head, 'I said that I should resign the Chairmanship of the Hospital if you were not offered the post ...'[135]

On the face of it this was an opportunity to realize his ideal for the advancement of medical science. In Fletcher's words, the proposed new Unit was to be: 'A wholly new departure ... in which the science of medicine, as such, was to be advanced by research and by higher teaching, in contrast with the present system in which the practical arts of medicine are handed on to students at the bedside'.[136] Initially Head did seem to entertain the proposal seriously and set out various conditions under which he might accept the offer. In particular, he demanded the right to appoint his own assistants. He further stipulated that, instead of 'chasing patients', these men should 'take one and half days off a week (apart from Sunday), for thinking and for reading'.[137] As we have seen, such relief from the constant pressure of routine work was something Head had himself craved earlier in his career.

These negotiations, however, soon ran into trouble. Head was concerned about the degree of autonomy he would enjoy as Director of the Unit, especially in the choice of staff. The real sticking point proved to be the number of beds in the Hospital that would be assigned for his use. Head demanded that no fewer than 100 beds be made available for this purpose; the Hospital felt able to offer no more than eighty, with a share of the beds of another physician.

Knutsford did his best to address these concerns. But by May 1919, it was clear that Head would not accept the directorship. He conveyed his decision to Knutsford, writing that 'I would gladly have taken up the task, and attempted to give the experiment a start on what seem to me right lines, and had already thoughts of the first class assistants ... But I was not content to sacrifice my time and strength to running a second rate department based on non-scientific lines, such as the staff desire'.[138] Head thus put the blame for the collapse of the proposal on the obstructionism of the clinical staff at the Hospital. Such resistance was a common occurrence.

Knutsford did not conceal his incredulity and exasperation at this turn of events. He told Head that it was 'a bitter blow ... to me that you refused the post'.[139] A lively exchange followed between Fletcher and Knutsford over where the responsibility for the failure to persuade Head to return to the London really lay. Knutsford insisted that: 'there were absolutely no differences between the Staff and him which could not easily be adjusted ...' He maintained that the matter of the number of the beds was also an issue that could have been resolved if Head had shown the least flexibility. When offered an initial allocation of eighty and a share in the beds of another member of staff, 'He just put this down at once, and refused in an emphatic way to entertain for one moment a lesser number than 100'. Knutsford went so far as to suggest that Head had been disingenuous in the account of the negotiations he had conveyed to Fletcher:

> I think Head has misled you. You tell me he 'would certainly have accepted about 90 beds'. Well, why the devil could he not have said so to me? He knows I am his friend, he knows I was deeply keen to get him, and he absolutely, and in the most emphatic way, refused to consider the post under 100 beds, and shut down any discussion on the matter at once when I spoke of Roxburgh's offer. 'Oh, no, that won't do at all'.[140]

These exchanges leave the impression that Head was in fact seeking pretexts to refuse an offer that might at one time have been highly attractive to him. Fletcher hinted at what may have lain behind this reluctance to be tempted back to the Hospital: 'On selfish grounds, he is far better as he is; he would have to give up both money and ease of mind to take up the work'.[141]

Head had been frank about the importance of financial considerations when in January 1919 he had explained his reasons for resigning from the London. His wartime experience had shown him that 'the work of the Hospital and Research are enough to occupy my whole working time'. Moreover, Head felt that: 'I cannot revert to the old methods of carrying on my duties, which existed before the War; and at the same time I cannot afford any longer to do no private practice'. Of the three main components that made up his professional life, he had therefore decided that hospital work was dispensable. There is also a hint that Head had also taken this decision in light of the deteriorating state of his health: 'I have accumulated many observations which I must put together during the few years of activity that remain to me'.[142] By 1919 the early signs of Head's Parkinsonism were becoming manifest; he knew better than any the likely prognosis. This was a further reason to refuse to take on the responsibility of the running of the new academic unit.

The Doctor As Mage

We have seen that Head's practice included a significant number of 'functional' patients who manifested psychic, or psychosomatic, symptoms that could be ascribed to no obvious somatic cause. To treat these cases, Head took on the role of a 'psychologist' who sought to understand this illness in terms of the personality and life history of the individual concerned. He sought to formalize this dynamic approach in a 1920 paper on 'The Elements of the Psycho-Neuroses'. When confronted with such a patient it was, he insisted, 'futile to waste time in considering whether he is a case of neurasthenia, psychasthenia, anxiety neurosis, or hysteria'. The only diagnosis that was:

> worthy of the dignity of our profession, is the laying bare of the forces which underlie the morbid state and the discovery of the mental experiences which have set them in action. Diagnosis of the psycho-neuroses is an individual investigation; they are not 'diseases', but morbid activities of a personality which demand to be understood.[143]

Thus when a female patient presented ostensibly because of a minor physical ailment, it quickly became clear to Head that her real problems were psychological. She suffered from suicidal tendencies that arose from a conflict between her career aspirations and a sense of duty to a parent (see p. 59 above). Having arrived at this understanding of the case, in this instance Head's solution was to advise the woman to abandon her metropolitan career and devote herself to the case of her mother – a prescription that was apparently effective in relieving her morbid mental state.

This was an instance of what might appear as a highly pragmatic, common sense, kind of psychology – albeit one clearly shaped by strong presumptions about preferred gender roles. Head was, however, versed in the latest trends in psychological theory, keeping abreast with the latest French, German, American as well as British literature on the subject. He was, for instance, among the first

British practitioners to take onboard Sigmund Freud's theories on the origins of the psychoneuroses. In his 1920 address, Head maintained: 'The importance of the mental conflict arising from the failure of automatic control was not recognized until Freud formulated his conception of the process of repression, and its bearing on the genesis of the psycho-neuroses'. In the same paper, Head also revealed a knowledge of Carl Jung's concept of regression.[144]

Freudian ideas even entered into the casual banter of the Head household. Head and Ruth Mayhew had throughout their early correspondence often recounted and sought to interpret their dreams. By 1917, Freud's influence was evident on the form this interpretation took. Ruth told Head 'I had had such an amorous dream. It would have delighted Freud ...'.[145] This dream proved to be a somewhat disappointing affair involving Ruth flirting with Henry's friend Viscount Knutsford. Head did not feel threatened by this revelation: 'how I laughed over your sex dream and its repressed complex'. However Ruth's revelation led him to recall a Freudian slip of his own:

> I must tell you an extraordinary example that occurred yesterday of the psycho paths of common life. [Henry] Lyon was talking in the callous way he sometimes does of the death of one's friends that must happen in middle life. Nothing would stop him and I called out 'Oh shut up, Rivers'. [Robert] Nichols at once saw what had happened and leaping from his chair made me analyze the causes of this curious lapse of the tongue. It evidently went back to 1908 when I was writing our paper on the arm along with the haunting dread that Rivers would never come back from Melanesia again.[146]

Head's persona as a 'psychologist' consisted of more than theoretical knowledge and clinical expertise. Others ascribed to him a personal charisma that made him formidable – and in some instances, even frightening – in this role. Thus Head's mother conveyed to Ruth Mayhew a view current in the family that: 'Harry has a terrifying influence over other people's souls'. Mayhew chose to make a joke of this teasing Head – 'Had you realized the dynamic quality of your mental gaze?'.[147] It is apparent, however, that Head could on occasion make use of these gifts of insight and analysis to devastating effect.

Not even his mother was safe. Head's previously close relations with his mother suffered during his extended courtship with Ruth. In June 1903, after he and Hester Head attended a performance of Ibsen's *A Doll's House* together, his patience with his mother finally snapped. The two

> as usual, bickered – She has a haughty way of laying down the law about marriage illustrated by entirely false instances from her own life that always drives me to cold blooded scientific truth – I develop a vile genius for dissecting her and her marriage – I know my observations are the truth by the paralysed struggles with which they are received but I wish I had been permitted to be silent ... All the many years that her jealousy attempted to tie my affection by slandering those around and deprecating

my father's wonderful work she has only talked to a horrible recording machine that registered and classified her as a psychological phenomenon – It is a terrible nemesis ...[148]

Head's representation of himself as a 'horrible recording machine' driven by 'cold-blooded scientific truth' is striking. His anecdote suggests that his 'vile genius' as a psychologist could be used tool to dissect and render helpless another person. Elsewhere he regarded psychological acumen of this sort almost as a form of connoisseurship: an ability to discern what was most striking and remarkable in a personality that could have artistic as well as scientific value.

In December 1913, Head wrote to Charles Ashbee: 'Your wife's letter is a wonderful document. She seems to have observed the conditions with the discrimination of a trained psychologist but, at the same time, has clothed them in language which makes her observations as interesting as a chapter in the finest realistic novel'.[149] In this letter Janet Ashbee had discussed a woman named Phoebe for whom: 'The sexual act is, & has always been "repulsive" to her – has no meaning or reality – she admits that Ron [her husband] has to "implore" her before she will consent at all'.[150]

Head added some analysis of his own: 'I see the situation clearly – A woman, the product of town life, "thinly" sexual, ready to flirt but not to give herself freely – A man, lazy, hating responsibility but sensual & heavy of body'. He concluded that the couple: 'are not pathological but the problem is the outcome of modern conditions'.[151] The particular 'modern condition' to blame appears to have been Phoebe's desire for a social identity separate from her marriage; Janet Ashbee had suspected that at the root of the problem was 'the taint of the typewriter & the "independence"'.[152]

Head suggested that having a child might help resolve some of the couple's problems: 'This will change her at any rate during the pregnancy and they may shake down together'. He also asked Ashbee to, 'Please let me know what happens'.[153] Head seems to have had no professional interest in this 'case'. His involvement was more recreational in character, consisting of a mixture of aesthetic appreciation for Janet Ashbee's letter as a piece of fine 'realistic' writing and psychological fascination with the details of the relationship under discussion. This conflation of literature and scientific psychology was a trait that Head also revealed in other contexts.

It is clear that Head held that the same criteria of psychological truth were to be applied to literature as to clinical practice. A passage he came across in his reading claiming that: 'the mind having learnt to abide unshaken in one trial will be more likely to abide unshaken in another and bring all the knowledge and art she has to bear upon it' provoked a vehement refutation:

This I absolutely deny. An officer who is oblivious to death & danger in the field is found to be as nervous as an East End Jew in face of Anti-Typhoid inoculation. I have seen brave men mount the operating table to take an anaesthetic, pale with fear – None but a trained person can face the refractory women airing court in an Asylum unattended – I have seen a young girl, an attendant in an asylum, who was negligently fearless in the refractory court jump onto the chair flying in terror from a supposed mouse.[154]

The various aspects of Head's persona as a psychologist are most clearly evinced in his relationship with Robert Nichols (1893–1944). Head first came to know Nichols as a patient. Their interaction retained a clinical aspect to the end. Because of their shared literary interests, however, the doctor/patient relationship was from the outset mixed with other overlapping strands of discourse.

We have seen that during World War One Head felt it his patriotic duty to devote himself to the treatment of casualties returning from the front. These included soldiers who had suffered wounds to the brain or other parts of the nervous system; Head was to put these 'organic' cases to good scientific use (see Chapter 3). Like his erstwhile collaborator, W. H. R. Rivers, during the War Head also encountered a large number of soldiers suffering from what was sometimes called 'shell-shock'.

These were men who manifested no apparent physical injury, but who had suffered some form of psychic trauma as a result of their experience of combat.[155] His encounters with these cases led Head to revise his views about the psycho-neuroses.[156] He regarded most of these hysterical patients as being unusually suggestible. They were best treated by psycho-therapy that relied upon the physician's powers of suggestion. As a rule, Head maintained, hypnosis was unnecessary.[157] What was required was a sensitivity to the individual personality of the patient. Thus one of Head's patients had a severe tremor of the hand that prevented him from performing most acts. He was, however, capable of playing the banjo perfectly. Head 'used this musical aptitude for effecting his cure'.[158] This provides a simple example of what Nichols – using a golfing metaphor – described as Head's aptitude for discovering the 'stance' best suited to a given individual. This was a relatively modest medical intervention in the personality of a patient: 'you have always tried not to correct a given stance but so to speak to rock him gently backward & forward & this way and that until he fell into a stance natural & useful to him …'[159] We shall see, however, that Nichols believed that Head was capable of effecting more radical change in a personality.

Head first encountered Nichols as one of these shell-shocked soldiers. Nichols had been an undergraduate at Oxford when war broke out in 1914. By his own account, he volunteered for military service out of a sense of patriotic duty coupled with a principled opposition to German aggression. In September 1915, after a particularly heavy bombardment, Nichols suffered what his commanding officer called 'a slight nervous breakdown', and was eventually evacuated back to England and committed to the Palace Green Hospital, an institution dedicated to the care of traumatized officers.[160] His early experience as a hospital patient in a 'dismal attic in Palace Green' was evidently an unhappy one.[161]

Nichols was thus overtly a victim of traumatic neurosis engendered by what he referred to as his 'little war'.[162] In later years, however, he ascribed remarkably little significance to his wartime experience in the aetiology of his neurasthenia. He saw this as more the result of heredity (his mother was committed to an asylum) exacerbated by some appalling childhood experiences. By his account, after having attended an English public school, the western front held few terrors.

While at Palace Green, Nichols attracted that attention of an acquaintance of Head who urged him to go to see 'a young friend of hers who was eating his heart out and getting no good from his doctors'.[163] Nichols left a graphic account of his first encounter with Head:

> On the whole the most important event in my life to date was the occasion on which Dr Henry Head F.R.S., a plump, bland slightly Mephistophelian figure pushed open the door in the hospital cell & sat down & asked me if I liked Conrad – the first sensible & honest question that had been put to me since I came out of France. You know, Henry, I took to you from the very moment you sat down. It was your way of putting your bag on the ground & settling into your seat. You set down that bag with such care – evidently you were 'on the job': I spotted the craftsman. You settled yourself down with such an air of relish & of not being hurried – of guarding yourself against any possibility of being rushed – for all the world as if you descried something in me at once & felt I might be worth not a cure, but a chat. The immense & subtle flattery of that![164]

This passage shows that Nichols regarded his first meeting with Head as a turning point in his life. It marked the beginning of a relationship that was to endure until Head's death in 1940. It also demonstrates that, from the outset, this was more than a 'patient/doctor' relationship in any narrow sense. Head's first attempt to connect with his new patient took the form of a reference to a shared passion for literature. We know from other documents that part at least of the therapeutic sessions that Head and Nichols undertook in the early days of their acquaintance took the form of intense discussions of the psychology of fictional characters, such as Conrad's *Lord Jim* (see Chapter 5, p. 216).

After his release from Palace Green, Nichols remained in regular contact with Head. They corresponded and when he was in London Nichols became a regular visitor to the Head household in Montagu Square. In the early years of their acquaintance, Nichols continued to suffer symptoms that could be ascribed to his wartime experience. Thus in May 1917, he wrote: 'Dear Headlet, I go on fairly well. Bad night last night. Kept on waking up. Slight fever. Bad dreams – dreamed that man took a bet to kiss a corpse – problem for Freud. Cannot say it was an inhibited impulse eh? Shall use the idea in a ghastly playlet'.[165] The letter is also notable for the playful appellation Nichols applied to Head (in previous letters he employed a more respectful 'Dear Dr Head'), and also for the casual reference to the theories of Sigmund Freud. As the years went on, Nichols was to acquire an ever more extensive acquaintance with psychological literature.

But Nichols was soon to depend upon Head for help with a much wider range of emotional and personal difficulties. An early example was the 'Daisy Kennedy business' when Nichols was rebuffed by a woman with whom he had become infatuated. He recalled that in the aftermath he had been:

> right off the rails & it culminated in my going to the Deuce...When that last happened I really thought of making away with myself & as I was walking along outside Scott's Restaurant I thought 'No', there's one chance left! There's Henry – the only grown-up who ever gave me a chance'. And I jumped into a taxi & went to him. And when I'd told him he sort of sank down in his chair & I could see tears in his eyes. He saw me through – you & he. I had a haven at last.[166]

Even after Head's retirement, Nichols continued to rely on him for help with a long series of emotional upsets resulting from his love life and from the reverses he experienced as a playwright. When he withdrew from clinical practice, Head had referred Nichols to the care of his protégé, George Riddoch. Nichols was at first appreciative of Riddoch's virtues as a doctor; after an initial consultation, Nichols reported:

> [Riddoch] went over me very thoroughly. I found him extremely sympathetic – an extraordinarily sympathetic person. His voice is beautiful & he has a nice Scotch accent. We talked a lot about you & Ruth – we're rivals in devotion! I felt much comforted by him.[167]

Subsequently, however, Nichols came to compare Riddoch in unfavourable terms to Head. Thus in April 1934, he opined that Riddoch was: 'very nice & I'm sure his heart is kind & good & I find every doctor entertains a profound respect for him. But he is not a genius like Henry'.[168]

In particular, Head's 'genius' as a doctor lay in the fact that he combined clinical acumen with aesthetic appreciation; Head was, in effect, 'an artist physician'.[169] Nichols maintained that it was this profound appreciation, not only of literature but also of music and the plastic artists, that made Head so effective a clinical psychologist. He had initially used a shared passion for the novels of Joseph Conrad to establish a rapport with Nichols and to seek an insight into his psyche. When it came to prescribing a remedy, Head's solution was again as much literary as medical.

In March 1925 Nichols exulted in his reading of: 'Great Goethe! Has there been a wiser man, more all-round wise? more generally wise? And I owe him to you'. He reminded Head:

> never has one, save you, told me to read Goethe. I had not known you six months when you began to insinuate Goethe into the conversation. Halting on the steps of number 4 [Montagu Square], with a sheaf of papers in your hand, you said 'Goethe is, I think, a mind that will eventually appeal to you with a peculiar force'. How did you know it? I had written next to nothing. I was in a chaos. I could not think five consecutive thoughts.

Nichols's other advisers had urged him to 'discipline' his mind; 'all the usual XVIII century stuff'. Head alone had understood that: 'with me building was the only possible discipline'.[170]

Nichols drew on metaphoric resources supplied by the war to try to convey the effect that reading Goethe had exerted on his outlook. He now felt distinguished from other men of his acquaintance:

> They are like generals commanding troops here & there & doing well in their several engagements, but the general battle over this terrain cannot be won by them ... while I ... having been up in an aeroplane, constructed & piloted by Goethe, have the scene <u>as a whole</u> in my mind, have now & again, some perception of the relations between two or three points, & thus am their superior – not through any peculiar gift of mine but simply & solely because a Dr Henry Head advised me to take the aeroplane ...

For him, indeed, Goethe was the 'First Modern' – a fact that Nichols had tried to impress upon other literary men, such as Aldous Huxley.[171]

In one of his few surviving letters to Nichols, Head replied:

> My dear Robert,
> Your letter filled me with pride and pleasure, for it delights me to think that it was I who turned your attention to Goethe. In my busy life, I am forced to give much advice which I fear is often beside the point. In many cases where I would gladly be silent, I am compelled to speak dogmatically, and repeatedly make mistakes, where had I been left to myself, I should have kept silence. It is therefore particularly delightful that a remark made so casually has borne such abundant fruit; for indeed Goethe is your man. He is the only genius who seems to me to have combined both the methods of scientific and poetical thought.[172]

Ruth Head added that Nichols's words had made her husband 'blub all over, but which was true every word of it, and the one to him about Goethe which pleased him even better, because he does love to think he put you on the scent of so great a man'.[173]

Nichols saw an artistic aspect to Head's activity as a physician because he not only 'cured bodies & mended minds astray', but also engaged with his patients in a truly creative way. In his clinical activity, Head had something of the removed perspective of the novelist: 'Many vile things must have been poured out in his consulting room. He remains untouched.' But when he came truly to engage with his patients, Head assumed more the aspect of the sculptor, remoulding the material he found: 'He has fortified hundreds in spirit & some, such as myself, he has <u>made</u> – physically, morally, philosophically, intellectually, even aesthetically. I am a changed man since I knew him.'[174] Among the most remarkable aspects of the correspondence between Head and Nichols is the degree to which both operated with the assumption that the self was a malleable entity capable in the right hands of being reshaped.

From their correspondence it is evident that for his part Head regarded Nichols, not only as a patient, but also as a kindred spirit. They were both poets with shared tastes in novels and music. Moreover, Head considered Nichols a skilled 'psychologist' in his own mould. This psychological acumen was manifest in Nichols's assessment of some of their mutual acquaintances. In 1926 Ruth Head told him that her husband: 'is also tremendously pleased with your masterly study of Frank Kidd and says you have analysed him to the life'.[175]

After Head's death, Nichols produced a series of dialogues designed to give the reader an idea of the 'psychological' conversations the two had over the years. Some of these concerned characters in literary works, such as Joseph Conrad's *Lord Jim*. Others involved the collaborative analysis of real individuals; the technique differed little between the two cases.

Thus Nichols recalled 'Henry explaining the conduct of a novelist, now dead, whom I will call 'X''. Nichols had remarked at seeing X in a theatre bar drunk and involved in a brawl. He asked Head to 'Explain his psychology'. Head's analysis centred on the fact that X had 'a taste for gay boys'. Head saw this as 'a craving with evidently deep psychopathic foundations'. Because he was wealthy, X enjoyed some protection from being prosecuted for his sexual orientation. X knew, however, that 'there will come a day when money will avail him nothing. Every morning when he rises he resembles in his realization of his predicament a man who has to enter a room containing twelve switches, the pulling of any one of which will electrocute him'. The switches corresponded to the twelve hours of the day that X was obliged to traverse. He perversely sought to deal with this constant fear of discovery by gravitating to locations frequented by heterosexuals of the kind: 'who would show little charity towards such an anomaly as his. He is seeking safety!' This was why X was so often found in bars frequented by prostitutes and their clients.[176]

Having analysed X's psychology to his and Nichols satisfaction, Head struck a compassionate pose. He urged:

> Let us be lenient toward X––. Let us encourage him to spend as many hours a day as possible at his desk & as few as possible at the bar in the promenade. If he gets into trouble let us not fling him into gaol but take him to a clinic. His life is more dif-

ficult than ours. Let us be merciful toward his weaknesses, keep possible victims at a distance and encourage him to produce works, which, queer as they are, yet possess a certain indubitable merit & which wouldn't be what they are or even perhaps exist were he not what he is.[177]

The only point such analysis seemed to have was to provide Nichols and Head with a kind of mental gymnastics. On occasion, however, Nichols tried to put the psychological expertise he had gained through his interaction with Head to more practical ends. This might take the form of acting as a kind of neurological scout who identified likely cases to be directed to Head. During the war years, most of these were servicemen whom Nichols diagnosed as suffering from traumatic neurosis. In May 1917, he advised Head of 'a case I am sending you'. The case in question was an airman Nichols had encountered:

> It is easy to see that the strain of continued fighting on three fronts & of – more particularly – flying & fighting is beginning to tell on him. He is slow of speech now (which I don't think he always was), a little contracted about the eyes which fall into a dead stare at intervals – which again I don't think they always did. But you will see the symptoms. I send him to you because you seem to specialize in flying men & you will do him justice: he has a fine record.[178]

Nichols also tried to persuade Head to help the Belgian socialist politician, Emile Vandervelde, who lived in London during the War. He remarked that Vandervelde 'is in rather a groggy state & unless taken in hand the worst may happen'. Nichols offered to arrange an apparently purely social meeting between Vandervelde and Head. By way of encouragement he added: 'I assure you not only would you confer a great benefit on the world at large but you would be deeply interested in the case if you looked into it.'[179] Most intriguingly, Nichols tried to persuade the novelist, D. H. Lawrence to see Head: 'but he wouldn't hear of it & that wasn't because of the money for I told him I thought that could be arranged'.[180]

After Head's retirement, Nichols often referred 'cases' to George Riddoch. But he also became sufficiently confident in his skills as a clinical psychologist to seek to advise neurotics himself. When assuming this role, he confessed that he took on something of Head's comportment. 'Henry', he claimed, 'would have laughed like hell for all the seriousness of the matter – the utter seriousness – if he could have seen my clinical eye on the lady and my Montagu Square manner'.[181]

He referred to the process whereby he attempted to alter a patient's attitude as 'moving his furniture about'. In this technique, he claimed to be a true student of Head, claiming: 'Everything H.H. has done with me ... is due to this habit – for it becomes habit – of moving the furniture gently & with love.'[182] Eventually, Nichols even presumed to interpret a dream of Ruth Head after Henry refused to attempt the task. He told Ruth: 'I think I can read it. But the reading is sad.'[183] Nichols interpreted the dream as a reflection of Ruth's growing despondency at the way of life she and Henry were obliged to endure in retirement (see Chapter 6).

Nichols was clearly unique even among the select group of officer-patients that Head acquired during the War. Over many years he and Head engaged on a variety of levels of which patient and doctor was only one. Nichols was therefore well placed to give an account of what he called 'the wholeness of [Head's] genius'. He sought various epithets to describe the qualities that he saw in his neurologist, fellow-poet, and fellow-psychologist. 'Old Clinical Eye' was a favourite tag.[184] Head was, on one occasion, 'the Quaker Zeus'.[185] On another, Nichols saw him as 'a sort of Prospero'.[186] But the soubriquet that finally seemed most fitting was that of 'mage'. According to Nichols, 'H.H. [is] the only fully blown Mage I know'. If a man attained this status,

> everyone becomes simpler in his presence, calmer & more sensible both in respect of commonsense & of uncommon sense as they certainly did in the presence of the scientist-poet Tchekov, as they do to a certain extent in the presence of [Arnold] Bennett when Bennett is at his best, as they seem to me invariably to do in the presence of H.H ...[187]

There was an element of whimsy in Nichols's choice of this title. However, the notion of the doctor as mage does capture something of the personal, charismatic quality that Head brought to his encounters with at least some of his nervous patients. Something of the magical potency that Nichols ascribed to Head can be discerned in the efficacy he ascribed to his former doctor's image. In 1930, he told the Heads: 'Henry's portrait now hangs opposite my bed – I put it there last night – & as I've one in my writing-room too he's with me most of the time'.[188] Even in the absence of a physical representation, Head's image could be powerful. On his way in a taxi to a dinner-party that he dreaded, Nichols felt 'as if somebody had slipped a silk noose round my neck & was pulling it tight'. However, 'just before we arrived I suddenly found myself looking at Henry's head floating in the darkness – 'pon my soul it was enough to make me believe in telepathy--& that got me into the house & up into the drawing room before I knew where I was'.[189]

After The War

In 1932, Nichols told Ruth Head that: 'the War came on you & Henry in the maturity of your powers when you were both able to render your maximum to the alleviation of distress'.[190] The War was, indeed, a turning point in Head's life. His immediate reaction was that of a patriotic Briton. After a visit to the Continent shortly after the outbreak of hostilities, he wrote proudly to Harvey Cushing:

> Q. 'Are we downhearted?'
> A. 'No!
> You should have heard that from 10,000 soldiers' throats as we did at St Nazaire in September. It was like the rattle of a machine gun as we steamed out of the harbour on our transport.'[191]

Head bitterly lamented the fact that he was too old to serve on the frontline[192] and sought to compensate by – at considerable personal cost – abandoning private practice for the duration so that he could treat wounded and traumatized soldiers. The experience he gained through his encounters with these casualties of war was to provide Head with the material for much of his later researches.[193] He also sat on a number of official bodies, including the MRC'S Air Medical Investigation Committee and National Health Insurance Committee. In recognition of these services, Head was awarded the rank of Captain in Royal Army Medical Corps.

During the war, Head on a number of occasions travelled to France on government business. Once hostilities had concluded, he continued his public service. For instance, Head went to Rome in 1919 to take part in the negotiations that led to International Convention for the Regulation of Aerial Navigation. Head chaired one of the sessions, observing that he: 'had no difficulty in understanding the Italian and in directing the proceedings in French'. But he lamented the fact that: 'Latins are not used to our strictness in procedure and our determination that each speaker should be heard.' The Italians, moreover, knew 'little or nothing about flying', and seemed more interested in hosting excursions for their guests than in conducting serious business. The British delegation, on the other hand, was 'bitterly determined to come to grips with the problems of Aviation'. Head remarked that this experience of diplomacy was: 'exactly like the Peace Congress and it fascinates me to play a part in this international drama.'[194]

The following day the City of Rome provided a lunch for the delegates at which Head gave a speech. He contrasted: 'this most ancient centre of civilization with the newest of Man's activities – Aviation and pointed out that it had absorbed so many places of culture it was fitting that it should be the first meeting place of those interested in Aerial Navigation.'[195]

The War also forced Head to abandon some of his most cherished beliefs and commitments. In particular, the dedication to German culture that had since his Halle days formed so important a part of his sense of identity was shaken to the foundations. He even came to challenge the premise that Germany was pre-eminent in the field of medical science. Writing to Bulloch in October 1914 about the contribution of German neurologists, Head declared:

> Everyone must have noticed the deterioration in the quality and intelligence of the work turned out in the last ten years. In my line some of the old workers like Edinger kept the flame going and some of the men of my generation like Forster were cosmo-

politans. But the others could not understand that the new work was not coming from Germany but from France & England.[196]

His initial view was that the impact of the 'this horrible war' on medical science could only be detrimental. In September 1916, he told Cushing: 'It is scarcely credible that all the men we knew so well in Germany can hold the doctrines of the Kaiser & his group. I am afraid it will throw back Science for this generation: for we are going to crush them if it costs us our last man and out last penny.'[197] By the end of the War, however, Head had arrived at a more positive assessment of the impact that this cataclysm might prove to have on medicine.

Indeed, Head came out of the War with a clear sense of mission that had before been absent. He set out in a series of addresses what was, in effect, a modernist catechism for medicine. As early as November 1918, within days of the Armistice, Head told the Neurological Section of the Royal Society of Medicine: 'The cataclysmic events of the last four years have shaken men's belief in the old order, and medicine has not escaped the universal demand for a restatement of current values.' The 'old order' had been characterized by a sterile reliance on diagnostic labels that were: 'but anodynes to the conscience. They explain nothing; they serve only to put to sleep the salutary feeling that further investigation is necessary'.[198] In place of a medical system that was content to restate problems in 'bastard Latin and Greek', Head proposed a new dynamic neurology based upon the application of evolutionary principles to an understanding of disease of the nervous system.

In the following March, Head expanded this polemic to clinical medicine as a whole. Again, he used the characteristically modernist trope of a fundamental bifurcation between the past and the future. The war, he declared,

> has drawn an indelible mark across the history of our times. Much that before seemed possible has now become incredible and our beliefs have been shaken to their foundations. As the ancient edifice came tumbling about our ears, we wondered at the structural enormities revealed in its ruins. With the passage of time additions had been made to the original structure, which were founded on the flimsiest basis, and we are filled with astonishment that they could have held together so long.

The great conflict had, however, only been the catalyst for this seismic event: 'In many departments of medicine the old order was tottering before the War.'[199]

Head once more concentrated his critique upon the inadequate – indeed, risible – linguistic practices of the old order. Head epitomized this as 'penny-in-the-slot' medicine that confused diagnosis with naming. He applied the ethnographic stance he had displayed in other contexts to the linguistic practices of his own professional culture. The most egregious embodiment of the diagnostic Dark Age was the *List of Diseases* issued for the use of Army medical officers during the War. This work, Head insisted, 'forms a relic of a past epoch in medicine, and merits a prominent position on the shelves of an anthropological library, as evidence of an extinct phase of human thought'. Indeed, the book manifested an 'acceptance of diverse and contradictory categories of belief, so common in all primitive cultures'.[200]

Head's alternative was a clinical practice firmly grounded on rigorously determined scientific fact. This was, in many respects, a modern laboratory-based medicine that aspired to quantitative precision and made use of such technological devices as the electrocardiograph. Head did not, however, deny a place for intuition at the bedside: 'True diagnosis is an orderly procedure in which all the faculties of the mind, logical and instinctive, play their part.'[201] Nichols's description of Head as an 'artist physician' comes to mind.

As well as making these programmatic utterances, in the post-war period Head began to show an interest in social issues that had been previously lacking. He joined the National Council for Mental Hygiene formed in 1922. The aims of this movement, as explained by Humphry Rolleston at the founding meeting, were to detect and seek to remedy early signs of mental disorder in the population.[202] In his own contribution to the meeting, Head insisted that: 'mental hygiene was the application to the life of the community of highly specialized scientific work, the results of which belonged to the most diverse categories of knowledge, and were gathered by workers in widely different fields'. He pleaded that these various sources of knowledge be coordinated for the 'mental health of the community'. Head defined mental hygiene as: 'the maintenance of that state of health in which the human being could respond normally to the calls made upon him by daily life'.[203]

Central to this endeavour was a recognition of: 'how inextricably mind and body were intermingled even in the simplest mechanical acts'. Head's wartime experience had brought this point home to him in a forceful way. He gave the example of an RAF pilot who had been referred to him because he had lost the mechanical skills necessary to making successful landings. On closer examination, it proved that this loss of bodily aptitude was part of a wider complex of symptoms that included anxiety and disturbed sleep. From such cases, Head drew the conclusion that: 'The hard and fast line so commonly drawn between organic and functional conditions was grossly fallacious.' Nor should the detrimental effect of afflictions that were deemed merely 'nervous' be underestimated. Head insisted that, 'So-called 'nerves' produced more individual and corporate misery than cancer'.[204]

Such 'misery' could impact upon the industrial efficiency of the nation. Head cited the work done in America and elsewhere on the effect of unnecessary fatigue on the worker's performance. Nervous disturbances might even have pernicious political consequences. Head 'considered that there was little doubt that much industrial unrest was due to the worry and fatigue induced by unsatisfactory working conditions ...' It was therefore incumbent on the state to address the mental as well as physical hygiene of the working class. Because of the importance and prevalence of nervous disorders in the population, Head insisted, moreover, that all doctors and nurses should have some knowledge of morbid psychology. Indeed, he maintained that even clergymen should be enlisted in the campaign for mental hygiene, 'for depression in an early stage not uncommonly assumed a religious form'.[205]

By the 1920s, Head had, through his work on the physiology of sensations and on aphasia, acquired an international reputation that gave weight and authority to his demands for a fundamental reform of medical thinking. When Abraham Flexner was undertaking his survey of medical education in Britain, he was advised by William Osler to seek out 'the ablest and most cultivated of English physicians, Sir Henry Head ...'[206] Head's advice was sought and his work lionized by physicians and scientists from around the world.[207]

We can only speculate where the campaign to revolutionize medicine might have led if Head had been able to pursue it. His Parkinsonism was by 1925, however, so advanced that Head was obliged to retire, and he spent his remaining years as an invalid. Ruth Head gave a poignant account of the end of her husband's career as a doctor. She told Janet Ashbee that the secretary that had previously helped with the running of the practice had been dismissed and she was now left to deal with the aftermath of Head's retirement:

> There are endless letters to write to would be patients and despairing doctors. I had no idea H. was so appreciated, poor lamb. But everybody in the profession is touched by the tragedy of a neurologist stricken down by one of his own maladies ...'[208]

This review of Head's practice has yielded a bewildering range of representations. He is the principled physician anxious to do the best by his patients. This Head even saw his own illnesses as an opportunity to improve his effectiveness as a doctor. After a bout of influenza, he told Mayhew: 'I have learnt a great deal by being ill that I hope to practice for the good of my patients when I get well ....'[209] But he was also a clinician who often cursed the same patients as distractions from the intellectual pursuits for which he craved. Head was the apparent anti-Semite who cherished his Jewish friends. He was the severe materialist who believed that all mental disorders would ultimately be shown to have a basis in the nervous system. Yet he was also an accomplished psychological healer. It is a futile exercise to try to reconcile these contradictions into any definitive depiction of the man. Instead we are left with an ever-shifting and elusive enigma of self-contradiction. The one thread that seems to run through all the various permutations of Head's professional life was, however, his commitment to the pursuit of science.

3 'THE GREAT HARD ROAD OF NATURAL SCIENCE'

The Scientific Self

When in November 1926 Head sought to find retrospective coherence and meaning in his life, he discerned one unifying theme. 'I cannot remember,' he wrote, 'the time when I did not wish to take Medicine as my career in life'.[1] It is apparent, however, that for Head it was *scientific* medicine that had this enduring appeal to him. It was only when clinical practice was combined with the elucidation and application of fundamental truths about the working of the body that medicine became a worthy vocation. As Head put it in 1909 he when gave testimony to the Royal Commission on Vivisection, in medicine, 'practical results were the by-products of a manufacture of which knowledge was the aim'.[2]

Head claimed that his mother's cousin, Marcus Beck, had been an early exemplar for him of this kind of medicine. Beck had been an assistant to Joseph Lister when the latter pioneered antiseptic surgery at the University of Glasgow. Antisepsis typified, in Head's view, the ideal relation between theory and practice postulated by the rhetoric of scientific medicine. First had come germ theory, which revealed that such pathological processes as mortification were due to the action of microscopic organisms. This advance in knowledge led to the invention and implementation of techniques designed to exclude these germs from wounds during and following surgery, thus reducing the incidence of postoperative sepsis.[3]

Head seems to have felt an almost familial allegiance to this conception of scientific medicine. Lister himself was another kinsman, and Head declared that he 'was brought up in an atmosphere of worship for the great man who was connected with my own people'.[4] In September 1900, Head reflected on the relationship between his scientific vocation and his ancestral religious background in somewhat more jaundiced terms. He speculated that what drove his quest for knowledge was 'simply the Quaker conscience gone sour'. Whereas an earlier generation of Quaker had: 'shunned delights that they might live in Paradise ...

their degenerate descendant lives laborious days to eat of that tree that was the cause of our first parents expulsion from the garden'.[5]

The claim that science formed a necessary and obvious foundation for medicine was by no means uncontroversial in the later nineteenth century. Even within the orthodox medical profession, there were those who doubted the value of such sciences as physiology to clinical practice. There was, moreover, some resistance to the notion that the doctor should identify himself primarily with science. Especially within the London consultant community where Head spent most of his career there were influential voices that insisted that the physician should form his identity around alternative, more humanistic, cultural resources.[6] Head's own humanistic credentials were strong, although he was more 'modern' than 'classical' in his literary tastes. As far as his professional identity was concerned, however, he did not deviate from the view that the doctor must, first and foremost, be a man of science.

In the nineteenth century 'science' was, however, a term with varied connotations.[7] It might for instance imply a natural historical, primarily descriptive and classificatory orientation. But by the time he began his undergraduate career in Cambridge, Head was committed to a laboratory-based, experimental approach to the pursuit of knowledge. His brief stay in Halle had constituted an initiation into this way of life. During these months he accumulated forms of cultural capital that he was to grow and expend in later life. As well as the technical skills he had acquired through the physiology and histology classes he attended, he gained dispositions that were essential to the life of the experimentalist. In particular, Head overcame his initial revulsion at the idea of vivisection and came to view animals merely as laboratory materials (see Chapter 1, pp. 29–30).

These inclinations were reinforced once Head became a cadet member of the Cambridge physiology school. Indeed, at one point Foster appeared to have persuaded him to abandon medicine entirely in favour of a career as a laboratory scientist. Cambridge also seems to have inculcated in Head the significance of being a member of a scientific community. He recalled fondly the spirit of fellowship that obtained among his fellow Natural Science undergraduates as well as the personal attention they received from more senior men such as Langley and Lee.[8] Head was to carry the notion of science as a collaborative enterprise, in which the support and recognition of one's peers was of crucial importance, into his subsequent career.

Head's enthusiasm for science tended to take an embodied form. He was inspired less by abstract principles than by individuals who represented scientific virtue in an exemplary fashion. During his student days, Foster, Hering, and, above all, Gaskell provided Head with models of scientific virtue to be emulated. Later in life he was to find other heroic figures among the London medical and scientific circles in which he moved. Head sketched some of these paragons in the 'Rag Book' he composed with Ruth Mayhew.

One of these was David Ferrier (1843–1928), who, like Head, was a clinical neurologist with a strong interest in the physiology of the nervous system. Head recalled in reverential tones his last encounter with Ferrier when a general practitioner arranged for the two physicians to hold a joint consultation. Head made his way to Ferrier's consulting-room to discuss the case, choosing:

> an hour when the ordinary Consultant ceases to be busy, but was shown into a waiting room overflowing with patients – many of whom had obviously been waiting some time – Envious eyes followed me as I was quickly sorted out & taken to a little back room by myself. The inner door suddenly opened and a brisk little man with side whiskers and gold spectacles took me by the hand & placing his left hand on my shoulder led me into his consulting room. His voice is a rich baritone with a somewhat pompous enunciation. The nature of the case was rapidly considered, details of time & place settled and I left him to continue his attempt to empty his waiting room.

This account stresses Ferrier's success as a practitioner and the almost religious faith his patients appeared to vest in him as well as his personal idiosyncrasies. But it was only on the following day that the true essence of the man was revealed:

> On the appointed day I arrived first at the station and reserved a carriage for the great man. He made his appearance brisk well groomed and business-like, settled himself into his corner drew a dust cloth across his knees and said 'I've heard enough about our patient – Now tell me what you are all doing'.
> During the journey we talked physiology only and I saw the ashes flame up once again. For in the old days he too held the fire from heaven.[9]

Here the religious imagery becomes explicit. The heavenly 'fire' that Head recognized still smouldering in the elderly Ferrier was the same fervour for science that now burned in him.

But even Ferrier's glory dimmed in comparison with that of a scientific deity so revered that Head hesitated to name him. He wrote in awed tones of his impressions of an encounter with this being:

> The door of a small house in the corner of Manchester Square is opened by a very old butler who says he will see if the Doctor can see me – 'he has not been well lately but is better now'. I am ushered into a large room in which my eye is at once attracted by two immense arm chairs. By each of them stands a low table with notebooks & pencils. Round about, in apparent hopeless disorder, lie innumerable books, pencilled journals & scraps of typewritten manuscript. From one of the chairs rises a white haired man in a wrinkled old fashioned frock coat. Standing with his head a little on one side like a great bird, he stretches out to me a most friendly hand. His words come thickly through the veil of his heavy white beard & moustache which entirely cover

his mouth. I draw close to his great chair ~~for he is somewhat deaf~~ and plunge into the middle of a scientific conversation. At first he is somewhat shy for his life during many years has been lonely and this loneliness has been increased by the onset of deafness. Although nearly 70 years old he is still full of ideas and has maintained a wonderful freshness of interest. But he has a curious and embarrassing habit of assuming, in his great modesty, that fundamental principles enunciated by him a quarter of a century ago are still unknown to me, to whom they have been elemental steps in intellectual training.[10]

It is left to the reader to identify this figure as the neurologist, John Hughlings Jackson (1835-1911). Head noted Jackson's physical frailty and infirmities as well as his eccentricities — for instance, the great man's taste for trashy novels. But these peculiarities serve only to highlight the grandeur of his hero's intellectual achievements. As we shall see, Head's mature conception of the workings of the nervous system did indeed owe a great debt to Jackson.

In the 'Rag Book,' Head also recounted the visit he and W. H. R. Rivers paid in 1903 to the now elderly Gaskell (see Chapter 1, p. 39). Head noted that Gaskell's 'life seems to be completely wrapped up now in his theories of the origin of vertebrates. To this key to all the mythologies he refers all new information put before him'.[11] Head's adulation of his old mentor is here tempered somewhat by a touch of wry amusement: there is an allusion to the obsessive scholar, Casaubon in George Elliot's *Middlemarch*. Gaskell's theories of the evolution of the vertebrate nervous system were, however, to have a major influence upon Head's early researches on the organization of the spinal and cranial nerves in man.

Ferrier and Jackson were two of the extensive network of professional acquaintances that Head developed after he settled in London. For Head, science remained an essentially communal enterprise albeit one that sometimes required solitary study and experimentation. He maintained that no scientist, however gifted, could operate to his full potential without the support and stimulation of this social matrix. Thus Head's evaluation of the Burnley physician James Mackenzie (1853-1925), with whom he shared some research interests, was: 'He is a man of great ability and energy but too far from the centre of knowledge to recognize the great general tendency of the things he had seen. His mind was fixed on the individual element in each case and he failed to see the full general bearing of his work on the structure and function of the nervous system'. Head reported that a paper Mackenzie had delivered at the Neurological Society was fatally flawed because 'It was entirely personal ... In Science the recitation of a Credo is entirely uninteresting unless it be that of a heretic who is ready to fight for his Antinomian doctrine'.[12]

Head maintained that a true understanding of and dedication to the scientific way of life was confined to a tiny elite even within his own profession. His comment on his rivals for a Cambridge fellowship for which he had unsuccessfully applied was that: 'The researcher's outlook, with its sense of perfect form and completeness of demonstration for the sake of completeness, was to them unknown'. The aesthetic dimension with which Head endowed the researcher's outlook is noteworthy: being confronted with these unworthy competitors: 'gave one the sense of aloofness that the artist experiences amongst those who are ignorant of art and its methods'.[13]

The pursuit of scientific truth moreover possessed a vital moral component. In November 1900, when complaining of the shortcomings of a pupil's MD Thesis, Head lamented that: 'He was perfectly satisfied with the most slipshod statements and I had the greatest difficulty in convincing him that it was morally wrong to make indefinite assertions about definable facts'. Such 'prostitution of the mind' was, he declared, 'the foulest of all sins'.[14] In such passages Head came close to admitting that for him science constituted a surrogate religion.

The intensity of Head's dedication to the scientific ethic could be frightening. Ruth Mayhew recalled a cloudy evening she and Head spent on Hampstead Heath soon after he had learned of the sudden death of a colleague at the London Hospital. This was a man with whom Head had planned 'to work for next twenty years'. As she sat next to him on a bench on the heath, Mayhew confessed: 'I have never felt so alienated from you as then'. It was:

> so horrible to me to see so clearly that for your friend as man you really were not mourning. You cared because he was working at your side for truth and righteousness. It is a noble Roman Senator's way to mourn one's dead, but to realize that <u>that</u> was, and always would be, your way was like a cold hand at my heart.[15]

Head was an active member of such professional and scientific organizations as the Neurological Society and the Royal Society of Medicine. A token of Head's growing esteem within the scientific community came in 1899 when he was, in recognition of his researches on the physiology of sensation, elected a Fellow of the Royal Society. His proposers included Gaskell, Langley, Ferrier, and Jackson.[16] It was, however, the surgeon, Victor Horsley (1857–1916) who came 'quite late on a Thursday night' in July to Head's apartment to inform him of his election. Far from deriving satisfaction from the news, Head felt: 'anxious & frightened. It is rather as if one heard one's own funeral oration'. Head recognized that this distinction had been awarded not so much for 'work already accomplished but on the promise for the future in the work already done'. He was all too aware that: 'this flowery promise is a most fallacious thing. An acute frost of worldly success or the slow blight of advancing age may prevent the greater part of that blossom from forming fruit ...'. Because of these misgivings, he was 'filled with present fear and dread for the future ...'.[17]

In the following twenty years Head regularly attended Royal Society meetings. His recollections are informative of his encounters with contemporary developments in the sciences. In February 1902, for instance, he attended an address on the electron theory by the physicist Sir William Crookes (1832–1919), from which Head gathered that: 'the hard boundary which was supposed to exist between Matter & Force can be upheld no longer'.[18] Four months later, Head heard his friend William Bateson's (1861–1926) exposition of Mendel's long-forgotten theory of heredity. He noted in particular Bateson's claim that: 'it will become evident that an experimental study of heredity pursued on the lines Mendel has made possible is second to no branch of Science in the certainty and magnitude of the results it offers'.[19]

These gatherings also enabled Head to observe and refine his views of the different types of scientist that comprised the Society. Thus a Royal Society dinner Head attended in December 1902 led him to discriminate in a judgmental way between the various kinds of personality that a career in science seemed to generate. He found himself next to the chemist Henry Edward Armstrong (1848–1937). Armstrong was:

> a man of imagination and has attempted to reconstruct our ideas about the ether. I drew him out about his conceptions of immunity and other conditions that seem to be forming a link between the chemical physicist and medical man. But next to him sat a foolish creature who interrupted each conversation by saying 'how serious you are getting'. When Armstrong had been put off his line for the second time I felt murderously inclined to this shallow fool. Then to my astonishment I found that this creature was the recipient of a medal for advances in our knowledge of vaccination and smallpox.

Head concluded that the 'curse of Natural Science lies in the fact that it may lead purely to technical aptitude of a high order which then masquerades as intellect ...' On the other hand, when science 'does educate its educational value is incalculable'. The Royal Society was composed of both scientific types; on the one hand were those:

> in whom knowledge is a vitalizing force impregnating them to great intellectual fecundity – On the other hand men like his neighbour (C[opeman].) know, act on their knowledge and produce results of practical or scientific importance remaining throughout entirely uncultivated. They resemble their own filters apt to strain off even the minutest particle of solid; but when this function is performed the cold water of life rinses them clean. They have imbibed nothing but matter which can be washed out again ...[20]

Such discrimination between lower and higher types was a recurrent motif in Head's writings. As we shall see, he regarded the power to make fine discriminations as a natural property of a highly evolved nervous system.[21]

Head's account of his own routine work as a scientist showed a due respect for 'technical aptitude' – when combined with true intellectual insight. Indeed, from his Prague days he enjoyed a reputation as an exceptionally skilful and ingenious experimenter. Even after he embarked upon a clinical career, he retained an association with laboratory life. For five years as part of his early researches into the physiopathology of sensation, he worked on a weekly basis at the pathological laboratory attached to the Claybury Asylum in Essex. Later, Head was to visit the physiology laboratories at Oxford to seek experimental confirmation of phenomena he had observed in the clinic. Thus in July 1902, following 'a rather dull meeting', he spent some time 'most fruitfully in the Physiological Laboratory'.[22]

Head's accounts of his times at Claybury have an almost rhapsodic character. This was in part due to the semi-rural location of the establishment; he recalled the walk he took one Sunday 'in the glorious autumn morning up that long country road to the Laboratory'. When the time came to quit the Laboratory, Head became even more lyrical, recalling that while at Claybury, 'I have watched the chestnuts flower and the Poplars break out into leaf; I saw the hay made and carried and the water come again onto the low meadow with the autumn rains – How much the merely physical surroundings of the place had been to me I did not know until today'.[23]

The fact that the Laboratory was often deserted when Head worked there was also part of Claybury's attraction. But it was the nature of the labour itself that most elated him; he looked forward:

> with delight to more hours of solid silence occupied only by the Mechanical Employment of [?] bodies, like the caps of a child's pistol, from one fluid to another. Many verses were running through my head, Wordsworth lay on the table, and I expected that something would [crystallize] out before night-time. However, I became so excited in my little caplike bodies when night came I was surrounded by many glass slips on which were fixed twisty purple bodies and not a thought had I given to anything else. You cannot conceive the excitement of getting a positive result. First came months of waiting – then the supreme moment when that material lies in your hands. Then the months of waiting till it is prepared for examination and the constant fear that after all that the whole will fail.

Head described the feelings of ecstasy that this mix of manual work and intellectual exhilaration produced as 'what the Germans call 'Wissensrausch' – an intoxication far exceeding that of [?] love – which is unfortunately usually followed, like the other intoxication, by a period of depression. This however I await philosophically. I have been gloriously drunk'.[24]

Head's account of the 'wonderful five years' he laboured at Claybury often gestures towards Romantic conceptions of scientific creativity.[25] Technical aptitude and tireless application alone did not suffice.[26] To be a true scientist, rather than a mere mechanician, required a certain genius – the 'fire from heaven' that Head discerned in his mentors and with which he believed he was also endowed.

One way the *ignis dei* manifested itself was through an aptitude for *discovery*: the ability to achieve truly original and profound insights into the workings of nature. For Head 'true, as opposed to fortuitous research',[27] had a teleological aspect. In retrospect it was clear how the end was foreshadowed in the origins of a successful programme of investigation. Although he scorned scientific claims that did not meet rigorous logical and technical criteria, it is clear that Head believed that the capacity to grasp a key germinal idea was not entirely rational in nature. This gift was contingent upon an intuitive grasp of the essence of a particular phenomenon. In September 1900, Head recounted one such flash of inspiration:

> I so well remember the moment at which I first got the idea which underlies the work I have been doing for the last ten years & shall continue, I hope, till I die. It was a beautiful summer Sunday afternoon and I then lived in the little room in Wimpole Mews. I was sitting at the window sill reading Matthew Arnold with a sense of well earned Sunday leisure – Suddenly it came apparently from nowhere & I spent the rest of the day looking up authorities to see if it worked – and it did work and I saw opening up before me the whole vistas of unexplored country.

This recollection came in the context of a lament at how the pressure of his routine medical work left him with ever less time for science. He complained that: 'I do not have those intuitions now – Not because I cannot get them but simply because the necessary idle hours are wanting – those hours in which the seed ripens into a living being'.[28]

These insights into the inner workings of nature sometimes took on an almost mystical aspect that enabled Head to overcome his frustrations with the mundane world. In August 1902, after a particularly vexatious bout of hospital politics at the London, Head spent four days conducting research at the Royal Army Medical Corps hospital at Netley. The experience had a transformative effect, which began when he alighted at the station: 'there came through the starlit night a breath from the sea and next morning in the sunshine the world seemed to laugh again – in comparison with the logical beauty of the work past worries seemed the fussy unrest of an ant-heap'. The following day, 'in the blue air, with the mild salt wind blowing over the beautiful Hospital grounds it seemed that for once we touched something that lay behind the surface ripples of the world'. Head saw with sudden clarity the logical structure that underlay the complex phenomena with which he had been wrestling. A guiding principle that had long been growing in the hidden recesses of his mind suddenly became fully manifest – 'For many months the idea underlying the whole has been present with me and I have known that ultimately another child will be born – But during these few happy days the work has taken form and I am as glad as the mother who has felt the moving limb of her unborn child'.[29]

In this letter, Head compared the complexity of the physiological problems with which he was wrestling to that of a 'Chinese puzzle' that had to be unravelled. In conversation with the ethnologist Brenda Seligman (1882–1965), he varied the metaphor somewhat while retaining the oriental reference, with its connotations of mystery and exoticism. Seligman recalled that Head had once told her that: 'to understand a Chinese picture one had to get "inside" it'. He also shared with Seligman a revelatory moment of scientific insight he had experienced:

> He had been working on his neurological research & had got to a stage when he felt stale – could not see the wood for the trees. He took a walking holiday & put it all out of his mind. Suddenly looking up at the sky he saw clouds that formed the image of a hand with out-stretched fingers; from that moment facts that had seemed isolated tumbled into place, & he knew the lines on which he was to work for many years to come.[30]

One assumes that this was the flash of inspiration for the experiments on his arm that Head was to conduct with W. H. R. Rivers.

In the 'Rag Book,' Head gave an account of a similar moment of revelation. In October 1901 he wrote:

> For five years I have recognized the existence of a scientific problem which it would be necessary to face and to investigate. But the progress of other work necessitated that only up till the present time all observations towards the solution of this question could only be perfunctory. Last week I began to consider the matter, read what others had done and reviewed my own observations – Suddenly when walking down the street this afternoon the answer came whilst I was thinking of indifferent matters. It came in the form of a picture or diagram which however revealed more than any one picture or diagram could ever do – I also seemed to at the same moment to see all the far reaching consequences of the hypothesis by changes in the form of this mental picture. No formulated reasons & nothing approaching a syllogism played any part in what I believe to be a statement capable of strictly logical proof.[31]

The last sentence in particular hints that such insights – while retrospectively amenable to rational justification – originated in subconscious levels of the mind that did not observe the rules of logic. This faculty was in many ways akin to artistic creativity. The German Romantic conception of *Anschauung* captures something of this intuitive capacity.[32] Such power was not given to every scientist; it was reserved to those who possessed true genius. Head felt no need for false diffidence in this regard. 'A genius,' he declared, 'must be immodest about his own work because he has a true conception of its value which is great.'[33]

Others were less convinced. Some of Head's critics – such as James Mackenzie – felt that the inspirational style of science that he practised was flawed. Pursuing research in accordance with some initial germinal idea led the scientist to find the results he has expected and to disregard those that failed to meet his preconceptions. Others, however, accepted that Head did possess a special gift. In 1934 Robert Nichols, during a stay in Germany, encountered the former editor of the *Lancet* Samuel Squire Sprigge (1860–1937). Sprigge expressed 'the greatest admiration for Henry'. He told Nichols: 'Head would frankly tell you "I don't know why this should be so, but it will be proved to be so". His intuition was phenomenal. He had that which marks out only the greatest – he saw into nature'.[34]

Head considered the researches in which he engaged as 'the permanent fruit of my life'.[35] His discoveries were, however, only of value if they could be communicated to the wider world. This involved writing scientific papers and monographs. Head found this aspect of the creative process trying. In 1898, however, he invested in an item of furniture intended to give structure and discipline to his literary activities. He asked Ruth Mayhew: 'Have you ever had the curious feeling that I experienced this week of expending your energy over something that is to last your life?' Most possessions were, he continued, of a transient nature:

> This week however I have received a Cabinet for manuscripts so designed that the various subjects in which I am interested will be kept in separate drawers and as each nears completion that drawer will grow fuller & fuller. When that piece of work is thrown upon the world the materials will remain, until I am gone, for reference & revision. Thus from now onwards whether I die next week or in 40 years I may add to the drawers but each drawer that is in use will contain the same type of manuscript till I pass to a land where the pen no longer moves. It has given me a curious feeling of settlement – of determination in the old meaning of the word.[36]

The solidity and permanence of this cabinet thus supplied a metaphor for Head's readiness to dedicate his life to the scientific enterprise.

Head articulated clear views about the value of science to humanity. He was scornful of any purely utilitarian understanding of the contribution of scientific knowledge to the human condition. In November 1900, Ruth Mayhew told Head of a sermon she had heard by the Bishop of Winchester ('a florid old fogey') in which: 'he lumped together the benefits Science had taught us during the last hundred years thus illuminatingly – "Telegraphy, Anaesthetics, and photography"'.[37] Head professed himself to be 'immensely amused with the Bishop's catalogue of the benefits of Science'. This display of banality did, however, lead him to some serious reflections.

For reasons of his own, Head declared, the Bishop:

> dared not explain that the real gift of Natural Science to the last half of the century is that it has shown us the way to the Tree of Knowledge and has taught us to eat fearlessly & to become as Gods knowing good & evil. For our speculative intrepidity, though it bring death in its train, amply compensates for the bestial timidity of the garden where a dominant form walked even by noonday.[38]

The Bishop thus dared not speak of the true significance of science in the modern world quite simply because to do so would be to admit that, thanks to the rise of the scientific worldview, Christianity had become redundant to the spiritual needs of mankind.

Head's denunciation of Christianity has Nietzschean overtones.[39] The true value of science to humanity had been to free the mind from a slave philosophy: science was, in effect, the high road to spiritual freedom. In a poem he sent to Charles Sherrington in January 1915, Head gave a remarkable representation of his scientific self. He compared his activity to that of:

> ...the spider in his patterned web,
> Based on immutable law,
> Boldly I spun the strands of arduous thought.[40]

Such a being – at once embedded in nature and yet through force of intellect unravelling her laws – was far from any Christian notion of the scientist as the humble, and largely passive, vehicle for the discovery of divine wisdom.

In the 'Rag Book,' Head expatiated further on 'the message of Natural Science to man'. The physical sciences had, he argued, shown that the universe had arisen 'by the aggregation of masses of flying dust or gas and that it will ultimately return to a similar condition to be grouped again in some new combination'. The fact that 'man has inhabited the surface of the globe for some, to us, enormous span of years will make no difference of any kind to the new system in space that will arise from the remains of this world'. Biology, for its part, 'teaches us that man has arisen from the lowest form of living matter by a laborious series of changes'. These changes were in no way the results of divine design leading to some 'higher ideal'. On the contrary, in the post-Darwinian era it was apparent that: 'an infinite number of variations occur and of their nature destroys all those unsuited to the survival of the animal in its environment'. Thus the: 'Gospel of Natural Science, compelling belief in every mind sufficiently trained to understand it, is on the physical side hopelessness, on the Biological side Cruelty'.[41]

To those who still clung to a more comforting Christian view of the universe and of man's place in nature, this might appear a gospel of despair that must occasion 'an intolerable oppression' from which the only release would be suicide. Weak individuals would succumb in the face of this realization and sink into a paralyzing morbid melancholy. Head sought, however, an escape from this apparent existential impasse by positing a form of psychological dualism. The human mind, he maintained, possessed both a generic and an individual component. Natural science 'appeals to the generic aspect of the human mind. Conduct is mainly determined by that side of the mind which is bound up with the sense of individual personality'. Thus both 'Jekyll & Hyde probably believed in the nebular hypothesis & in evolution; but the Jekyll-Hyde individual was the prey of two opposite personalities & two opposed lines of conduct'.[42] In other words, it was, Head maintained, possible to divorce one's theoretical understanding of the world from the life choices one made within it.

In effect, 'exalted minds overcome this paralysis by a denial of the absolute reality of physical laws, whereby they escape the inertia generated by that sense of the resistance of the external world'.[43] That is, when confronted with the implacable reality into which he or she was thrown, it was open at least to the 'exalted' individual actively to construct meaning and liberty. Head suggested a number of ways in which this 'denial' might be accomplished – for instance, through religion and art. But there was also a third path: namely, 'by the use of the Scientific method'. He recognized that this might at first glance appear 'a contradiction in terms;' it was, after all, science that had created the vision of a physical world bereft of value against which the exalted individual was obliged to react. However,

> the study of philosophy, history or natural science by the Scientific method assumes a certain bulk of phenomena into the personal mind and so reduces the resistance of the external world. Thus to me the greatest personal happiness of which I am capable results from the investigation of some psychological or biological aspect of the world. But I can get no comfort from the stars – The heavens on a starlight night might be a sheet pricked with holes for all I care.[44]

Thus for Head, the pursuit of science – and in particular of biology and psychology – provided the 'sense of freedom' that Walter Pater had claimed was the chief requisite of the human spirit 'in the face of modern life'.[45] Head had, in effect, resolved the antinomies of modernity by making an existentialist choice. The same scientific method that had created a universe without intrinsic meaning also supplied him with the means to 'live and work and play'.[46]

The psychological and biological aspects of the world to which Head devoted most attention as an investigator fell into two main categories: the physiology of sensation and the cerebral mechanisms of language. These division, however, were both aspects of a unified endeavour to understand mind's place in nature – and nature's place in the mind.

The Evolution Of Feeling

In October 1925, while 'turning out some old papers,' Head came across a document that had been composed 'by certain members of the Neurological Society about the year 1880'. It had been passed to Head in the early 1890s by Armand de Watteville (1846–1925). This document was 'a parody of [John Hughlings] Jackson's doctrine':

WHOSOEVER will be saved, before all things it is necessary that he hold the Neurological faith, which faith except everyone do keep whole and undefiled, without doubt he shall perish everlastingly.

And the Neurological faith is this: that Unity in the Nervous System, and the Nervous System in Unity, is to be recognized. Neither confounding the Function, nor dividing the Structure. For there is one portion of the Brain, another of the Cord; and another of the Sympathetic.

But the structure of the Brain, of the Cord, and of the Sympathetic is all one; the cells similar, and the fibre identical.

....

So likewise the Brain action is reflex, the Cord action reflex; and the Sympathetic action reflex.

And yet there are not three reflex actions; but one reflex action.

....

The Cord is made of none; neither made, nor designed.

The Brain is of the Cord alone; not made, nor created, nor designed but evolved.

....

So that in all things, as aforesaid, Unity in the Nervous System, and the Nervous System in Unity, is to be recognized.

He, therefore, who will be saved; must thus think of the Nervous system.[47]

Head described this exposition of the Neurological Faith as a '*jeu d'esprit*', which he forwarded for Sherrington's amusement. But he added that the document: 'certainly shows how closely Jackson had arrived at the sort of position we hold to-day'.[48] This creed is also revealing of the disciplinary postulates of the scientific community that Head joined in the 1890s. Fundamental to the assumptions of this group was the doctrine of evolution – and in particular the version of that theory expounded by John Hughlings Jackson. On this view, the structural and functional unity of the nervous system was an unquestioned dogma. The human brain, for instance, was 'of the Cord alone' because it had evolved from the lower, more primitive parts of the system. Head would, moreover, have already been indoctrinated in this creed through his training in the Cambridge school of physiology and, in particular, through his devotion to Gaskell – whom he described as 'my spiritual father'.[49]

Of particular relevance to Head's early researches was the work Gaskell published in the 1880s on the evolution of the vertebrate cerebrospinal axis. Gaskell argued that: 'the evidence that not only the nerves but also the groups of nerve-cells from which these nerves arise are arranged not in a continuous chain but metamerically, points directly to the conclusion that the nervous tissue of the animal from which the vertebrates took their origin was arranged in a distinctly segmental manner'. This was true both of the spinal cord and of the cranial nerves, which were 'built on the same plan as the spinal nerves, it is clear that the different groups of cells from which they arise are arranged segmentally'.[50]

The researches that Head conducted during his first decade in London served, for those who adhered to it, to confirm this doctrine in man taking advantage of the 'natural experiments' that disease had performed. During his time in Prague, Head had acquired from Hering an interest in sensation, a liminal topic that encroached upon the interests of both the physiologist and psychologist. It was also a theme that required the use of human subjects and could thus only be explored in a clinical context.

In particular, Head saw the phenomenon of 'referred pain' as a means to elicit important truths about the organization of the human nervous system. His work in such institutions as the Victoria Park Hospital for Diseases of the Chest acquainted him with patients suffering from a variety of visceral diseases. He noted that in many cases as a result of these pathological conditions in the internal organs, pain was *referred* to the surface of the body. Furthermore, Head maintained that the areas of the skin affected in this way conformed to a fixed pattern: he 'found that these sensory disturbances also followed definite lines'.[51] In a letter to James Mackenzie, Head revealed that he had in fact obtained the post of House Physician at Victoria Park with the specific aim of facilitating these researches: 'I found it was quite hopeless unless every note on the patients condition was taken by myself and the patients absolutely under my charge day & night'.[52]

In 1893, Head described how he had mapped these tender areas on the bodies of his patients using nothing more sophisticated than a pin. He advised readers wishing to repeat his procedures:

> Travel over the surface of the abdomen using the blunt end only, in the same way as if you were using the point to test for analgesia. In a favourable case the patient does not complain until the limits of the tender areas are reached, when he at once complains that he is intensely sore and he will even cry out that you are pricking him. Thus in a favourable case the increased sensibility is so great that the contact of the blunt head of a pin is mistaken for a prick of the point.

Head took care to quote the patient's own accounts of the sensations they experienced during these investigations, noting that it was: 'wonderful how unanimous patients are in their description of the sensation produced by touching such a tender area'. The patient would say that he felt: 'sore and bruised, as if he had been beaten about the body or back'.[53] Another method to identify these tender areas, which was 'clumsy but theoretically interesting', was to apply a warm sponge or a test-tube containing warm water to the surface of the body. As soon as 'the tender area is reached the-patient experiences a feeling of great heat and flinches under the application'.[54] Cold would produce similar results. With some refinements, these were the techniques that Head was to use in all his subsequent work on the physiology of sensation.

Head also felt the need to devise a *literary* technology adequate to the needs of this line of research.[55] In his letters he frequently wrote of the pains he took to convert the raw material gathered in his investigations into a form that would be intelligible and persuasive to his scientific peers. While he found it relatively easy to expound these notions in lectures, expressing them in writing was much more challenging. As Head proceeded with his work on sensation, the linguistic resources that had served previous generations of physiologist seemed to him ever more inadequate. In 1901 he confided to Salaman that: 'I am troubled by the inadequacy of the English language and am trying to work out a simple style to express reiterated scientific statements'.[56]

To give added credibility to his claims, Head appealed in his paper to the familiarity with pain common to members of his audience: 'Many of my readers will have at one time or another suffered from these tender areas, and will be able to bear out my description'.[57] Such appeals apparently had some rhetorical force. In 1927, Head wrote to Charles Sherrington: 'I too have not forgotten the moment when you were converted to the theory of referred pain by that most drastic of all methods, personal experience'.[58]

The chief outcome of investigations undertaken in this manner was that:

> it is evident that these areas do not overlap one another, and it is a peculiar characteristic of these areas of painful sensation that wherever they are situated on the surface of the body, whether they be marked out by hyperalgesia or analgesia, their limits never materially alter or overlap.[59]

This was a 'both curious and unexpected' result given that Sherrington had shown there to be a considerable overlap of the areas that took their nerve supply from the spinal roots.

To try to explain this apparent anomaly, Head decided to take the disease known as Herpes Zoster, or shingles, as an experimental model. This was a condition in which painful eruptions on the surface of the body were thought to be the result of irritation of the posterior roots of the spine. Head saw this disease as, *inter alia*, an ideal opportunity to map the distribution of the sensory nerves emanating from the posterior roots. The presence of skin lesions in Herpes Zoster provided visual evidence lacking in other forms of referred pain. In the manner of a natural historian, he began to 'collect' cases of shingles in order to establish the distribution of the cutaneous eruptions that formed part of its symptomatology. He was:

> astonished to find that they agreed in an extraordinary way with the areas of tenderness I had observed in visceral disease, for not only did the actual areas occupied by the eruption agree with the areas of tenderness, but I found that if the herpetic areas were carefully marked and measured, they were never found to overlap. Moreover the advent of the eruptions is mostly preceded by pain and tenderness of the skin. In three

cases in which I was fortunate enough to see the patient before the eruption was developed, I was able to mark out an area on the skin which exactly corresponded to that seen in visceral disease, and I was unable at the time to say whether this area was due to reference from visceral disturbances, or was the precursor of a herpetic eruption.[60]

On the basis of this evidence, Head proceeded to produce a diagram, which purported to show the supply of nerves from determinate segments of the spinal cord.

Head was to extend his research on Herpes Zoster in the later 1890s relying chiefly on patients he encountered at the London Hospital. He was eventually able to compile a total of 450 cases that had come under his own observation. He added to this 'collection' by mobilizing his network of contacts at other hospitals – including Guy's, Great Ormond Street, Bart's, and Queen Square – to provide additional material.[61] In this research, Head was especially anxious to acquire a comprehensive visual picture of the disease. He marked the location of eruptions on pre-printed schematic body outlines and made extensive use of photography in an effort to fix the clinical aspects of the condition.

Moreover, because Head's theories had initially elicited a great deal of scepticism, he 'felt that if a *post-mortem* examination, could be obtained on a case of herpes zoster in which the distribution of the eruption had been carefully drawn or photographed, the question might be settled one way or the other'. Since herpes was not a fatal disease this raised some practical problems. To circumvent these, Head decided to concentrate upon occurrences of the disease in Poor Law Infirmaries and in Lunatic Asylums, 'a considerable proportion of whom remain in the institution until their death'.[62] To facilitate access to such cases, Head collaborated with Alfred Walter Campbell, Pathologist to the County Asylum at Rainhill in Lancashire. (This practice of finding ad hoc collaborators in a particular project became typical of Head's later research.) With Campbell's help, Head was able to accumulate 14 of the 21 autopsies that were presented in their joint publication.

These early papers on sensation are for the most part overtly empirical in nature. Page after page of clinical, pathological, and anatomical detail, complemented by diagrams and micrographs, aims to convince the reader of the incontrovertible truth of Head's claims. Even the sections of the papers labelled 'Theoretical' scarcely deviate from this register. Throughout, however, Head adhered to tacit assumptions about the nature of the nervous system consistent with the neurological 'creed' to which he subscribed. Thus when, after describing the areas uniformly affected by visceral disease, he asked 'to what level of the nervous system do they correspond?',[63] the question implied the hierarchical views articulated by John Hughlings Jackson.[64]

The word 'evolution' does occur in another paper in which Head extended his analysis to pains located in the cranium and neck. The term is, however, used in the old, epigenetic, sense. When he goes on to speak of the way in which the current form of sensory areas had arisen by the modification of 'a primitive band-like arrangement',[65] it is unclear whether he is speaking of embryonic or phylogenetic change.

Head's papers on referred pain amounted, however, to a corroboration of Gaskell's views on the origins of the human nervous system. He noted 'that the sensory supply of the viscera bears a very curious resemblance to Gaskell's scheme of their motor and inhibitory supply'. Gaskell for his part confirmed that the conclusions that Head had reached had been independently corroborated using anatomical evidence.[66] The fact that the nerve supply to the surface of the body took the form of discrete 'bands,' corresponding to spinal 'segments', was a vestige of the evolution of the human spinal cord from the chain of ganglia found in arthropods. Head believed that he had, in effect, discovered in the diseased bodies of his hospital patients vestiges of human ancestry (see Figure 1).

The importance of this deep commitment to the evolutionary creed in shaping Head's observations is evinced by some critical remarks made by the Burnley physician James Mackenzie on these researches. Mackenzie had addressed the subject of referred disease independently of Head. Indeed he complained that although Head had 'got the idea' from one of Mackenzie's papers, 'save for a passing reference in his first communication, he never alludes to my work in this connection in the many articles he has written'. Mackenzie did not question the greater 'brilliancy' of Head's treatment of the topic. He *did*, however, cast doubt on Head's claim that the areas of cutaneous tenderness were strictly demarcated in correspondence with the affected spinal segment. In Mackenzie's view his own researches had demonstrated that: 'no visceral disease can give rise to a "segmental" hyperaesthetic area ... But all these facts Head has studiously ignored and has gone on maintaining that the cutaneous heperaesthesia is segmental and that these areas do not overlap, ... and so has imposed 'Head's areas' on the world ...'.[67] For his part, Head viewed Mackenzie as a talented individual, but one whose scientific contribution was curtailed because of his intellectual isolation (see p. 102 above). It might be added that, unlike Head, Mackenzie was sceptical about the heuristic value of laboratory science to medicine.

Head noted that his researches on referred pain: 'opened up the whole question of sensation in its various forms'.[68] As he and Campbell prepared their 'vast paper' for publication, Head confided to Ruth Mayhew the hopes he had for these researches as well as his emotions as these laborious investigations drew to a conclusion. The process had, Head revealed: 'filled me with unrighteous exaltation that I know will end in dire depression. For we seem to have touched upon the fundamental & final instead of the hypothetical & temporary'.[69] Head was to explore various other aspects of the subject in the following two decades. Throughout Head retained his commitment to explaining sensation in man within an evolutionary framework as the royal route to attaining 'fundamental & final' truths.

These early papers also presage other themes that became prominent in Head's later writings. There is, in the first place, a strain of what might be called sensory elitism. The relatively crude methods that Head used to explore altered sensation in his patients have been outlined above. In his 1894 paper dealing with referred pain in the head and neck, he elaborated on the protocols that he employed in such cases:

> In every case it is best to use at least two methods, and to state which form of stimulation has produced the results figured on the recording chart. The method I adopt in working out a case is as follows: First the patient is asked to point out the situation of the pain. Then I rapidly test the skin, beginning from a part of the head over which sensation is unaltered, and working towards the hyperalgesic patch. Marks are made to show the points at which the tenderness begins. Then the history is taken or the patient engaged in conversation. After a time a second estimation is made. If the case be one of disease of the eye, ear, &c, I make a third estimation after the patient has been subjected to refraction testing, ophthalmoscopic or laryngoscopic examination, &c.[70]

The methodology remained relatively simple – especially when compared with the elaborate techniques that Head had devised in his experimental studies of respiration. The study of sensation did, however, depend upon the input

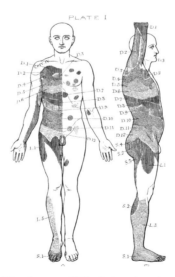

Figure 1: Distribution of 'Head's Areas' on the surface of the body.

of one singularly complex and often unreliable registering device: namely, the human subject upon whose reactions and verbal contributions the investigator depended. The management of this essential component of the exercise involved the deployment of skills and tact of a different order from anything demanded of the laboratory scientist. It was, Head observed, 'very important neither to tire the patient, nor on the other hand, to concentrate his attention on the patch of tenderness'. The latter point was of special importance; there was always a danger that in the course of these interrogations the researcher would unconsciously lead the subject in a particular direction. When dealing with affections of the head and neck, Head noted that it was fortunate that: 'marks made on the skin are ... invisible to the patient himself', thus obviating one source of error.[71]

As his researches came to focus on ever more subtle aspects of sensory appreciation, Head became increasingly concerned with the reliability of the human component to these investigations. He came to believe that just as there was an evolutionary hierarchy within the nervous system, so there was a hierarchy of sensory appreciation within the human race. Some individuals and classes of individual were therefore markedly more reliable as witnesses to their sensations than others.

There is a hint of this kind of elitism in the 1894 paper. Certain diseases were socially selective. Thus Head noted that: 'I have not yet seen a true case of megrim [migraine] amongst hospital patients. All my patients were medical men'. This was because this was a disease that in particular afflicted those 'engaged in intellectual work'.[72] From the scientific point of view, this was an advantage because: 'Sufferers from true megrim are of such a high order of intelligence that their own observations are very valuable'.[73]

From the outset, moreover, Head realized that the programme of research upon which he was embarking had implications that went beyond the limits of clinical medicine or physiology narrowly conceived. In his 'theoretical' discussion of why pain was sometimes referred, Head pointed out that: 'the localization of sensation is not a physical but a psychical phenomenon'.[74] His researches thus encroached upon the area in which the mental and the corporeal were commingled. In particular, the phenomenon of referred pain was an example of the dependence of the mind's understanding of the body – and indeed of the world in general – upon a complex of bodily phenomena, most of which under normal conditions lay outside consciousness. It was only when this mechanism was impaired through disease that the degree of this dependence of consciousness upon manifold nervous processes and contrivances became fully evident.

Head expatiated on these themes in the Goulstonian Lectures he delivered in 1901 at the Royal College of Physicians of London. He sent Mayhew an account of the ecstatic state that the composition of these lectures produced in him:

> I have been writing steadily & slowly at my Goulstonian lectures and rejoice in the prospect down fairy avenues of shadowy trees to a horizon that seems to grow clearer every moment – I leave my room feeling a prophet, jostle against some daily acquaintance and a sympathetic word of enquiry brings upon him the whole flood of what I have seen. I watch the look of incredulity on his face and to me in the foggy daylight I seem to talk nonsense. Then I creep back to my tower a worm with the certainty that I am launched on a colossal mistake. Gradually the records scattered about my room and the tables hanging on my walls work their will upon me and again I am away in the land of my dreams.[75]

The theme of these lectures was the psychic changes that accompanied visceral disease. He thus went beyond the effect of such disorders on sensation narrowly conceived to turn to how the higher mental functions might be altered as a result of the abnormal workings of the nervous system. Now that 'disease of internal organs has been shown to be accompanied by pain radiating around the surface of the body and by tenderness of its superficial coverings it becomes', Head argued, 'necessary to examine in how far the intrusion of such stimuli upon the nervous system are accompanied by changes in consciousness'.[76]

He had begun gathering observations on this topic in 1893 while he was employed at the Victoria Park Hospital. This gave Head the opportunity to observe such phenomena in patients suffering from a range of thoracic affections. He later supplemented this information with cases where there was some kind of abdominal disorder. Head noted that most of the patients belonged to 'the usual hospital class'. In this instance, he did not seem to see their social origins as detrimental to their status as sources of evidence. He added, however, that: 'I have excluded all those with an hereditary taint of insanity, and all Hebrews, owing to the recognized tendency of this race to functional neuroses'.[77] The reason for these exclusions was Head was trying to isolate instances of psychic confusion that did not fall within existing categories of mental disease.

Head went into some detail about the methods he had employed in this investigation. It was precisely because the patient might fear that he or she was going mad that it was difficult to elicit information about their abnormal mental states. Head therefore took pains to gain the patient's trust, because 'it is only after his confidence has been gained that the full extent of the mental changes can be estimated that underlie an apparently normal behaviour'.[78]

In order to assuage such fears, Head asked these patients no questions about the existence of any family history of insanity or neurosis until he had come to the end of his examination. He also realized that it was unlikely that he would gain the patient's confidence if he were accompanied by 'a crowd of students' or even by a nurse: 'for men are peculiarly adverse [sic] to exposing their emotional weakness before a woman, and women do not trust another woman not to tell "how silly they have been"'.[79] Head did, however, feel the need to have some companion present during these examinations 'in order to control my observations'. He accordingly explored the mental state of these patients in the company of his house physicians. He noted that, although at first reluctant to reveal the delusions from which they were suffering, for many of these patients, 'it is a relief to talk freely of phenomena that may have puzzled or frightened them'.[80]

The phenomena in question included inexplicable feelings of elation or depression. The patient might also experience a sense of dread or delusions of persecution. But the most striking of the symptoms occasioned by visceral disease that Head encountered were hallucinations that could afflict almost all the senses. Thus a man suffering pulmonary tuberculosis 'imagined he saw someone standing at the foot of his bed or walking across the room'. This apparition struck fear into the patient. Head found that the appearance of such a figure was the typical form that visual hallucinations took in these cases: 'Patients are unanimous that this figure is unlike anything they have ever seen and they are peculiarly definite in describing it as "draped", "wrapped in a shawl", "wrapped in a sheet", and in two instances volunteered that "it was dressed like statues are dressed and not in proper clothes"'.[81]

Such descriptions of course recalled many of the conventional features of the spectres found in folklore – especially when accompanied by 'sweating, heart-beating, and "goose-flesh"'.[82] The point was not lost on Head. But he observed: 'Curiously, not one of my patients spoke of the hallucinations as a ghost, possibly owing to the almost entire ignorance shown by the London-bred population of ghost stories or fairy tales'.[83] This was an interesting ethnographic observation. Head clearly associated belief in supernatural beings with the culture of rural, pre-modern, societies such as the one he had encountered in Bohemia. Patients of the 'hospital class' were presumably not familiar with the literary genre of the ghost story that flourished in the Victorian period, of which a leading exponent was Head's Cambridge contemporary and friend, Montague Rhodes James (1862–1936).

As well as these visual hallucinations, Head found evidence of auditory and olfactory delusions among these patients. One of them, for example, was troubled by 'a noise resembling loud breathing or inarticulate whispering under her bed'.[84] Another complained of: 'a foul burning smell that made him feel sick'.[85] Head had no doubt that hallucinations of taste also occurred; these were, however, more difficult to establish.

At the conclusion of these lectures, Head made an attempt to explain the phenomena he had described. This proffered explanation took for granted the truth of the theory of evolution. Head argued that: 'The lower we pass down the animal scale the more purely is the life of the individual determined by visceral impulses'. The 'aim of human development', however, was 'to keep such impulses in the background in order that the mind may be ready to receive impressions of all kinds in an ever shifting environment'. For this reason, in health visceral activity was 'pushed out of consciousness' so that the mind could focus its attention upon the external world. When the effects of visceral disease became disseminated by means of the nerves, however, those sensations 'crowd into consciousness, usurping the central field of attention'. As a result, images and dispositions of which the mind was normally unaware now became inescapable.[86]

In short, the referred pain accompanying visceral disease brought about a kind of psychological atavism; the lower body regained a degree of psychic prominence that had been wrested from it in the course of human evolution. As a result, the individual lost some of the self-possession that was the chief distinction of refined human nature:

> The patient's character appears to be altered, for the content of his consciousness is changed. He will become moody, at one time unduly exalted, at another depressed without cause. Reason is displaced and he is the victim of each passing wave of feeling-tone, and he will have lost control over the expression of his emotions and of his temper.[87]

Head's explanation of the clinical phenomena he had observed thus revealed a wealth of assumptions about what was most valuable in human nature. Reason, restraint, and self-control – the key bourgeois virtues – were shown to have a biological basis. Evolutionary history became in effect a narrative of the progressive triumph of these characteristics. This was, however, a fragile achievement, constantly threatened by more primitive impulses. These values were implicit in the 'evolutionary creed' with which Head operated.[88] It therefore comes as no surprise when he declared that the 'barrier which the normal mind sets between conscious life and that of the viscera' was 'less stoutly fixed'[89] in women than in men. Women were naturally more 'visceral' beings than men due to the greater prominence of the reproductive organs in their physiology.[90]

In his final paragraphs, Head at last allowed himself to indulge in what he – somewhat defensively – described as 'fantastic speculation'.[91] The autonomic (within which he included those parts of the central nervous system connected with the viscera) was the oldest part of the human nervous system. Its age was evinced by the fact that 'it still retains traces of a primitive segmentation lost elsewhere'. Indeed, 'the plan on which this nervous system is laid down is related to that of invertebrates more closely than to the cortical system of man'. The behavioural counterpart of this morphological antiquity was that the autonomic: 'is essentially the system of animal, as opposed to human activity'.[92]

The tendency of some of his patients to withdraw into depression or to be suspicious of those around them was explained by the fact that: 'when an animal is wounded or ill, he is killed by his fellows without mercy. A wounded animal therefore crawls away either to recover or to die in his hiding place'. However, 'an instinct that is salutary for an animal or for the lowest savage is entirely opposed to the highest development of the human mind'. For this reason, the mental states Head had described in his lectures appeared to his patients as 'an intrusion from without, an inexplicable obsession that can neither be controlled nor subjected to logical analysis'.[93]

In his Goulstonian Lectures Head thus finally made explicit the evolutionary premises upon which his explorations of sensation proceeded. He also demonstrated how readily this discourse could pass from anatomy to psychology and embody a complex of cultural assumptions and biases. A similar architecture can be discerned in his later researches in this field.

Increasingly, however, Head felt frustrated with the experimental subjects with which he was obliged to work. The 'typical hospital patient' had his or her uses. They could, when properly managed, be relied upon to identify areas of tenderness or anaesthesia on their bodies. They were sufficiently compliant to allow themselves to be photographed in various states of undress. A clever and conscientious investigator such as Head might even be able to elicit reliable accounts from them about abnormal mental states. But reliance upon subjects of this kind also had serious drawbacks.

Head was particularly interested in what could be learned from the phenomena of *dissociation* of nervous function: that is, from what could be elicited from dysfunction in cases of nervous injury or disease. Head maintained that 'these dissociations of function give the clue to the complex activities of the nervous system'.[94] In a letter to Robert Nichols, Head tried to clarify his concept of dissociation using a sporting metaphor:

> When some organized activity of the nervous system is disturbed, the loss of function does not reveal the elements out of which it is constructed, but shows how man faces a new situation with imperfect tools. Lawn-tennis with a broken racket may resemble the player's normal game, but the defects are not an integral part of his normal method of behaviour playing the game. In every instance there is first of all the mechanical act and secondly the situation under which it must be performed; the reaction which follows may be varied, either by changing the stimulus or by altering the circumstances in which it has got to act.[95]

Head maintained that: 'these dissociations of function give the clue to the complex activities of the nervous system'.[96]

As it was forbidden to inflict such damage deliberately on human subjects, the sensory physiologist was obliged to rely on clinical cases to supply research material. Head's post at the London Hospital provided him with a steady stream of such cases, which was supplemented by helpful colleagues who gave him access to patients suffering from various nervous disorders who came under their care. A city with the population of London offered a large pool of 'interesting' subjects.

Head, on occasion, supplemented these metropolitan resources by going farther afield – travelling for instance to the Royal Army Medical School Hospital at Netley in Hampshire to take advantage of the opportunity to study nerve injuries among casualties from the Boer War. This visit was arranged 'privately & unofficially', largely it seems through the good offices of Almroth Wright (1861–1947)[97] whose hospitality Head enjoyed during his stay at Netley. Head brought with him a battery and all his 'usual instruments'. He reported that: 'Everyone was extraordinarily kind. Small little men in dark uniforms were seen flying all over the building & grounds and shortly 15 cripples in coarse blue jacket & trousers with every form of military hat were paraded before me. From them I selected a few and started my investigation'.[98]

The visit also provided Head with an occasion to display his acuity in discerning social distinctions. Occasionally members of the RAMC staff would approach him to make polite conversation. Head felt awkward in the presence of their aristocratic grace: 'Their friendliness was apparent and their manners had that pleasant ease which always charms me in the best kind of soldier – I always feel a rough rude creature in the presence of so much ease and such an obvious desire to please'. When, however, he mistook civility for genuine intellectual interest, he was disappointed. If, 'I offer to share what I have learnt with them', he saw 'the obvious boredom on their faces as they are forced to listen politely for a few seconds'. Only one of the medical men he encountered at Netley understood that: 'I wanted to work after lunch and even after tea and did more to remove the restrictions of military routine than anybody'. When it came to assessing the patients he was examining, Head's sense of social superiority was restored. One of these men, Head noted, was 'the son of a workman in Margate'. This soldier 'talked with that curious precise drawl that this type assumes with no mistakes in grammar or pronunciation'.[99]

In contrast to the consideration he received at Netley, the conditions under which most of Head's researches on sensation were undertaken were far from ideal. In 1918 he finally gave vent to his frustration in a letter to Viscount Knutsford:

> I have been at the Hospital for nearly 22 years, but have never had a room that I could call my own, where I could examine a patient or leave my instruments and records.
> Sometimes I was beholden to the charity of the Registrars, the Residents, or the Sisters. During the war I annexed one of the Sick Rooms. For the last fortnight one of the Residents has been ill and even this refuge was taken from me.
> Today I have seen my patients in the bedroom kindly lent me by one of the House Physicians.
> Who but a lunatic would carry out clinical research under such conditions?[100]

Moreover, given the numbers of patients involved and the extended time period over which Head wished to study them, serious logistic problems arose. 'To those who have not worked in a town like London', Head averred, 'it may seem an easy matter to examine a case of nerve injury at regular intervals from the date of the accident up to complete recovery'. But, in reality,

> any systematic attempt to carry out such an investigation is hampered by innumerable difficulties, due solely to the conditions of life among the working population of this huge city. Firstly, the original wound may have been treated at some other hospital, or by a private practitioner. Often the state of the wound and the extent of the injury can then be inferred only from the patient's description. Again, after the nerve has been successfully reunited he may find it more convenient to attend some other hospital; or may leave his hand entirely untreated, and thus render useless the careful investigation at the time of the injury, the exploration of the wound at the time of suture, and the observations made during his stay in hospital.[101]

The patient might, moreover, suddenly move away without leaving a forwarding address: 'In spite of the help of an assistant, skilled in tracing the movements of hospital patients, and in spite of the fact that compensation on an ample scale was given for traveling expenses, and loss of time, many cases disappeared entirely, often at the most interesting period of recovery'.[102]

The principal limitation of the 'typical hospital patient' as a subject of scientific study was, however, intrinsic to his or her nature. They were, to use another of Head's favoured terms, insufficiently 'intelligent' – or more bluntly still too 'stupid' – to give reliable accounts of the fine gradations of their sensations. 'At best,' Head maintained, 'he answers "Yes" and "No" with certainty, and is commendably steady under the fatigue of control experiments'. But: 'such patients can tell little or nothing of the nature of their sensations'.[103]

It is apparent that Head saw himself as a scientific psychologist whose remit was to subject all aspects of mental activity to rigorous scrutiny. He was to write that: 'the day of the *a priori* psychologist is over as far as sensation is concerned. A man can no longer sit in his study and spin out of himself the laws of psychology by a process of self-examination'.[104] Head was in fact keenly interested in his own sensory universe: he often recorded particularly striking experiences in his letters and in the 'Rag Book'. However, to avoid the charge of solipsism, it was necessary to check such personal experience against that of others.

Thus Ruth Mayhew was encouraged to report on the more remarkable aspects of her sensations. She wrote in the 'Rag Book':

> I have realized lately that there is for me a connection between the sense of touch and that of taste. I was tonight cutting the leaves of a volume of my new Pater: the paper is thick and a little woolly. In order to cut very accurately I took a steel paper knife, not my favourite ivory one. The feel of ivory I love; it has smoothness for the touch and coolness, also colour for the eye. As I went on cutting, the contact of the steel with the paper brought to my tongue the taste I knew as a child when we loved

to put our disobedient lips to the iron railings of the Parks at home or liked to suck our pennies...[105]

Head also encouraged colleagues to share similar experiences of coenaesthesia. One fellow-worker, for instance, supplied him with a report of a 'tactile Dream' in which 'he dreamt he was at sea steering the ship in which he served as apprentice. He woke and for an appreciable interval felt the wheel in his hands'.[106]

Head seems indeed to have solicited reports of sensory experiences from acquaintances in a semi-formal way. For instance, the Oxford don, Adolph Paul Oppé (1878–1957), reported to Head not only on his own modes of visualization, but also on tests Oppé had conducted on undergraduates. The procedure was to show different individuals a picture of a landscape and to note different interpretations of various aspects of this image. Oppé informed Head:

> Two cases I have observed in men here give this: the sheep go thro' a gap in a hedge but the hedge is on the crest of a hill, i.e. there is nothing beyond it. (I see beyond the hedge another field, trees & a hill – sheep in both fields.) This gives the visual picture determined by the counting interest & is an intermediate step between mine & that ideal one of which I spoke to you in which the hedge – a great undetermined high thing with a gap – & the sheep – formless units – form the whole picture.[107]

The point of the exercise was thus to demonstrate how visualization was an active process governed by the different 'interests,' or inherent dispositions, of each observer.

Such sophisticated investigations required test subjects who were sufficiently 'intelligent' to provide detailed and reliable accounts of their own inner world. For some of the psychological experiences with which he was concerned, Head was able to draw such subjects from his circle of acquaintances. But in the case of the phenomena involved in the dissociation of the normal workings of the nervous system, he was presented with a quandary. Head was obliged to wait for such states to arise as the result of accident or disease. Moreover, even when such opportunities arose, they tended to occur in 'typical hospital patients' who were, for a myriad of reasons, profoundly unsatisfactory experimental subjects.

After 1900, Head devised a twofold solution to this problem. The historical accident of the First World War was to provide him with a class of patient that he did regard as a worthy collaborator in the grand scientific enterprise in which he was engaged. In the first instance, however, he arrived at a radical solution to the shortcomings of the human material with which he had hitherto been obliged to deal. Head concluded that: 'Introspection could be made fruitful by the personal experience of a trained observer only'.[108] He in effect decided to make himself his own experimental subject.

A Human Experiment

On 25 April 1903 Head underwent a procedure whereby the radial and an external cutaneous nerve in his left arm were severed by the surgeon James Sherren and then sutured together, 'with fine silk, and the wound was closed with silk sutures'. The limb was then put in a splint, 'with the forearm flexed at the elbow, and the whole hand left free for testing'.[109] Between May 1903 and 13 December 1907 Head underwent a series of extensive tests designed to chart the stages of recovery of sensation in the affected limb. In all, 167 days were devoted to the experiment.

These researches were undertaken in collaboration with W. H. R. Rivers, and took place in his rooms at St John's College, Cambridge. Head generally travelled up to Cambridge on Saturdays, 'after spending several hours in the out-patient department of the London Hospital',[110] and devoted the remainder of the weekend to these investigations. During these visits, he stayed at his old college, Trinity. Head saw these trips as a withdrawal from the hurly-burly of London life, with all its tedium and demands, to an idyllic setting in which he could pursue his true intellectual interests. It was, above all, a retreat from the demands of the world into the inner life of Head's own sensations. In a set of rooms on the top floor of the second court at John's, 'here, absolutely quiet and undisturbed, free from the petty worries of a busy life, H. gave himself over entirely to examination'.[111]

His collaborator matched his enthusiasm: 'Rivers', Head reported, 'is inexorable'.[112] The two sat at:

> a large table with my left arm exposed between us. My eyes are closed and he touches or pricks it in various ways recording his stimulus and my answer. Otherwise no word passes between us and the silence is only broken by the scratching of his pen and my ejaculations. After a while I open my eyes & rest at the same time dictating a consecutive account of what I have experienced. After a short rest away we go again. We cannot work for much more than an hour at a time. For though I am not conscious of any difference in myself my answers are less certain and it is obvious that the feeble sensibility of the affected part is giving out.[113]

Head kept Ruth Mayhew informed of the progress of these investigations, providing in November 1903 a further account of the method he and Rivers employed but now with an emphasis on the accompanying mental states:

> I sit with my arm bared & my eyes closed and Rivers carries out all manner of tests of which the general idea has been laid down in consultation but the details are unknown to me. With my eyes closed I try to let my thoughts flow by like clouds on a windy day. No one thought must occupy attention permanently and I must entirely detach myself from the idea that experiments are in progress. Suddenly in this flowing sea of thoughts there appears a flash of pain a wave of cold or the flicker of heat – It

Figure 2: Henry Head and W. H. R. Rivers.

should appear with the suddenness of a porpoise, attract attention and again disappear leaving the untroubled sea to its onward flow. Such is the most perfect condition for psychological investigation: this state I can now assume at will.[114]

It was this degree of psychological self-possession, which only the 'trained observer' could attain, that was supposed to guarantee the objectivity of these investigations.

These intensive sessions of experiment were punctuated by ludic episodes when Head and his collaborator allowed themselves to regress to the undergraduate or even schoolboy state. In February 1904, Head told Mayhew that: '... Rivers & I ran up and down the John's wilderness like boys ... Had you been here I should have been again the youth who had just passed his "little go" & was beginning then to know the great hard road of Natural Science ...'.[115] In the summer months, Head sometimes relaxed by boating on the river, as he had done in his student days. After a morning's work, Head might also cycle out to visit Gaskell and stay for dinner. As he returned from one of these excursions, 'The moon was up and I rode home in the clear autumn air singing ...'. Following another 'good morning's work on my arm', Head and Rivers 'took horses and rode for many hours – He took me to grassy roads of the existence of which I was totally ignorant ...'.[116] A fortnight later, Head made the trip to Grantchester where he 'dined with Bateson and his wife ... We talked heredity and other things in his line interspersed with discussions of French plays & French novels ...'.[117]

There was, however, no doubt about Head's determination to follow through fully the consequences of the ingenious self-mutilation that he had devised. Even his relationship with Mayhew took second place to this quest for knowledge. He informed her on one occasion: '... I am afraid Sunday is impossible – I must go up to Cambridge – Things in the arm are becoming exciting & we fear to lose any phenomena of recovery'.[118] As the process of recovery advanced, it became, however, ever more difficult to do justice to these phenomena and the personal cost correspondingly greater. In July 1903, Head revealed that:

> The arm is beginning to recover and the returning sensation is most aggravatingly difficult to test. I prick a portion of skin with a needle or pull a hair and at the end of an hour a number of points are found and marked where pain can now be felt. After suitable rest Rivers tests these places and finds that all the sensitive spots have now become insensitive and the only places where I can feel are unmarked. It is trying, wearisome work and all the time we are haunted by a fear of forgetting to test some important point until the suitable period for such testing is past. I shall know a great deal about pain by the time this experiment is over.[119]

News of the experiment soon spread among Head's acquaintances. Some took the matter lightly. Redcliffe Nathan Salaman indeed made it the subject of an extended joke when he sent Head New Year greetings in January 1906. Salaman wrote: 'As I fully expect to see you made a peer in today's list of honours I am sending you a ready made coat of arms'. The arms in question consisted of such elements as 'Dext. a much tried hand insensate to finger tips' and 'Compasses marking the borderland'.[120]

For her part, Ruth Mayhew felt a mixture of pride and anxiety over the course on which Head had embarked. She had visited Head a few days after the operation; as she entered the room, he claimed that the affected arm 'glowed and tingled'.[121] As the experiment proceeded, Mayhew tried to match Head's courage; but she admitted that: 'fears grow up hydra-headed and I cannot calm them until I see you again with both your dear arms free and strong and every morsel of that brave left hand able to respond to the lightest touch of my littlest finger'.[122] In July 1903, she begged to know: 'Is your arm recovering really? Will it soon be well? May I before long sit on your left side as happily as on your right, without the chill feeling that the arm I am so proudly holding is only half-sensate?'.[123]

The reaction of Head's mother to the experiment was in striking contrast to this solicitude. According to Ruth, Hester Head had opined: '... I am not half so miserable when I think of what Harry has had done to himself [i.e., to his arm] as when I think of the poor dogs and animals who are tortured without the chance of refusing'.[124] Mayhew was appalled by this apparent lack of maternal compassion. Head, for his part, was probably not amused by his mother's apparent display of antivivisectionist sentiment. By this time, a serious rift had opened between the two largely because of Head's attraction to Ruth Mayhew (see Chapter 4).

For Head, what justified the pain he suffered – and even the anxiety he inflicted on Ruth – was the significance of the results that he and Rivers obtained by these means. These results were in large part dependent upon the peculiarities of Head's mental constitution. These were scrupulously detailed alongside the experimental procedure when he and Rivers published their conclusions. They noted that:

> H's mental processes are based upon visual images to a remarkable degree. Every thought is in some way bound up with internal vision, and even numbers, the days of the week and abstract ideas, such as virtue and cowardice, are associated with images of varying tones of white and black. He cannot recall musical sounds, except by seeing the notes or attaching the sounds to words which are clearly visualized.

The answers Head gave to Rivers's enquiries throughout the investigations of his arm were profoundly influenced by 'these vivid mental images'.[125] Since the tests were primarily of tactile, non-visual, sensations, his peculiar psychological type was indeed something of a drawback. He was often unable to recall a particular mental image because it had no visual aspect, a fact that 'greatly hindered its introspective study'.[126]

Moreover, because Head was 'at the same time collaborator and patient', it was necessary to take 'unusual precautions to avoid the possibility of suggestion'.[127] Thus Rivers was careful to give Head no clues – such as the clink of ice in a glass or the sound of a kettle being placed upon the hob – as to what stimulus he was about to apply to the arm. Any such lapse 'tended to prejudice [Head's] answers and destroyed that negative attitude of attention essential for such experiments'. Rivers was accordingly careful to make his preparations before the subject of the experiment arrived: 'the iced tubes were filled and jugs of hot and cold water range within easy reach of his hand, so that water of the temperature required might be mixed silently'.[128]

In an attempt to give precision to their results, a grid of one-centimetre squares was drawn on the arm and hand. The limb was then photographed with pains being taken that on each occasion it remained in the same position. Occasionally, Head was obliged to retain these markings for extended periods – for instance, between 28 January and 12 March 1906 – 'so that, whatever its faults, we might be certain we were photographing the same field'.[129]

Head was obliged to bear these stigmata even while on vacation. During a cruise to North Africa in April 1903 on the SS *Orotova*: 'My hand could not be hidden owing to the mass of square & dots with which it was covered'. These markings attracted the attention of Head's fellow passengers, who eagerly speculated that the letters he was writing to Mayhew on the voyage were: 'part of a "magnum opus" of which the hand is part. If they only knew what poor feeble things they are and how lamentably they fail in their object'.[130]

In later years, Head was to attempt to gain further credibility for the objectivity of his results by seeking to replicate them in alien contexts and with the help

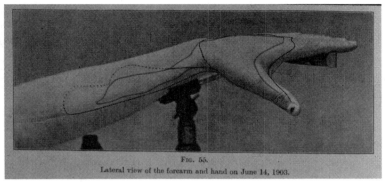

Figure 3: Lateral view of Head's forearm and hand.

of outside observers. Thus in April 1908 he travelled to Germany to enlist the aid of the physiologist Max von Frey (1852–1932) in this project. Von Frey was a recognized authority of the subject on cutaneous sensation; his endorsement of Head and Rivers's results was thus especially desirable. Head was delighted with the results of this meeting. He reported:

> As to the work it goes along famously my hand is in excellent form and I have been able to let Frey work out for himself many of the more striking experiments in favour of our views. He is evidently much struck with the simplicity of the observations on the abnormal as compared with the normal hand. More I am pleased to think that our work is based on such certain grounds that a stranger can work it out for himself on my hand.[131]

This intellectual satisfaction provided Head with some compensation for his disappointment with the philistine character of the von Frey household.

The chief conclusion of these laborious researches was that in the healthy individual peripheral feeling depended on the operation of two distinct nervous mechanisms each of which generated a distinctive mode of sensation. The first of these modes of sensation was the 'protopathic'. This was characterized by diffuse feeling of an all or nothing nature. The protopathic was also marked by excessive response; even a mild stimulus would cause extreme and widespread reactions. The second form of sensation Head and Rivers called the 'epicritic'. This was, in contrast, distinguished by being localized, graduated, and proportionate in character. It was thanks to epicritic sensation, for instance, that it was possible to make fine spatial discriminations between sensations.

In the healthy individual these two forms of sensation were conjoined in a specific fashion: the epicritic mode inhibited the workings of the protopathic so as to moderate the more extreme aspects of its characteristic form of feeling. 'Under normal conditions', Head and Rivers insisted, 'there are no "protopathic" or "epicritic sensations"'.¹³² In the case of the experiment on Head's hand, however, this normal relationship was disrupted. As sensation gradually returned to his hand, it was the protopathic that returned first. He was therefore able for a time to experience protopathic feeling in its raw unmodified form. As late as the autumn of 1908, epicritic sensation had still not returned to some parts of his hand. Head drew an analogy between this form of disseminated sensation and the referred pain due to visceral disease he had studied in his earlier papers.¹³³ Both phenomena were only explicable by reference to the evolution of the human nervous system.

Head and Rivers insisted that the existence of these two modes of sensation in man and the relations that pertained between them should be understood in evolutionary terms:

> We believe that the essential elements exposed by our analysis owe their origin to the developmental history of the nervous system. They reveal the means by which an imperfect organism has struggled towards improved functions and psychical unity.¹³⁴

Protopathic sensation represented the forms of feeling characteristic of the human species' primitive ancestors. Epicritic sensation was a more recent acquisition that had evolved because it endowed the organism with a wider range of ways to appreciate and thus respond to its environment. In particular, epicritic sensation provided a far more qualitatively refined form of feeling and the ability to make spatial distinctions unavailable to the protopathic animal.

Head and Rivers use of the word 'struggle' in this context might appear to allude to the Darwinian 'struggle for existence'. But it is clear that they were operating with a quite different conception of the evolutionary process. Theirs was a teleological understanding of human evolution. In the course of phylogenetic history, man had strived not merely for survival, but for ever greater perfection. Indeed, the true end of evolution appeared to be the achievement of the highest possible degree of 'psychical unity'. If Darwin's conception of the mechanics of evolution reflected a Malthusian vision of a bleak and brutal human destiny,¹³⁵ then Head and Rivers's narrative of nervous development clearly reflected a different cultural context in which the struggle for life had been superseded by a quest for psychical wholeness.

Moreover, a close reading of Head's papers reveals the presence of an implicit evaluation of the relative worth of the two modes of sensation he and Rivers had discerned. For instance, Head often writes of the epicritic as, not merely in evolutionary terms a more recent kind of feeling, but also as embodying the '*higher* forms of sensibility'. Epicritic sensibility was 'higher' because it embodied: 'all the finer and more delicate sensations'. In contrast, in the protopathic animal, 'all the finer and more delicate sensations involving discrimination and differentiation are wanting'.¹³⁶

In passages like this the social relations involved in the production of this knowledge can be seen re-inscribed within the finished product. Head undoubtedly saw himself as a more 'epicritic' being than the typical hospital patient whose limited ability to experience human feeling in its fullest extent and richness was so painfully obvious. 'Fine', 'delicate' sensations, and the ability to discriminate, were the reserve of the man of feeling – the defining characteristics of the aesthete.¹³⁷

The gradual triumph of the epicritic over the protopathic in human evolution could also be seen as a metaphor for the triumph of civilization over savagery in human history. This analogy is particularly brought to mind when Head declared in one of his papers that: 'Removal of epicritic sensibility exposes the activity of the protopathic system in its full nakedness'.¹³⁸ The image of the 'naked savage' was commonplace in Victorian and Edwardian culture.¹³⁹

Head followed up his studies of peripheral sensation with extensive research into the workings of the sensory nervous system at the spinal and encephalic levels. He once again made use of the method of dissociation – that is, trying to establish the physiological principles through the study of pathological cases. These later enquiries were greatly facilitated by the outbreak of war in 1914.

During the course of the war and in its immediate aftermath Head was afforded the opportunity to study a large number of soldiers with wounds to the spinal cord and brain. In particular, many wounded officers came under his care. I have described elsewhere how he found these subjects especially well suited to the exploration of the nature of the language disorders that followed injury to the brain.¹⁴⁰ They were no less valuable in his work on sensation. Because of their social background and level of education, Head was prepared to ascribe to these patients a competence to report the nature of their sensations reliably that he denied the 'typical hospital patient'. Thus 'Colonel M.', who came under Head's care after he had on 4 October 1914 been shot in the head by a rifle bullet, was described as: 'an unusually intelligent man of intellectual tastes and an excellent subject for examination'.¹⁴¹

In a letter to Ruth Mayhew, written while he was completing his paper on 'Sensation and the Cerebral Cortex', Head revealed the depth both of his emotional attachment to these wounded soldiers and his personal enthusiasm for the research that they made possible. 'It is', he declared, 'beautiful work and the things these dear young men said about how and what they felt in answers to my various tests are too astonishing for the world not to believe that I made them up'.¹⁴²

In his treatment of sensation in the central nervous system, Head remained faithful to the principle of functional unity adumbrated in the 'Neurological

Creed'. In essence, he maintained that specialized organs displaying characteristics analogous to the protopathic and epicritic systems he had discerned at the peripheral level were to be found in the brain. This was to be expected if the 'higher' structures and functions of the human nervous system had evolved from the more primitive.

Writing in 1911, Head maintained that 'We believe that there are two masses of grey matter, or sensory centres, in which afferent impulses end to evoke that psychical state called a sensation'.[143] One of these two centres was situated in the optic thalamus; the other was located in the cerebral cortex. The thalamic centre was particularly concerned with the 'affective' aspects of sensation. The cortex, on the other hand, contributed the 'discriminative' and relational dimension to feeling. Head insisted that these 'two centres of consciousness are not co-equal'.[144] On the contrary, a clear hierarchy existed between them. In the normal state, the 'activity of the thalamic centre ... is dominated by that of the cortex'.

The cortex in effect acted as a source of moderation upon the thalamus by inhibiting the more extreme manifestations of the affective side of sensation. This influence became apparent when, due to injury or disease, this cortical control was removed. The so-called 'thalamic syndrome' was characterized by extremes of pain and pleasure unknown in the normal state. Thus,

> a highly educated patient confessed that he had become more amorous since the attack, which had rendered the right half of his body more responsive to pleasant and unpleasant stimuli. "I crave to place my right hand on the soft skin of a woman. It's my right hand that wants the consolation. I seem to crave for sympathy on my right side". Finally he added, 'My right hand seems to be more artistic'.[145]

Head had himself experienced a somewhat similar state during the experiment on his hand. He told Ruth Mayhew in May 1903 of the effect that the operation had had on the sensations accompanying his creative activities:

> ... When a lyric comes into my head I think of it the last thing at night and as I am dressing it begins to form itself ... I have always suspected that the lyrical stage in the early morning was physical. The proof has come with my dead arm. For during the period on Saturday morning when I knew both by my internal feelings and by the fact that the lines were coming rapidly that my mood was lyrical the left arm glowed and tingled exactly as it did when you came into your room on the Wednesday after my operation.
> On the other hand the pleasure it has given me to see the finished lines on paper leaves my arm quiet and is obviously only intellectual.[146]

There was clearly something more primitive, more 'protopathic,' about the thalamic aspect of sensation.[147] Head indeed maintained that the thalamus was the more ancient part of the brain, the cortex a more recent addition. The protopathic was a relic of an old form of feeling that was adapted to automatic defensive reactions to noxious influences. Such responses had their use in the past, 'but they hamper voluntary action by the uncontrolled movements they evoke ...' It therefore became 'necessary' that 'these modes of reaction should be brought under control of centres which allow of choice and determine movements of the whole animal. The chief of these centres is the cerebral cortex ...'[148] Indeed, 'the aim of human evolution is the domination of feeling and instinct by discriminative mental activities'.[149]

Because the more primitive sensory centres had in the course of evolution been supplemented rather than supplanted, this control had to be actively maintained even when the cortex was in a normal state. There was, in effect, a 'struggle between the discriminative and affective aspects of sensation, which forms so important a factor in the activity of the human mind'.[150] Head had therefore identified the physiological basis of the dichotomies in human nature between the emotional and the rational, the corporeal and the intellectual, and the civilized and the savage. Head's conception of the bodily mechanisms underlying sensation thus incorporated polarities that had long preoccupied western culture.

Head's writings on sensation spanned several years, involved collaboration with a range of individuals, and resulted in a series of lengthy research papers. These were collected together in the two-volume *Studies in Neurology*, which appeared in 1920. Taken together these researches constituted a remarkably consistent and unified research programme. Thus in his 1918 paper on 'Sensation and the cerebral cortex', Head was able to show the significance of his early enquiries into referred visceral pain for the overall conception of the workings of sensory nervous function that he was unfolding. The internal organs, he maintained, 'are supplied from the protopathic system; there is no controlling epicritic mechanism. Any stimulus capable of exciting pain produces a segmental response; the sensation radiates widely and is referred to remote parts, including the surface of the body'.[151]

Moreover, Head made strong claims for the originality and wide-ranging implications of the theory of sensation that he and his collaborators had produced. 'If the view I have put forward in this paper is correct', he declared, 'all previous theories of cortical localization, as far as sensation is concerned, must be put aside'.[152] There was, he insisted, no *via media*: if his conclusions were correct, all previous understandings of the mechanisms of sensation had to be abandoned.

What underlay and necessitated this revolution was the 'universal belief that man has been evolved from the lower animals'.[153] Despite widespread acceptance of the theory of evolution among scientists, the doctrine had nonetheless persisted that in man: 'each psychical act of sensation is associated on the physical side with certain distinct processes, which, starting in the peripheral end-organs, pass unaltered to the cortex of the brain'. Such a view: 'was not unreasonable in an age ignorant of the gradual evolution of the human nervous system. Man was thought to have been created perfect, armed with apparatus to receive and conduct the special processes, which underlie each fundamental sensory experience revealed to him by introspection'. But now that 'this conception of man as a created being has long passed away', such a conception of the workings of the nervous system was no longer tenable.[154]

A conception of nervous function that was compatible with an evolutionary understanding of human origins eschewed the facile simplicity of the old model. Instead it emphasized the complex physiological processes that intervened between the stimulation of the sensory apparatus by some physical stimulus and the generation of a psychical event. Moreover, this evolutionary neurology did full justice to the contingent nature of the human nervous system implicit in the theory of evolution. But for the accident that our ancestors had been metameric worms, human beings would perceive the world in entirely different ways. Indeed, the world would *be* different.

Head betrayed a modernist sensibility in thus demanding a complete break with the physiology's past. The language with which he sought to express the truth of the sensory mechanism also situated him historically. In his account of the sensory nervous system, Head deployed a variety of metaphorical resources. When trying to explain the selective character of the sensory receptors within the spinal cord, for instance, he drew upon his musical interests:

> It is as if the gallery of a concert hall were fitted with a series of resonators, each of which was tuned to a certain note. Each resonator would pick up a peculiar tone, whether it was produced by the strings, the brass, or the wood-wind.[155]

In an analogous way, Head argued, the spinal receptors gathered together afferent impulses of a particular quality regardless of their point of origin.

A master metaphor does, however, pervade these studies in neurology. In order to convey his conception of the true nature of the nervous system, Head had repeated resort to metaphors of fabrication and distribution. The model of nervous function Head sought to supplant postulated a relatively simple transit by impressions received by the sensory end organs through the central nervous system to the cerebral cortex where they became the occasion for sensation. For Head, on the other hand, this process was far more complex and characterized by continuous refinement, reorganization, and transformation.

In places he compared the workings of the sensory nervous system to the sifting of raw materials in a mining operation. 'Epicritic and protopathic impulses', he wrote, 'and those associated with deep sensibility, travel along the fibres of the posterior columns like rubble over a graduated sieve. Stones of small size drop through at once, those that are larger pass on further, and some travel the whole length of the sieve to fall into a heap at the end'.[156] These selected impulses might at a later stage in the process require further 'sifting'.[157] But as well as these relatively crude forms of sorting, the nervous system was capable of performing actions analogous to the processes of 'transmutation and recombination' found in a chemical laboratory or factory.[158]

These processes of selection and refinement took place at different points along the cerebrospinal axis. The modified nervous materials were then conveyed to the next stage in their journey. Head characterized this centre variously as 'end-stations', 'relays', and 'terminal stations'.[159] At these 'junctions' nervous impulses were modified and regrouped before being transmitted further. The pathways of the central nervous system were thus akin to a sophisticated transport network.

The final 'end-stations' in this system were found in the cerebrum. The thalamus served both as a centre at which certain impulses terminated and as a final relay that conveyed others to their ultimate destination in the cerebral cortex. The cortex was, however, no mere passive receptor of impressions. Although Head occasionally used the conventional metaphor of the cortex as a 'repository' or 'storehouse' for mental images, he regarded the static implications of this figure as misleading.[160] The lower nervous centres supplied: 'the afferent materials out of which the cortex manufactures those forms of sensation for which it is responsible'.[161] The cortex was thus a further active participant in the 'production of sensation'.[162]

In short, in Head's formulation the sensory nervous system is composed of factories, rail networks and termini: the basic components of an early twentieth-century industrial economy. All these elements collaborated in the great process of 'the production of sensation'.[163] Like the economic infrastructure of Edwardian Britain, this was, moreover, not a system that had been created *de novo* from some pre-existing blueprint. It had, on the contrary, evolved gradually through accretion and adaptation over time. In consequence, the human nervous system was characterized by certain redundancies, idiosyncrasies, and anomalies. It was in many ways an admirable contrivance, but one that bore the unmistakeable marks of its adventitious origins.

Head's revised vision of the nervous system also had wider implications for understanding man's place in the universe. On this view, the Cartesian notion that the self was a given entity that passively contemplated the universe was a delusion. On the contrary, selfhood was an active construct and man's knowledge of the material universe – the body included – dependent on a complex of physiological processes of which the mind was normally unconscious.

In Head's words, 'Man perpetually builds up a model of himself, which constantly changes'. Knowledge of the body and of its relations to other objects was dependent upon these constantly updated models, which Head called 'schemata'. These schemata were inherently dynamic in character: 'Every recognizable change enters into consciousness already charged with its relation to something that has preceded it'. These preceding states were not, however, themselves psychical in character. To make the point, Head had resort to another metaphor, which would have been unavailable to a psychologist of an earlier era:

> Just as on a taximeter the measured distance is presented to us already translated into shillings and pence, so the final product of spacial [sic] changes rises into consciousness as a measured postural appreciation.[164]

The physiological processes underpinning consciousness were, moreover, responsible for more than simply the mind's ongoing sense of posture. They provided the true source of the Kantian categories of time and space. Head noted that in the course of evolution the relatively autonomous and egalitarian metameric model of organization had been supplanted by a tendency for the 'headwards parts' to become 'increasingly dominant'.[165] One aspect of this progressive cephalization was the development of specialized 'distance receptors,' such as the eye that were peculiarly concerned with the processing of what Head called 'projected sensations'. These 'projected' sensations were what common sense regarded as the outside world – that is, those parts of the universe that lay beyond the body.

'From the physiological point of view', therefore, 'an external object is a group of functional events'.[166] Later in the same paper, Head elaborated that 'an "object" might be defined as a complex of projected responses; it is said to have characters such as size, shape, weight and position in space, which distinguish it from all others'. The recognition of such features supposedly inherent in the object itself was, however, dependent on: 'physiological activities, the product of certain definite centres in the cortex'. If these processes were altered due to disease, 'the "object" disappears …'[167]

These physiological processes located in the cerebral cortex were, moreover, 'characterized by a strict dependence on past events'. This diachronic element was inherent in the organization of the schemata that underlay postural awareness. It followed that: 'the sensory activities of the cortex are not only responsible for projection in space, but also ensure recognition of sequence in time'. Damage to the cortex could therefore lead to a loss of temporal awareness, as well as a loss of spatial cognition. It was thus: 'the projected elements in sensation to which we owe our conceptions both of coherence in space and time'.[168] Head's rewriting of the sensory nervous system was thus both a reflection of and a contribution to the new understanding of the nature of time and space that arose around the turn of the twentieth century.[169]

As we shall see, the broader philosophical implications of these researches were not lost on Head's contemporaries. More immediately the extended series of investigations into the physiology of sensation that he undertook between

1893 and 1918 gained him recognition within the scientific community to which he ascribed so much importance. One token of this appreciation was the fact that in virtue of his work on sensation Head was nominated on four occasions by his peers for the Nobel Prize in Physiology or Medicine. On the first of these occasions, Head expressed his reaction to Ruth Mayhew. He noted that: 'The prize in medicine will probably be worth about £10,000 pounds – I have not the slightest chance of obtaining it but what has pleased me so much is that when it came to selecting a name to represent a group of London scientists they, having watched me at work for years, should select me'.[170] Ruth shared in Henry's satisfaction in the news: 'No gladder prouder woman than I ... has to-day walked the gloomy streets of England'.[171]

Aphasia And Kindred Disorders[172]

In January 1901, Head had written to Ruth Mayhew of an out-housed hospital patient he had encountered who, 'in consequence of a blow on his head has entirely lost the [visual] element in speech ...' In this instance, however, Head appears to have been less concerned with the details of the man's pathology than with the fact that the rustics with whom the patient was boarded had failed to provide him with adequate clothing and occupation. Head's conclusion was that, 'Country people seem to have learnt or rather kept a passive experience that they and the beasts learn from their struggle with nature, totally unsuitable for survival in any other struggle'.[173]

Around 1910, however, Head began to develop a special interest in the disorders of language arising from injury to the brain collectively known as 'aphasia'. This interest was heightened after the outbreak of war when, 'patients began to pass into my care with wounds of the head, who suffered from defects of speech slighter and more specialized than any I had seen before'.[174]

What seems in particular to have spurred this new line of research was Head's 'discovery' of John Hughlings Jackson's papers on the subject. These suggested ways of approaching the subject of aphasia in what seemed to Head novel and exciting ways. Head maintained that Jackson's views had suffered from an unjustified neglect largely because of the obscurity of his style. He admitted to Constantin von Monakow (1853–1930) that: '[Jackson's] writings are so difficult, even for an Englishman, that they have not received that attention due to them'. Even the eminent French neurologist, Pierre Marie 'told me he had often tried to read them but had never succeeded'. Head sought to remedy this alleged obscurity by adding an introduction, which he hoped Monakow found 'sufficiently clear and easy to read'.[175]

Head felt that Jackson's contribution to aphasia studies had also been diluted because his writings on the subject were so widely scattered. Head used his position as editor of *Brain* to remedy this state of affairs by in 1915 publishing a selection of Jackson's papers on aphasia in a special issue of the journal.[176] In 1920, when he delivered the John Hughlings Jackson Lecture, Head chose 'Aphasia: an Historical Review' as his subject. In the same year he took the opportunity of another public lecture to expound for the first time the results of his own investigations on the subject of aphasia.

Head's idolization of Jackson as an embodiment of scientific virtue has already been noted. In his papers on sensation Head on several occasions acknowledged his intellectual debt to this master. Thus he acknowledged that his key concepts 'on the sensory side of the nervous system' mirrored what Jackson had written thirty years before about the regulation of motor function.[177] Head now saw the possibility of bringing similar conceptual resources to the explication of the enormously complex phenomena evinced in cases of aphasia. He told Monakow that his new research on aphasia would proceed 'on the same lines' as what he had already published on sensation and the cerebral cortex. Head would, moreover, rely on the same class of experimental subject in this project, drawing upon 'observations on the wonderful material provided by the war'.[178]

Many of these subjects were drawn from the 'Special Hospitals for Officers' that were set up in London in 1915. Head's friend and mentor Viscount Knutsford was a prime mover in the establishment of these institutions. Cases of 'shell shock' were treated at 10 Palace Green, while the Empire Hospital in Vincent Square was where officers suffering from 'organic' injuries were directed.[179] It was in this latter establishment that Head was to find much of the 'wonderful material' from which he derived his understanding of aphasia.[180]

Head stressed that these officer patients were 'intelligent' and well educated. They tended, moreover, to come from a similar social background to his own. He was therefore able to relate to them as equals and to treat their testimony with some confidence. He often reproduced their own accounts of the psychological nature of their disorders at some length, remarking that these 'highly intelligent officers ... were able to give introspective information of the greatest value'.[181] Head took pains to insist that their essential personality remained intact despite the damage to the powers of expression that followed from the injury. This respect was in marked contrast to the mixture of condescension and scorn he showed to the 'typical hospital patient'.

For their part, the patients Head encountered at the Empire Hospital appear to have been ready to join wholeheartedly in the investigations that he wished to undertake on the neurological disturbances resulting from their injuries. This was despite the tedious nature of the battery of tests to which they were subjected. Indeed some of them continued to return voluntarily after discharge so that their condition could be further assessed. Head noted that he had been able to examine one patient over a period of over seven years. As a result of such active cooperation, Head was able to gain a far better sense than had his predecessors in the field of how these disorders progressed in the long term.

It is apparent that these aphasics were willing – indeed enthusiastic – participants in this research. Head remarked on the 'euphoric' mood of his consultations with these men.[182] While George Riddoch recalled that these patients: 'could not resist the glamour and excitement of [Head's] work on themselves'.[183] In other words, Head had succeeded in transforming these patients into willing collaborators rather than merely passive experimental subjects.

The affinity that Head clearly felt for these wounded officers was especially significant in the case of language disorders. Because of their shared background, he and his patients spoke the same language to a much more marked degree than was the case when a middle-class physician met a proletarian patient. As a result of this sense of familiarity, Head was able to engage in forms of social interaction with these convalescent officers that would have been unthinkable with a hospital patient. Sometimes these casual encounters yielded valuable insights into their condition. For instance, having tea one day with 'Captain C'. and another guest helped Head better appreciate the former's linguistic capabilities.[184]

As with his work on sensation, Head saw his contribution to aphasia studies as marking a fundamental break with the past. Indeed, he rejoiced in the effect this iconoclasm would have on his more orthodox colleagues. He wrote gleefully to Robert Nichols that: 'the paper I published at the end of last year, seems to have stirred up a hornet's nest of contradiction and acclamation. It has put the whole question of cerebral localization into the melting pot again ....'[185]

Head was highly critical of the methods, assumptions, conclusions, and above all of the aesthetic pervading classic aphasiology. In particular, he drew attention to an excessive zeal among his predecessors to impose a spurious clarity upon the subject. They had attempted to attain a tidy classification of the various forms of aphasia that conformed to the notion that language could be analysed into a set of sensory and motor operations localized at determinate sites, or centres, which 'starred' the cerebral cortex. Each of these centres was supposed to possess as well as a definite location a specific function: 'like the push of an electric bell, it is fixed at one spot and produces a constant and determinate effect'.[186]

This approach glossed over the nebulous nature of the phenomena in question: 'The question is asked, "Can the patient understand what is said to him, can he speak, read or write?" and he is placed in one or other group, according to the answer'. Such empty terms as 'alexia' or 'agraphia' were then coined to characterize the spurious pathological entities that had supposedly been identified.[187] Head instead insisted on the necessity of asking in what ways and to what degree was a patient's ability to speak, write, or read affected.

Head also found the forms of representation characteristic of classical aphasia studies profoundly repugnant. He dubbed the exponents of this school the

'Diagram Makers' because of their insistence on the possibility and indeed the necessity of providing schematic depictions of the cerebral mechanisms that supposedly underlay language. These diagrams were the most aggravated symptoms of the tendency to strive for a Euclidean lucidity and simplicity in a field that was inherently obscure and complex. Driven by these imperatives, previous aphasiologists had done serious violence to the clinical evidence presented to them. Head alluded to the myth of the Procrustean bed to drive home the point. Case histories were either stretched on the rack of casuistry or ruthlessly truncated to make them conform to some preconceived schema.

These diagrams also had an important role in pedagogy; as Head knew from personal experience, they provided an all too convenient resource for both the lazy lecturer and the indolent student. 'Teachers of medicine', Head declared, 'could assume an easy dogmatism at the bedside and candidates for examination rejoiced in so perfect a clue to all their difficulties'.[188] For eighteen years, he recalled,

> at University College Hospital, [H. Charlton] Bastian had demonstrated to generations of students a man who had been seized with loss of speech in December 1877. On each occasion the famous diagram was drawn and we were told what commissural fibres were affected, and why the visual centre must be intact, although that for hearing was in a state of lowered vitality.[189]

Head added with some satisfaction that when an autopsy was eventually performed on the patient in question, Bastian's predictions were falsified.

Early in Head's career Bastian and Jackson had alike been his mentors: both had, for instance, been among those who proposed him for the Fellowship of the Royal Society. But by the 1920s, Head portrayed the two as representing polar opposites in the history of neurology. Bastian served to highlight all that was supposedly vicious in the state of aphasia studies; he had: 'set the points on the catastrophic road to schemas and diagrams'.[190] Jackson, on the other hand, occupied a lonely and isolated island of scientific virtue. The mark of his greatness was that Jackson's attitude to aphasia was entirely foreign to any conception put forward in the history of aphasia.[191] Chronology notwithstanding, Jackson – 'that great empiric philosopher'[192] – was thus the first true modern in neurology.

In marked contrast to the obsession of the Diagram Makers with anatomical localization, Jackson took a dynamic, functional approach to language and its disorders. Moreover, he was perhaps the first medical writer to grasp something of the complexity of language. The Diagram Makers operated with a simplistic model that assumed that *words* were the basic linguistic units; it was these that were stored and recalled in the various centres or 'depots' supposedly situated among the convolutions of the cerebral cortex. Jackson, on the other hand, pointed out that from a functional point of view the *proposition* is the true element of speech. This insight was rich in implications. It drew attention to the intentional aspect of language – its goal of communicating thought or feeling. It also showed an appreciation lacking in most early discussions of aphasia of the structural complexity inherent in language, which was resistant to any attempt to reduce its elements to simple sensory and motor components.

While he followed Jackson in maintaining that the sentence, rather than the isolated word, was the true unit of language, Head felt it necessary to elaborate on this notion. He maintained that the essence of aphasia consisted of a reduced capacity for symbolic formulation and expression. These disturbances could occur at any stage in the process, and might affect syntax, comprehension, or operate at the level of particular words. Loss of symbolic competence might also affect the aphasic's ability to perform certain apparently non-linguistic actions.

Any adequate account of aphasia would need to take account of the normal linguistic performances distorted or suppressed in these cases. In short, the aphasiologist must make a far more detailed and sensitive assay of the deficits of his patients than had hitherto been the rule before he could attempt to refer these disorders to damage to specific areas of the brain: 'Before we can determine anatomically what parts of the brain are responsible for these manifestations, it is essential to discover the nature of the disorder itself'.[193] Such an undertaking necessitated, however, the development of a vastly expanded range of methods of examination and assessment.

Head devised an extensive battery of tests designed to assess the nature of aphasic disturbances. This preoccupation with technique and method is another distinctively modern aspect of his approach to the subject. These tests, which were far more elaborate and systematic than any employed by his predecessors in the field, were intended to test all aspects of the capacity for the use and comprehension of language. Some of these were relatively simple, asking the patient, for instance, to name and appreciate the relations between various images. Others assessed non-linguistic performance such as the ability to read a clock face and to mimic the movements of another. The most famous of these was the 'hand, eye, and ear' tests in which the subject was placed in front of the observer and asked to copy his movements. Head supposed that a verbal element was normally involved in performing these movements; hence they would be affected in cases of aphasia.

Head also asked his patients to make ground plans of the quarters they occupied to assess their appreciation of spatial relations as well as their ability to represent those ideas. Moreover, he placed great stress on gauging patients' aptitude for such familiar games as draughts, dominoes, and chess; he noted that many aphasics who were able to play these successfully failed at card games. He also tested their ability to complete jigsaw puzzles – noting that this had been a common pastime during the war.[194] He showed patients cartoons to assess their ability to appreciate visual humour. The inclusion of these forms of examination within the test battery is an indication of Head's insistence that cases of 'aphasia' would, when properly scrutinized, be shown to involve far more extensive disorders of cognition and behaviour.

While insisting on the essentially inchoate nature of the condition and its resistance to any hard and fast system of classification, Head thought it possible to distinguish four principal forms of aphasia. He designated these: verbal, nominal, syntactical, and semantic. In the first of these the leading symptom consisted of difficulty in enunciating words. Nominal aphasia involved inability to comprehend the relationship between words and the objects to which they referred. Syntactical aphasia was characterized by disturbance of the normal flow of speech leading to a kind of jargon. In semantic aphasia there was a failure to comprehend the function of words within propositions. Head acknowledged that Michel Bréal had invented this term in his pioneering study in 'the science of the ultimate meaning of words'.[195] This is one of Head's few specific references to developments in linguistics.

To some, these conclusions belied Head's stated ambition of effecting a revolution in the state of aphasia studies. Were not his 'verbal' and 'nominal' aphasias merely new names for the motor and sensory complaints that had long been recognized? Such criticisms betrayed a failure to grasp the true nature of the conceptual revision that Head was attempting. Instead of a spurious attempt to analyse a condition into the supposed elements of language implicated in a particular case, he was attempting to understand aphasia in terms of language considered as a function. The pathological form of speech manifested by the aphasic represented an attempt by the organism to adapt to a reduced level of functional efficiency resulting from some insult to the brain. Thus the phenomenon of paraphrases constituted an effort to operate at a lower level of symbolic competence.

Head was, in fact, loath to consider linguistic performance in isolation, insisting on situating speech in a wider behavioural context. He maintained that language must be regarded as a biological trait that had evolved in the context of more primitive forms of behaviour. 'Speech, one of the highest and most recently acquired functions,' Head wrote, 'was from the first concerned with formulating and expressing man's relation to the world around him for the purposes of action'.[196] Language – or the capacity for 'symbolic formulation' more generally – was, in Head's view, merely one of the 'available resources' that a highly evolved organism relied upon to deal with the world with which it was presented.[197]

By 1920, Head had turned to the daunting task of turning the mass of research materials he had accumulated over the preceding decade into a monograph. In December 1923, he excused a lapse in his correspondence with Nichols on the grounds that: 'I ... have put in every moment on my book'. Cambridge University Press had agreed to take the work. At his first interview with his publisher, Head was, however, 'horrified ... with their demands; for I feel it will be quite unsaleable and will only be read by the very few. However it will be finished in time and in twenty-five years perhaps somebody will see what it means'.[198]

Head pursued this task even while on holiday. During a three week stay at Malvern, Ruth Head reported: 'The Oracle has been in so far lucky that he has written every morning from 10 to one at his great work on Aphasia. Then regardless of storms he has tramped the hills waving his wet tail in the wet wild woods from 2 till 4'. She added that Head resumed his writing in the evening.[199] By 1922 a note of exasperation had entered into Ruth's accounts of this endeavour. She told Nichols: 'My Henry is really a Devil for work' who laboured a 'whole morning each day at his never ending Aphasia book'. Head's idea of recreation was to spend 'all the rest of the indoor hours "swotting" at medieval history so as to be up to all the actors in the history of our surroundings'.[200]

In a more serious vein, Ruth confided to Janet Ashbee that there was something obsessive about her husband's devotion to this project. She voiced her concern at 'Harry's absorption in his work which increases as the years go on'. She thought she understood Head's compulsion: 'Now he hears perpetually the horse-hoofs of eternity and I dread his being disturbed at all. There isn't a soul in the world who could make the least use of his mountain of material'.[201] By this date, Head was indeed aware of the deteriorating state of his health and determined to complete the book before Parkinsonism rendered him a complete invalid. During the latter stages of the book's composition, Head could 'only walk for short distances, as he gets quickly tired'.[202]

Writing to Charles Sherrington in 1924, Head confessed that: 'the work has been long and arduous and sometimes I have almost lost courage'. He felt especially guilty about: 'how much pleasant travel I have deprived my wife on account of the necessity of taking our holidays in a place where I can write. Poor thing!' Ruth, according to Head, 'wishes "la petite Aphasie", as her French friends call it, would be less sluggish in making its entry into this world'.[203]

By July 1925, the task was largely complete. With evident relief, Head told Nichols that:

> Four out of the five parts of my book have gone to the Cambridge Press and proofs are beginning to arrive. Even the last theoretical part requires the addition of a few sentences only and some minor alterations before it too departs for Cambridge and the printer. Thus, I have entered upon that lull which preceded the stormy period of proofs, corrections of references, and the making of the index.[204]

The two volumes of *Aphasia and Kindred Disorders* were finally to appear in 1926. When Janet Ashbee visited Head the following March, she noted: 'His gigantic "Aphasia" Book "(which will not be well <u>recognized</u>" he says "For 30 years, as it is so revolutionary; but wasn't it lucky I just got it done in time?" with a twinkle) lay before me, beautifully printed & got up by the Cambridge Press'.[205]

Ramifications

It is clear from his remarks to Ashbee that Head did not expect the revolution he was attempting to have immediate or general success. His works on the physiology of sensation and of language have indeed received a mixed response from neurologists.[206] However, Head's views struck a chord with many of his contemporaries.

Some of these were fellow scientists who saw the applicability of his work to their own fields of research. Thus when he heard that Head was preparing his monograph on aphasia, the psychiatrist and neurologist Adolf Meyer (1866–1950) wrote to tell him that: 'my attitude toward the Aphasia problem coincides to quite an extent with your own … '[207] In particular, Meyer approved Head's insistence that aphasiology should shift from being 'a study & science of words as entities of an independent & detachable reality' and become 'a study of less conspicuous <u>functions</u>'. Head's approach to the subject was, in Meyer's view, 'not an aphasia theory, because it implies much more than speech and aims definitely to treat speech only as an incident'.[208] Meyer also shared Head's conviction that the study of aphasia was the most promising way to resolve some of the most intractable neurological and psychological problems. He wrote that: 'The problem of aphasia & apraxia is and will be the main entering wedge into the problem of cerebral activity of psychological order, and it cannot be sacrificed to clinico-anatomical convenience'.[209]

The American physiologist Walter Bradford Cannon (1871–1945) found aspects of Head's account of the mechanisms of sensation pertinent to his own research. In particular, Cannon considered Head's hierarchical conception of the nervous system a useful resource in explaining the phenomena of emotional expression and repression in animals and humans.[210]

Head's former collaborator, W. H. R. Rivers, extended the concept of nervous dynamics that he and Head had elaborated into the field of psychology. Rivers argued that the distinction between 'protopathic' and 'epicritic' action could be used to illuminate a wide range of phenomena in both normal and morbid mental operations. In particular, Rivers drew an analogy between the protopathic and the instinctive, affective side of the mind.[211]

The impact of these ideas, however, extended beyond medicine and the life sciences. Because Head's researches on sensation and language touched upon the way in which the human mind acquired knowledge of and interacted with the world, his work was also of interest to philosophers, linguists and social scientists. Head anticipated these wider ramifications when in 1925 he told Robert Nichols: 'I have got a new reading of the Mind Body problem'.[212] Head evidently hoped that others would recognize these wider implications of his seemingly abstruse researches. Following the publication of *Aphasia and Kindred Disorders*, he wrote to Stanley Cobb: 'I am delighted to hear from you that the work has excited interest in America, for I have attempted to knit my observations into a coherent conception of the nature of neuro-psychical processes of wider application than simply to the phenomena of defects of speech'.[213]

Indeed, in Head's view, 'The fact that speech can be disordered by injury to the brain is one of the most wonderful means placed in our hands for investigating the relation of mind and body'.[214] He expatiated on this theme at a Congress of Philosophy held in Oxford in September 1920. Among his auditors was the French philosopher Henri Bergson (1859–1941) who had himself addressed the topic of aphasia as a means of exploring the relations of mind and body.[215] At the meeting, Bergson expressed: 'his profound admiration of Dr. Head's researches on the question of aphasia; they appeared to him of capital importance for psychology, and even for metaphysics'.[216]

Although he had no expertise or training in the subject, Head took an interest in contemporary developments in philosophy; he evidently felt that the subject had some claim on the interest of every medical man. Thus Head was among the signatories of a letter calling on members of the medical profession to support the British Institute of Philosophical Studies. The letter maintained: 'that what are termed generally philosophic studies would be of advantage to many medical men, both by expanding and defining their outlook in relation to general or individual experiences and by affording assistance in the solution of practical difficulties'.[217] Head even ventured in 1919 to correct on neurological grounds Bertrand Russell's usage of such terms as 'painful'. Head explained to Russell that neurologists 'have given up the use of "pleasure and pain" as contrasted terms; pain is a quality of sensation, like heat, cold and touch'.[218]

In 1919, Head contributed to a paper to a symposium on 'Time, Space and Material', which gave him the opportunity to draw out some of the wider implications of his work on sensation. Among the other speakers was his friend Alfred North Whitehead, who elsewhere maintained that: 'No doctrine of sense-perception can neglect the teaching of physiology'.[219] Although he never seems to have directly cited Head's work in his own writings, Whitehead's later philosophy shows that he had grasped the metaphysical import of these doctrines.

In his contribution to the meeting, Head insisted that: 'human conceptions of space, time, and material rest ultimately on the nature of the spatial and temporal elements in sensation'. The complexity of the physiological apparatus underlying sensation belied all simplistic philosophies that sought to derive the contents of mind from 'sense data'. 'This complexity', he maintained, 'of the intervening processes becomes comprehensible, if we recognize that sensation, as known introspectively is the product of evolutionary changes'.[220] A review of the Symposium in *Nature* found Head's paper of 'particular interest'. Indeed, the arguments he had presented amounted to: 'the complete scientific refutation ... of all psychological theories which build up knowledge out of original sense-material'.[221]

George C. Campion was another philosopher who in the interwar period sought to address the 'Psycho-Neural problem'. Campion collaborated in this effort chiefly with Grafton Elliot Smith (1871–1937), the Professor of Anatomy at University College London. Smith had championed a new vision of his discipline, which sought closely to associate anatomy with clinical neurology and anthropology.[222] Campion saw aspects of Head's neurology as conducive to the revision of some basic epistemological assumptions. In particular, he maintained that Head's theory of 'schemata' could underpin a revolutionary new understanding of the nature of basic concepts underlying thinking. These were now seen to be neither the result of the processing of 'sense data' nor of the existence of transcendental categories. They were, on the contrary, part of the mental apparatus that the human species had acquired in the course of evolution.[223] In the early stages of this project, Campion hoped to enlist Head as an active participant and asked Elliot Smith to arrange an introduction.[224]

The linguist, Charles Kay Ogden (1889–1957) also tried to recruit Head to take part in his own intellectual enterprise. Ogden sent Head proofs of his famous book *The Meaning of Meaning* (1923) for comment. This collection of essays sought to present a radical new understanding of how meaning was generated through language. Head's invention of the category of 'semantic aphasia' provided a point of contact between his neurology and this line of enquiry. 'Dr. Henry Head's contributions on the subject', according to Ogden, 'have already become classic'.[225] Ogden appears to have asked Head to contribute an essay – an offer that Head declined on grounds of pressure of work.[226] Head did, however, read the proofs 'with the greatest interest', and he arranged a meeting with Ogden and his co-author, I. A. Richards, to discuss the book.[227]

Head's scientific theories also achieved a certain wider public currency. In October 1921, *The Times* newspaper carried an article entitled: 'Sensation Controlled by the Brain' in which the main results of 'Dr. Henry Head's Researches' were explained to a lay audience. These readers were told: 'A hierarchy of nervous controls has been discovered designed for the advantage of the organism. It is of such a nature that the highest, or latest acquisition of the race or of the individual, is the most discriminating'.[228] In the previous year, the same newspaper had published a somewhat facetious account of the philosophical conference held in Oxford, at which Henri Bergson and Head had both spoken. After complimenting Bergson on his 'histrionic art,' the author concluded: 'Evidently M.

Bergson is not a 'semantic aphasiac'. I cull this delicious name from a paper read by Dr. Henry Head ...'. Despite its 'formidable' title,[229] Head's paper: 'proved to one lay hearer, at any rate, more intelligible than anything else at the congress'. In jocular vein, the writer added that while listening to some of the other presentations at the congress, he had felt himself afflicted by 'semantic aphasia'.[230]

Some 'hearers', however, went beyond such superficial acquaintance with Head's views; they found an unexpected relevance and resonance in these seemingly obscure and technical writings. As early as 1902, Head was somewhat perplexed to discover that Almroth Wright was: 'writing a book on the "Physiology of Belief" which he says is based on and suggested to a great extent by my work'. Wright read parts of this treatise to Head who confessed that he 'could not have believed that my innocent observations could be used as a basis for such theories'.[231]

Later, others were to urge Head to produce his own popular exposition of his views. One of these was Violet Paget (1856–1935), who, publishing under the name of Vernon Lee, was a prolific writer of fiction (especially ghost stories) around the turn of the twentieth century. Paget also made contributions to the field of aesthetic theory. Around 1920 Paget became interested in the psychology of musical apprehension and appreciation. She corresponded with Head on the subject and in 1932 published a book in which she classified different responses to music using the protopathic/epicritic distinction.[232]

Paget also urged Head to publish a popular work that would expound the ethical and social implications of these ideas about the physiological underpinnings of mental processes. She argued that:

> Such a book is altogether more needed & useful than anything I could do of my own. –– [?] are the needs of people requiring to have their basic notions of the relations of the bodily & the 'spiritual' cleared up. For instance to be told what Dr Head could tell them about consciousness (that all-important question of its lapses) and of the physiological plane on which so much of the 'mental' life takes place – also the distinction between the protopathic & the epicritic, which strikes me as at the bottom of all ethics, let alone all education.
>
> We need a méthode de bien conduire son esprit if we are to survive; and we must get some scientific notion of what our esprit is and how protopathically & epicritically it can and does conduire itself. That is what I want Dr Head to do![233]

In a letter to Robert Nichols, Head maintained that he was not unresponsive to such urgings. He revealed that Paget had: 'so worked upon me the last time she was here, that I have actually denoted one of my boxes in the office with the heading "Essays". Whether anything will come of it or not, I cannot say; but I could not help jumping when I read your letter and found that the advice you both gave me almost exactly corresponded'.[234] Head also intimated an interest in the work of the American economist, Simon Nelson Patten (1852–1922), noting 'his ideas seem to run somewhat on the lines we have followed in our attempts to show the adaptive evolution of the functions of the nervous system'.[235] Whatever plans Head may have nurtured to draw out the cultural and social implications of his neurological ideas were, however, thwarted by the rapid advance of Parkinson's Disease.

Nichols – a poet and playwright with no scientific training – studied Head's neurological writings in some depth. Indeed he appears to have internalized some of his former doctor's ideas about the workings of the nervous system as part of an active process of self-fashioning. Thus in June 1933, Nichols wrote to the Heads: 'I have been reading Henry's Conception of Nervous & Mental Energy again. It is right on the spot about the mechanical side of poetry'. Nichols was especially interested in the concept of nervous 'vigilance' that Head had introduced. This referred to the state of functional efficiency the nervous system displayed under various conditions. Nichols declared that:

> Henry's conception of 'vigilance' interests me enormously. I'd like to know where I can find more about it. You see there is something very odd about the state of 'vigilance' during the composition of poetry. I presume that the vigilance is lowered when a man is tired or when he has drunk a good deal of beer. Yet Wordsworth wrote best when tired after a long walk & A.E. Housman has just made public that beer threw him into the right condition for composing The Shropshire Lad.

Nichols's own experience was that:

> for composition the sort of vigilance that exists after a couple of cups of black coffee is excellent for revision & for 'architecture' (knitting together patches, seeing what joins to what) & not good for conception. In conception there is a 'floating sensation' ... The state is slightly somnambulistic & yet one part of the brain is absolutely knotted in effort, an effort which on introspection is felt as having much in common with the state of trying to hear a tiny sound.

He had apparently taken onboard Head's contention of the continuity between merely physiological and psychical events, declaring that reading poetry was 'tremendously physical'. Nichols declared: 'If ever I win a Derby Sweep or make any real money I shall found a Henry Head Chair in the University of Cambridge & propose that it devote itself to an investigation of the Psychology of Creation more particularly in the domain of Poetry'.[236]

Elsewhere in the correspondence, Nichols tried to make sense in terms of Head's neurology of the feelings of claustrophobia that he and others experienced. He maintained of the panicked reactions of the claustrophobic:

> Surely this is Henry's regression-in-flight. The passage prevents flight – hence the motor-system as it were over-excites itself. For the currents that should be used up in flight – i.e. setting legs & arms going, run back into the central switchboard. The panic in the nerves excited the limbs, the limbs strike the wall (actually or metaphori-

cally) & the panic as it were bounds off the wall & back into the limbs & back into the nervous system. And note the result – after a violent struggle the body suddenly falls quiet (I've seen a prisoner in the guard-room make a bolt for the only door, struggle with the military police & suddenly collapse entirely motionless): in a word tries the <u>other</u> regression to the hare's immobility.

On the basis of these observations, Nichols went on to identify certain shortcomings in the penal system that were, in his view, the result of the law's ignorance of such psychological processes.[237]

Nichols also tried to make sense of his own everyday sensory experience in terms derived from Head's writings. In September 1927 he wrote to Head that he had: 'made an observation not without interest to you I think last night'. Nichols had been entering his bedroom with a lamp (which he sketched in the letter) in his hand. As he entered the door the light was extinguished leaving Nichols in the dark. He tried to make his way to the bed but found that he had taken the wrong path. Nichols included plans of the room with dotted lines showing his intended and actual course to make the nature of the error in orientation clearer. This seemingly trivial incident led him to ask 'Why?' such an aberration should have occurred. Nichols suggested to Head that the confusion had happened: 'Because the weight of the lamp in my hand seemed to upset my sense of direction. Perhaps this has been investigated. If not, it seems to link up with your complex of senses – I mean the idea of "a synthesis" of sense impressions guiding us'.[238] Nichols's understanding of the event was thus shaped by Head's notion of the sensory nervous system as a mechanism for the processing of diverse sensory inputs into a coherent whole that could guide the organism in its interactions with the world.

Nichols went on to propose that the phenomena in question:

> might be worth an experiment. I think it important that the weight should not be in the palm of the hand because there the <u>known shape of it</u> acts as a sort of compass. It should be a weight held by a handle.
>
> My idea is that there is a synthesis of sense-impressions from which certain deductions are unconsciously made & acted on & that the 'synthesis' is very delicate.

Such an experiment would, in Nichols's view, have to take account of both the objective and the introspective aspects of the experience.

Head's researches on the physiology of sensation and language thus had an impact that went far beyond the fields of medicine and life science narrowly conceived. It is conventional to account for such ramifications in terms of the 'influence' that a set of theories might have upon those working in other fields or even upon lay understandings of the world. Head's work was certainly 'influential' in this sense – even if he himself believed that its full impact would not be felt for decades. A more dialectical understanding is, however, needed if the full complexity of the relationship of his work to his historical situation is to be appreciated.

Much of the resonance that Head's ideas had derived from the fact that he was addressing the issues and concerns that engaged bourgeois intellectuals of his time. The turn of the twentieth century saw a questioning of many of the certainties of Victorian positivism. In particular, the notion of an unproblematic relationship between the mind and the physical world was increasingly challenged. A defining aspect of the modernist sensibility that arose in this period was an insistence that consciousness was not constituted by the passive reception of impressions from some independent outer reality. On the contrary, the mind was an *artificer* that actively created the world with which it interacted.[239]

Head's conception of the nervous system as an elaborate instrument for the processing of the raw materials into psychical content was clearly consonant with this cognitive modernism. His exploration of dissociation also revealed the precarious nature of the unified subject that had since at least the seventeenth century been at the centre of western metaphysics. The fictive nature of this entity was manifest when injury or disease revealed the human mind's dependence on a mass of neurological contrivances that were responsible for the sense of continuity of personal identity in time and space.

The fact that the nervous apparatus that generated man's knowledge of the world was the product of evolution also had epistemological implications. The human subject was no longer the detached observer of classical epistemology.[240] The mind was a historicized aspect of the world about which it strove to acquire knowledge. It was, more particularly, a set of devices evolved to enable the organism to relate successfully to its environment. Human knowledge therefore had of necessity an accidental, contingent character. The mind created the reality it purported merely to represent. As Campion, among others, pointed out, a recognition of this contingency cast doubt on some of the excessive – and dangerous – claims to absolute truth to which humanity was sometimes prone.[241]

There is no evidence that when Head set out in 1890 on 'the great hard road of natural science' he in any way foresaw the far-reaching implications of his researches. The resonance that his views of the mind's relations to the body and thus to the material had are, however, an indication of the degree to which his project was shaped and guided by the problematic of the modern era.

What is clear is that from an early stage in these researches, Head fashioned himself as a self-proclaimed rebel and revolutionary who showed scant deference in the face of authority. The sense of alienation he experienced in the face of so many aspects of society, his profession and the urban environment in which he was obliged to live and work, extended to the scientific heritage that he confronted. His critique of the body of established knowledge that he wished to displace was primarily cognitive and technical in nature. Indeed, Head's stress on technique is another symptom of his modernism. But this critique was also aesthetic. There was, in particular, something inherently ugly and vulgar about the seductive simplicity of the 'Diagram Makers' who had for so long dominated the field of aphasia studies. In this respect the distinction between Head the scientist and Head the man of taste and feeling melted away.

4 RUTH AND HENRY

Among the Head papers is a perplexing letter written in the autumn of 1898. It reads:

> Dear Miss Mayhew,
> You will wonder how I a stranger, venture to address you on a matter of so private a nature, but believe me, it is because I too have suffered that I feel compelled to speak to hinder perhaps greater suffering for you.
> It has come to my knowledge that you receive visits from, that you correspond with, a friend, and that these visits, these letters are approved of neither by your parents nor his. Oh, I beseech you, pause and consider, before it is too late. You occupy a responsible position. Your girls look up to you. Do not lower yourself in their eyes.
> Then, oh forgive me if I pain you – but is not this friendship, in his life, at most a pleasing episode? But may it not become in yours, an absorbing interest?
> Renunciation is hard, I know it – but dear friend believe me, it is by renunciation alone that we can attain to Peace.
> My husband and I would welcome you at Richmond. Perhaps if you would let me talk to you a little face to face, I could help you, for indeed, I feel for you and long to help you.[1]

The letter is signed 'Dorothea Ladislaw'.

A previous reader of this document dismissed it as the work of a 'nosy parker'. The letter is, in fact, written in the hand of the purported addressee. The existence of a draft version shows that Mayhew took considerable trouble in its composition. Moreover, 'Dorothea Ladislaw' is the name of a character in George Elliot's *Middlemarch*. In the novel, Dorothea developed an attachment to Will Ladislaw whom she marries only after many misadventures and difficulties. Ruth Mayhew (1866–1939) was thus addressing herself in the person of a fictional person with whose story she evidently felt some affinity. The fact that Mayhew should employ such a device in order to articulate her own condition says much about the centrality of imaginative literature in the way she interpreted her own place in the world.[2]

The letter also reveals that by the latter part of 1898 Mayhew felt that she had entered upon a relationship with a man that transgressed Victorian notions of propriety and might bring scandal down upon her and her family. The man in question was a London physician named Henry Head. After science, Ruth was to be the great passion of Head's life. They were eventually to marry and to share the remainder of their lives. But first they were obliged to undergo a long series of trials.

A Beautiful Period Of Probation

Ruth Mayhew was the daughter of Anthony Lawson Mayhew (1842–1916), Fellow and Chaplain of Wadham College, Oxford and Jane Innes Mayhew (1843–1915). Ruth was the eldest of eight children, the youngest of whom was born when she was seventeen. Despite the frequency of these pregnancies, she recalled that: 'I never noticed any change whatever in Mother's figure! But of course I was exceptionally unobservant'.[3] One of her brothers – Arthur Innes Mayhew (1878–1948) – was to attain some distinction as an educationalist and Colonial civil servant. Ruth had warm relations with her brothers. But she confided to Janet Ashbee that: 'My sister and I hated each other till the age of 14. Then I began to admire Ethel and rather worshipped her'.[4] Despite the respectability of her family, Ruth was to recall a childhood dominated by a penny-pinching mother: 'I was', she lamented in 1917, 'brought up poor'.[5]

Literature played a special place in Ruth's childhood. In 1916, she remembered that she had been a: 'bookish girl whose world was sharply divided into books which were delightful, and life which was at best pleasant only in streaks …' In her teens and twenties the writings of Henry James were to have a particular impact on Mayhew's sense of self; indeed through their influence: 'books were to me no longer a world apart in which I happily wandered, but an integral portion of my individual life, and the two worlds became indistinguishably mingled'.[6] It was less the plot details of James's stories that entranced her than his characters' 'mental capacity for receiving impressions'. In Ruth's later years, 'such impressions were to be always the one thing of importance which life would throughout hold for me'.[7]

When she was thirteen, Ruth and her younger sisters attracted the attention of Charles Lutwidge Dodgson (Lewis Carroll) (1832–98), who obtained permission from Mr and Mrs Mayhew to photograph the girls. In a letter of May 1879, Dodgson maintained that the photograph he had taken 'of Ruth as "Comte de Brissac"' was 'decidedly good'.[8] Dodgson later thanked Anthony Mayhew for permission to photograph the younger Mayhew girls in a state of undress, noting somewhat ruefully: 'I hardly dare hope that it includes *Ruth*'.[9] When, however, Dodgson sought parental consent to photograph the children naked, he was refused, and a rupture in relations with the Mayhews evidently ensued.[10] Any unseemliness was, however, apparently lost on Ruth. Years later

Head's mention of a poem made her remember that: 'Mr Dodgson had quoted it to me in my little petticoat and bare ankles, a quarter of a century ago ...'.[11]

From 1875 to 1883 Mayhew attended the Oxford High School. She subsequently studied Modern Languages at Oxford and as a young woman spent extended periods on the Continent enhancing her command of French and German. When she visited France and Italy, she viewed these countries through the: 'gorgeously appreciative spectacles' of a devotee of Henry James's writings. Ruth: 'never lifted the heavy leather curtain of some old Church, or smelt the air of some bright prosperous Paris morning, or crossed the shining floor of some great picture-gallery sumptuously housed, but it was with the ardour of a disciple treading a pathway whose windings had long before been indicated by a Master's hand for my enjoyment'.[12]

On returning to England, Mayhew was obliged to devote herself to the more mundane business of earning a living. She worked as a temporary lecturer in German at Holloway College in London in 1887 before beginning a career as a schoolteacher. Her first appointment was as an Assistant Mistress in Oxford. There she laboured under a colleague drawn from 'the Spinster class', who was apparently obsessed with the fear that the pupils might form Sapphic attachments.[13] During this time, she continued to live with her parents. In 1899 Mayhew gained a novel degree of independence when she took over as Head of the High School for girls in Brighton.[14]

Ruth Mayhew was thus something of a *fin-de-siècle* 'New Woman', determined to make her own way in the world. Her bluestocking credentials were further affirmed by a taste (much censured by her mother) for 'advanced French novels',[15] as well as by a penchant for smoking cigarettes in public. At the same time, she could articulate deeply conservative views about women's maternal destiny, and was ever ready to give voice to her 'hatred of lady-doctors'.[16]

Ruth and Henry first met in Oxford on 3 August 1895. The circumstances of their encounter are not clear. However, it seems that Ruth was acquainted with some of the Head women before she met the eldest son of the family. Years later Mayhew admitted that at first Head's talk: 'half mystified and half alienated me'.[17] It seems that Henry, in contrast, from the first found Miss Mayhew fascinating. Despite the equivocal initial effect that Head had upon her, Mayhew also felt some attraction. She confessed in 1902 that she had 'kept in a box in a drawer in my secretary a little local old photograph ... done one Christmas at Buckingham in 1895'. The image of Henry Head preserved in this photograph was: 'young and jaunty and critical ...'[18]

The two had much in common. They shared a passion for literature that was to supply a foundation for their early interactions – although Head cared more for Flaubert than for James. They found, for instance, a shared enthusiasm for the celebration of aestheticism found in the works of Walter Pater (1839–94):

the two took long walks on which *Marius the Epicurean* was the main topic of discussion. Ruth's command of German meant, moreover, that she could converse with Henry about his favourite poet, Heinrich Heine. 'Poetry', she later told Thomas Hardy (1840–1928), 'has always meant a very great deal to us both …'. But she also confided to Hardy that her tastes in poets had differed from Head's more than she had cared to admit in the early stages of their relationship. Mayhew 'noticed from the first, that though he was usually pleased with the poems that aroused my highest enthusiasms, I was frequently baffled by those he selected for my admiration'. She ascribed these divergences to: 'the different way in which men and women appreciate literature, especially poetry'.[19]

Both Ruth and Henry were, moreover, ardent and energetic cyclists: while on a tour of the Rhine region in 1899 Mayhew described herself as 'a creature so brown and hard and strong, who rides 50 miles every other day on a strong little wheel'. Like Head she carried books of poetry with her on these excursions, maintaining that: 'There is nothing in the world so refreshing after long riding in sun and stress as to sit on the ground in the shade and read and slowly take in the delicious words'.[20]

There was also a congruence in worldview. Although she came from a religious family, Mayhew had, like Head, lost faith in Christianity at an early age. Both her father and her brother Arnold were clergymen and her mother was conventionally pious. While she remained under the parental roof, Ruth was therefore obliged to make a show of religious observance. Thus during a family visit to Nijmegen she complained about being obliged to attend 'a dreary service at the French Protestant Church'. She added that: 'the French Pasteur yesterday condemned most of my friends to hellfire in a kind of wailing bellow which really made me feel uncomfortable'.[21] While she was obliged to hide her feelings from her family, Mayhew felt from an early stage in their friendship able to confide her true views on this sensitive subject to Head. She even allowed herself the mischievous – and entirely frivolous – suggestion that her disgust with Protestantism might set her on the road to Rome: it would be 'a charming diversion to allow myself to be converted by a stuffy village priest wouldn't it?'[22]

Once she had left home, Ruth was less coy about expressing her views. When her clergyman brother became excessively 'priestly', she was goaded 'into an assertion that I hated Religion, which he certainly for the moment did make me feel'. She advised Arnold that: 'it was indelicate in a priest to give advice on my spiritual state as it would be for a doctor to prescribe my bodily needs, unsolicited …'. Creeds, Ruth declared, were 'detestable … Can't one love truth and justice and charity and leave those who will to quarrel about the apostolic succession and the real presence as much as they will'.[23] She took it for granted that Head would sympathize with this humanist sentiment. In March 1901 he indeed confided in her that: 'I have come to the conclusion that no sick bed can be happy without complete purity from the vice of revealed religion'.[24]

Mayhew liked to present her sceptical, ironic self as a reflection of the spirit of the French Enlightenment. 'By taste and temperament', she declared, 'I am pure Eighteenth Century'. She therefore found it ironic that she should find herself drawn to 'exchange ideas once a week with you most modern of moderns'.[25] In fact, Mayhew and Head drew upon similar sources in shaping their attitudes to Christianity. Ruth may have been unaware of the writings of Charles Darwin; indeed, she confessed to being ignorant of natural science in general. But both she and Henry were acquainted with modernist approaches to religion that denied Christianity any special status and insisted on understanding it as a cultural and historical phenomenon. When in January 1902 Mayhew toured Berlin museums in the company of her friend, the art historian and classicist Adolph Oppé (1878–1957), they were fascinated to find among the Greek vases symbols that seemed to anticipate the Christian cross (the accompanying illustration is of a swastika). The cross, Oppé informed Ruth, 'is one of the very earliest forms of decoration and he is working at a theory that it was adapted first by the Emperor Constantine as his symbol and became later the symbol of Christianity ... We found an excellent example of Constantine's symbol on a clay Roman lamp'.[26]

In the course of the same visit, Mayhew revealed that both she and Head were familiar with James George Frazer's *The Golden Bough* (1890) – a work that had scandalized so many of their contemporaries. Ruth and Henry, on the other hand, saw the book merely as a source intellectual fascination.[27] When a new edition of the book came out in 1901, Head told Ruth that he was planning to read it: 'for it to help me over the many languid hours I see before me'.[28] Frazer had caused particular outrage among the orthodox by drawing parallels between the New Testament narratives and aspects of pagan myth. Head had analyzed the early-Christian iconography he found in Rome from a similar perspective.[29]

Ruth and Henry thus shared a common social background and cultural interests. Moreover, to a remarkable extent they manifested a shared worldview: one that was secular and modern. Their rejection of conventional Christian belief did not, however, lead either of them to a surrogate religion such as socialism; they were indifferent to political creeds of all kinds. A low opinion of the working classes was matched by scorn for the philistine and materialist tendencies of so many of their own social stratum. They found their own substitute for a discredited religion in a dedication to the cultivation of the inner life through a pursuit of aesthetic virtue that matched in intensity the devotion of any religious zealot.

By March 1899, Head felt able to divulge to Mayhew some of his most intimate thoughts and sense of alienation from the common run of humanity. 'At one time', he confided:

> I doubted whether it was abnormal to have feelings so remote from those of the world around. Of late years I have found similar feelings in so many whose minds are attractive that I can only look at one's own mental State as a normal variation in the great genus of mankind. This sounds pedantic & priggish but youth always fears it is abnormal when standing apparently alone. I have no doubt that the first of our Simian ancestors who took to walking on their hind legs only did so in the privacy of their ancestral woods until emboldened by finding some fellow creature who also practised the same eccentric mode of locomotion.[30]

Mayhew's was undoubtedly one such exceptional mind that Head found attractive.

Such shared interests and views drew the two together. Potent social conventions as well as practical exigencies, however, threatened to separate them. There was in the first place the question of where and when they might meet. Head was obliged to spend most of his time in London; Ruth worked in Oxford and later Brighton. Mayhew at first remained ensconced in a parental home where she was subject to rigorous surveillance by both her mother and her sisters. She made occasional visits to Buckingham, the Head's country residence in Sussex; but here it was Henry's female relatives that kept careful watch. Even when meetings were possible, privacy was difficult to attain. In 1897 Head made a joke of his sense of frustration at these hindrances: 'We certainly have up till the present behaved much like the man and women of the weather glass. But is not this the English ideal of the relation of the sexes? And do you not feel a burning sense of righteousness that your behaviour has been so discreet?'[31]

In the same letter, Head suggested a weekend meeting in London, where, he assured her, they would be under 'efficient chaperonage (behind the curtains)'. This raillery at the conventions intended to regulate the interactions between the sexes became a theme in later letters. In February 1899 Mayhew suggested that they should take walks in locales regularly patrolled by the police and thus 'implore the protection of the State as Chaperon'.[32] Head took up the joke, but his reply also showed genuine scorn for the bourgeois standards that trammelled his relations with Ruth:

> the State as Chaperon? – Men and women who would willingly forgather (not as lovers for they are shameless and neglectful of all observation) but for pleasant communicating one with another are hindered by the necessity of a chaperon. For either the chaperon is bored with such things as please the two companions under his charge & terminates their conversation at the most unsatisfactory periods or no suitable person is obtainable. Now of all chaperons for such intercourse a policeman is the most perfect. He is not bored and he in never sufficiently interested to listen and thus destroy the perfect sense of free play of wit. Now the State provides such chaperons free, gratis, in certain large rooms in the town …[33]

This disdain for Victorian propriety was to translate into increasing frequent meetings between the two. Mayhew kept a careful log of these encounters. In almost every case a cultural event, or shared appreciation of some work of art, provided the occasion. Thus in September 1896, Ruth and Henry attended together a performance of *Cymbeline*, their 'First Shakespeare Play'. That Christmas they spent at Buckingham reading poetry. Further meetings followed in early 1897, including one in West Kensington devoted to the work of Robert Bridges. In May, Mayhew and Head showed their avant-garde credentials by attending one of the early London readings of Henrik Ibsen's play, *John Gabriel Borkman*. In November followed *Hamlet*, their second Shakespeare play. The next month they visited the Rossetti Exhibition at the New Gallery in Regent Street. 1898 saw a similar pattern of readings, plays and exhibitions.[34]

In the spring of 1899 Head and Mayhew embarked upon a new, still more daring, departure; they planned to hold a joint holiday in Germany. Head was to travel ahead to the Continent for a cycling tour of the Loire valley before rendezvousing with Ruth in Berlin. The prospect evidently excited him; he looked forward to 'another bright meeting and in hopes of further delight under foreign skies and in a guttural tongue'.[35] Their main goal was apparently to visit the Sans Souci Palace in Potsdam in order to admire Jean-Baptiste Pater's decorative work. Ruth had developed an interest in Pater's master, Antoine Watteau, perhaps through her reading of the Goncourt brothers.

Around the same time, both Ruth and Henry underwent significant changes in their lives. Head – by now a physician at the London Hospital – left his digs in Wimpole Street to move for the first time into a flat of his own. At the time of his departure for his holiday, the lease had yet to be signed. Head confessed that: 'I dread the necessity of furnishing but perhaps mother & Hester will help me'.[36] In the event, Ruth, while staying at Buckingham, also took a hand in relieving him of these domestic cares; on 3 April 1899, she wrote: 'All this morning your Mother and I have been sitting with Stores Lists on our laps, adding up and looking out your kitchen pots and pans. It was very amusing work, but the Matthew Arnold afterwards was on the whole more inspiring'.[37]

For her part, Mayhew was about to undergo a yet more important transition. She had applied for the post of headmistress of the High School for Girls in Brighton. She forwarded the testimonials she had garnered to Head who appraised them in encouraging fashion. He assured her that: 'I do heartily hope that you will be successful'.[38] By 16 March 1899, Ruth was able to advise him that she was 'a selected candidate', and due to be interviewed the following Wednesday.[39] When he heard that Mayhew had been offered the job, Head assured her that: 'The tidings filled me with joy that you may now pass on to a higher & fuller work'.[40]

A major implication of the new appointment was that Ruth could at last leave the parental home in Oxford thus gaining a greater degree of independence. Ruth was surprised by the reaction of her family. She had not imagined that 'my people would mind and oh they do'. Ruth herself felt some misgivings at this break with her past. She told Henry: 'It seems to me I have never once felt warm since I realized what leaving Oxford meant'.[41]

She took some comfort from the farewell presents she received. Her pupils gave her an illustrated book on art history. Mayhew's fellow mistresses for their part presented: 'two little bits of old silver. The dear unobservant things had never noticed the singular inappropriateness of the symbol on the caddy-spoon, a very charming little Cupid with bow outstretched'.[42] Writing from his hotel in Blois, Henry was: 'amused ... with your account of your metamorphosis'.[43]

In Berlin, Ruth stayed with the Rogge family in their house on Priesterstrasse. She had made their acquaintance on one of her earlier trips to Germany a decade before. This return occasioned 'a curious dream-like feeling back, with everything outwardly so unchanged and everything in myself so different'. She recommended that Head stay at the 'Hotel Einsiedler' – 'a charming-looking old inn quite near us ...'[44] She added the hope that he would not, however, live as an 'Einsiedler' [hermit].

There was no need to worry. No sooner had Henry arrived in the city than he wrote to Ruth proposing an itinerary of cultural activities for the following day. Head was, however, willing to accommodate himself to Ruth's plans, professing himself to be: 'quite at your disposal but let us meet somehow'.[45] He was even prepared to accompany Mayhew to church (presumably out of deference to the habits of her hosts) if that was her plan.

Ruth and Henry had a week together in Berlin. Head was obliged to return to England first. He at once composed a letter to tell her how much he had enjoyed their time in Potsdam:

> The Frühstück by the lake and the hot afternoon beneath the bust in the wood – the bright restful town days when Berlin showed us only its cleanest holiday face – Above all the exquisite Sans Souci mornings ('Sans Souci' must be our name in the future for a perfectly successful unruffled Expedition–). And all this was rendered possible by the admirable friendliness of a German family to whom I was an absolute stranger.

Members of the Rogge family had presumably played the role of 'chaperon' deflecting at least some of the outrage that Mayhew and Head's assignations might have provoked. Even so, Head admitted that the holiday had been a 'somewhat risky experiment'.[46]

With almost comic solicitude, Head sent Ruth detailed instructions on how best to negotiate her journey home. He gravely advised her that it was prudent to secure her sleeping berth when the train arrived in Oberhausen, as it: 'saves the scurry and delay on arrival at the boat'. Mayhew was also instructed on when to take nourishment: 'You will have had your evening meal at about 7o'clock (Berlin time) and it will be well to have a bowl of Bouillon or a roll from the women who go through the train at Goch about 9p.m. (London time or 10 o'clock Berlin time)'. Upon reaching English shores, 'You had better have cup of tea & bread & butter on the boat & a proper breakfast at Victoria'.[47]

It is unknown how closely Mayhew followed this prescription. She was, however, glad that Head found the time to meet her train at Victoria Station. 'You have no idea', she told him, 'how much I appreciated that your face was the very first thing I saw as we sailed in and it seemed like an incredibly pleasant dream. And I never even stopped to think how horribly unappetizing a female is apt to look after 22 hours of travelling ...'[48]

Even before the trip to Germany, Ruth and Henry's growing intimacy had attracted censure from both the Mayhew and the Head side. Most of their friends and acquaintances also looked askance at the relationship. Long afterwards, Mayhew recalled that: 'the Whiteheads were the only people who never tired of us nor shut the doors on us "because of what the servants might say". (This quite true of the 90s)'.[49] Head's undergraduate companion, Alfred North Whitehead and his wife did indeed allow Ruth and Henry to meet regularly at their home, the Mill House at Grantchester. Whitehead apparently found these visits refreshing. After one such stay he told Henry: 'how much he had enjoyed himself and I know by his manner he had spent one of those evenings on which he is at ease'.[50] At the Mill House Mayhew and Head were able to mingle with other broadminded friends of the Whiteheads, including Bertrand Russell and Sidney Webb. Such enlightened tolerance was in stark contrast to the attitudes of their own families.

In November 1898, Ruth warned Head that 'I am being sore pressed by my powers to give up my letters, and I cannot make up my mind how far I am right in resisting ...'.[51] The identity of these 'powers' became clearer in a later letter when Ruth revealed that she was preparing herself for 'mother's possible prohibition of my letters to you ...' Jane Mayhew – Ruth's 'stern Mamma'[52] – apparently saw Head's letters as an 'unsteadying influence' on her daughter.[53] Many years after the event, Ruth recalled that her mother: 'used to allude to H. during our stormy "friendship" period as "That Devil" and blame me pitilessly for 'running after him' in a way no man liked at heart!'[54] Quite why Mrs Mayhew took such exception to her daughter's friendship with Head is unclear. She may, however, have been aware that, because of his financial circumstances, he was in no position to marry Ruth, who thus risked scandal by continuing to associate with Henry.

In the event, Ruth did find the strength to resist this pressure. Her mother, however, made no secret of her continued disapprobation. At noon on 3 March 1899, just as Ruth was laboriously counting up her pupils' marks on her fingers:

'Mother comes in with her severest "How can you-be-so-indelicate-as-to-have any dealings-with-that-man-face and plumps down your letter"'.⁵⁵ This hostility did not diminish with time. Almost two years later, Mrs Mayhew could not 'conceal the horror that my friendship with you inspires in her'.⁵⁶

Ruth's departure from Oxford – her initial misgivings notwithstanding – relieved her of much of this pressure. On the anniversary of her move to Brighton, she rejoiced: 'it was this wonderful Wednesday last year that gave me my School, my home and my liberty and I have been thinking about it all day with much searching of the heart …. But it was a great risk'.⁵⁷ The risk was worthwhile now that she could freely receive and reply to Henry's letters. Mrs Mayhew was not entirely done with her daughter; but she now confined herself to advising Ruth of 'the crushing need of my saving every penny I could against my needy old age'. Such perorations served merely, however, to make Ruth: 'thank my stars I am really free at last, at last'.⁵⁸

A more serious and persistent threat to the relationship came from Henry's female relatives. Ruth appears to have known Head's mother and sister Hester from the outset of their friendship. As her focus shifted to Henry, she began to voice anxieties about how Mrs Head, in particular, would react. After thanking Head for a Sunday outing, Ruth observed that she should also write to his mother, 'who will, I fear think I have been very bad and mad. Do you think she will have the heart to scold me all across the seas, when she can't make it up with the kindest kisses the moment she thinks I am looking miserable?'⁵⁹ By September 1898, she confessed that she was 'stirred up' by the letters she received from Buckingham, 'which read to me very tragic indeed and make me rue the day I was born …'⁶⁰

The grounds of Mrs Head's opposition to her son's growing intimacy with Mayhew are also obscure. By July 1899, however, Ruth was left in no doubt of 'how glad our two mothers would be'⁶¹ if she and Henry were to cease their correspondence. Mrs Head found more sophisticated means than had Ruth's own mother to convey her hostility. She treated Ruth in a 'poor deluded child' manner implying that Henry was far superior to her.⁶²

Head's response to these shows of disapproval was to side with Ruth; as a result his previously close relations with his mother suffered. While indignant at the way that Ruth was treated, he could not refrain from deriving some psychological insights from these events: 'How is it', he asked:

> that women who fail to understand correctly the simplest problems in the life of those around them, who walk among their fellows in impermeable blinkers can yet so exactly touch the spot that hurts another woman vitally – By what curious gift can my mother, who has no conception of the least little movement of your mind, invariably touch that part of you which hurts the most?⁶³

Head was subsequently to employ 'the dynamic quality of [his] mental gaze'[64] to analyse his mother's behaviour in a ruthless manner that left her distraught.

During the summer of 1899 there appear to have been attempts to reconcile Mrs Head to the liaison. Ruth invited Head's mother to attend a poetry reading at her school in Brighton and the two later lunched amicably enough together; 'never since you were a cause of contention between us', Mayhew reported, 'have we had so pretty and peaceful time together'. Hester Head apparently: 'immensely admired my Spectator in 8 volumes which Arnold and I fished out of a beautiful secondhand shop here for 5/'. By way of reciprocation, Ruth was invited to stay at Buckingham while her new accommodation in Brighton was being prepared. Head coached Mayhew on how to conduct herself while under his mother's roof:

> When you are at Buckingham be your own bold self and shun those interminable personal discussions which delight my mother & which she can get none of her family to listen to. Choose rather some non-controversial book and above all things shun Dorothy or you will be overwhelmed with 'I have found it wisest' 'I have learnt by experience' and the like from one whose wisdom varies with the mood of the hour and who never observed but in lightning flashes, the worst illumination for a formulated judgement.[65]

Such attempts at appeasement were, however, to prove futile. By the end of the year, details of Head 'family politics' gradually filtered down to Ruth. She was 'shocked to find, what, in my obtuseness I had never realized, how all your people loathe and detest my small friendship for you ... I wonder they trouble themselves about something so small, so private, so personal a matter, but it is horrible to think of, and I feel I must see your faces no more'.[66] With Henry's encouragement, Mayhew persevered in her attempts to make herself acceptable to the Head clan staying at Buckingham again during 1900. But she confessed to being reduced to tears after one such visit when she discovered 'in your House Beautiful for the first time an atmosphere of positive hostility. It would be silly to brood over it, but believe me it was palpable ...'[67]

Mrs Head's resentment of Ruth was highlighted by shows of affection to her brother. Mayhew informed Henry that his mother had: 'fallen entirely in love with Arthur and I believe that if she had not gone away she would have adopted him. He treats her like a dutiful son and after her terrible family she repays him with endless affection. On Friday night, I was the guest and Arthur the son of the house'. Ruth concealed any pain she may have felt at this asymmetrical treatment. She chose to regard the matter as a 'delightful Comedy', and noted that Mrs Head was 'as gay as a young girl'.[68] Later, however, she revealed that the gay Mrs Head had used her acquaintance with Arthur to bring various 'railing accusations' against his sister.[69] Henry sought to console Ruth by assuring her that:

'My mother's reasons are as you know invariably false for she does not know really why she does things; she acts on some impulse that passes like the wind and the cause of both good or bad actions are forgotten for they rest only on the emotion of her moment'.[70]

Nonetheless, the daunting presence of the Head matriarch invaded even Ruth's dreams. The night after Head's parents set off on a voyage, she had 'a curious dream':

> It consisted of three distinct sensations. I saw nothing. I only felt. First your father said goodbye and pressed me so kindly I fell sobbing. Then your mother came and spoke to me without listening and looked at me without seeing, and kissed me without feeling, just as she had that last sad night at Buckingham. She drove away into the darkness and I felt myself walking on crying as I walked. Lastly I heard you calling and like Eve after the Fall I was ashamed, because I was crying. I remembered you must never see me cry. I did not stop but you were there and you walked beside me and you talked as you always do pleasantly and I grew comforted and I woke to realize that it did not matter who sailed away if you stayed behind.[71]

Ruth's siblings took a more sympathetic – if still sceptical – view of the liaison. During a visit to Brighton in July 1899, Arnold Mayhew noted that his sister: 'certainly does not pine, tho' she most certainly is in love'. Ruth herself seemed doubtful whether Head fully reciprocated her feelings. She: 'declares that when his present attachment shall as Hester & I prophesy, sink into cold friendship and later into indifference, she will bear it with equanimity, and I almost believe her, so strong she is'. When Arnold met Henry – apparently for the first time – he was pleasantly surprised: 'I confess I liked him. He was not a bit clever – there was no straining for effect – but he was pleasant and amusing, & his reflections on the East End, of wh. of course he has a wide knowledge, most interesting'.[72]

Later letters continued to show Ruth's emotional fragility in the face of the frostiness emanating from Buckingham. Yet there are also signs of a novel resilience and resistance. The sadness she felt in the course of lonely walk on a frightfully cold and windy Brighton day was mitigated by pride in the recognition that the 'grosse Head'sche Werk' had received in the scientific community. One day, she assured Henry, 'German men of Science will doubtless visit your birthplace and besiege the local stationer's stop for postcards commemorative of your Ursprung'. Ruth whimsically proposed a design for these cards:

> I should suggest a small portrait of you with the dates of most importance in your life and to left and right of you portraits of your Father and Mother, most interesting to you learned Physicians. I have composed the 'Verschen' to be printed below, but I should prefer to remain anonymous 'From Father I've only lazy being, and power to sleep all day. From mother I've my ready tongue, and domineering way!'[73]

She assured Head that: 'your friendship is my star of diamonds'. However, Ruth could not 'wear my bravery on my breast in the face of the world. I bear it under my bodice and sometimes the points cut into my flesh'. Even Head, she lamented, could cause the points to cut into her when he seemed to suggest that: 'you are only come to see me from Buckingham when it will not be noticed by your people'.[74] Henry, for his part, remained steadfast that his attachment to Ruth meant more to him than his mother's opinion of her suitability as a companion for her son. He also lamented that Ruth could be so upset 'by a mental state in your mother that has only pathological & no other significance'.[75]

Ruth and Henry finally declared their love for one another on a Sunday afternoon in July 1902. They were respectively thirty-six and forty years of age. Afterwards Ruth wrote that after Head had left her: 'I took my chair and sat at my balcony and saw the stars throb into being and it seemed to me I was drinking in great draughts of peace ...'[76] Henry's reply was: 'That I brought you peace has given me a joy beyond any joy that has ever fallen to my lot'.[77] In a later letter he drew (perhaps infelicitously) upon his experience of laboratory life to find a metaphor for the steadfastness of his love: 'My fondness for you, my dear, burns as steadily as the little lamp before a shrine or like the flame burning day & night beneath our great incubators from which come a multitude of growths in everlasting succession ...'[78]

By September 1903, it became apparent that Ruth and Henry were no longer merely friends but betrothed. Henry had declared that: 'The inevitable & preordained close of our friendship has come in this kiss in which we pledged ourselves'. They were entering upon a new stage in their relationship: 'Now come what may, it may be through many years of waiting, you are mine & I am yours not in friendship but in love'.[79]

Even after this declaration, however, marriage remained out of the question. What kept Ruth and Henry apart for so long was less the disapproval of their families than financial exigency. As Mayhew lamented in April 1903: 'Oh Harry if my tears could but be pearls and precious stones, how easily rich we should be and what ... possibility for a joined life we could purchase'.[80] Head's devotion to scientific research however kept him from more lucrative work in private practice, and thus obliged him to remain a bachelor.

For some time after their declaration, they were therefore obliged to keep their romance a secret from the world. Head took some satisfaction in this secrecy; whenever he met Ruth: 'in general company I have a colossal inward warmth to think of all the joy we have made together, we who seem to them amicable strangers – I hug my pleasure as a child holds some treasure hidden beneath its frock'.[81] Mayhew, on the other hand, found it ever more difficult on such occasions 'to play the part of the amiable stranger'.[82]

Despite these efforts at secrecy, Head's mother, however, sensed that something had occurred and sought to extract the truth from Mayhew. Ruth:

> fought hard. But it was no good. She asked the direct question that I could by no possibility evade. I had either to answer truly or to lie. But I cried to her to spare me and I said over and over again Write to him: let him tell you. And she would not let me go. I know now why you always said she must not be told.[83]

The episode had left Ruth with a sense of 'humiliation'. She took some comfort in the reaction of Head's father who, while advising continued secrecy, 'has been perfect to me'. Indeed, after his wife had left the room, Henry Head senior held Mayhew in his arms: 'and said very solemnly "I love you, Ruth" ... although he feared it would never be as a daughter-in-law'.[84]

Head sought to allay Ruth's fears. 'The world', he wrote, 'has been let into our Citadel and we must sit down & face the consequences'. He had, however, urged his father to remain discreet. While he excused Mayhew of any indiscretion, Henry's judgment of his mother's actions was damning: 'Her conduct is inconceivable did one not know that women are differently constituted in these matters than men'.[85]

Learning of this new state of affairs, Hester Head wrote to her son to assure him of his parents' love:

> for you and dear Ruth. She has always been very dear to us both, and I only wish you had told us at once of your new bond of union as we should have liked to have placed her in a still warmer place in our hearts ... We knew such a close friendship as yours and Ruth's could end in one way – for you have become necessary to each other – and we have long expected it. I never felt words were more inadequate but you know you have always the truest love and sympathy of your mother.[86]

This letter made Head 'so angry that I fear I may have answered by drawing the lash a little to [sic] strongly'.[87] In particular, he was furious at what he saw as the devious fashion in which his mother's had extracted news of the engagement from Ruth. In response, Hester adopted a tone towards her Harry that was at once piteous and reproachful: 'I do not think you can conceive a Mother's feelings or you would not have written as you have done'. She claimed that she had already guessed the secret from the fact that Ruth had begun to address Head by his first name, and maintained: 'I thought we ought to know, I was desiring to give the love that your Mother had to give and to help and cheer her in various Motherly ways ...' Mrs Head did not forbear to allude to the scandal that Ruth and Henry's attachment had supposedly generated; there had, she maintained, 'been much talk about you and Ruth on all sides ...' Her son was allegedly unaware of 'the trials I have passed thro' on the subject'.[88]

It was paradoxically Head's commitment to science that ultimately forced the issue. In the spring of 1903, Head felt that: 'I could not submit myself to the operation I then underwent without telling her how dear she was to me'.[89] Henry chose to propose to Ruth after they attended a London performance of J. M Barrie's play, *Quality Street*. More than thirty years later, she recalled to Janet Ashbee how Head had: 'made up his mighty mind to espouse me poor creature. He did it in his hired brougham by solemnly affixing a pair of earrings to my blushing lobes as we descended Constitution Hill on a rainy Nov: evening. So you see I don't remember much of the Play that first time'. *Quality Street* had clearly retained a special significance for the pair. They went to see a revival in 1922. On this occasion, however, the performance took place the day before Ruth was to undergo major surgery. She had nonetheless 'felt quite cheerful but I knew H. did not so again the Play missed its mark a bit'.[90]

Even at this juncture, however, his financial circumstances did not permit them to proceed with the wedding. It was only when in December 1903, Head's father – one of the 'two old people' he was later to scorn – offered his son a generous gift that marriage at last became possible.[91]

Anthony Mayhew, for his part, took the news with good grace. The announcement of the engagement was, he claimed, a source of 'gratification and satisfaction' both to him and to his wife. Jane Mayhew's cankered view of the relationship notwithstanding, Ruth's father averred that he regarded it as a good match: the two were 'bound together by similar intellectual and artistic tastes.' For some time, he told Head, 'you have been an inspiring influence in her intellectual life'.[92]

Alfred Whitehead, who had done so much to facilitate the pair's meetings, wrote a letter of congratulation to Ruth that combined warmth with insight:

> Dear Ruth
>
> My warmest congratulations to you on your engagement to Harry – you know how deeply I love him – we were boys when we first met, and a continuous friendship of twenty three years, always productive of unclouded happiness and the warmest feelings of affection, gives rise to a strength of feeling which I cannot put into words and sentences.
>
> With what pride we have watched his untiring energy and self-devotion, and what a joy to us his splendid triumphs! I feel like watching some tall argosy sail into port.
>
> One great personal happiness to us is that he is marrying one whom we may already claim – may we not? – as a close and warm friendship.[93]

Although Head maintained outwardly cordial relations with his parents, it seems that he never fully forgave them – and his mother in particular – for their treatment of his beloved. He wrote to Ruth in June 1904:

> My darling, it was sweet to see you yesterday and to hold your dear hand and kiss your sweet lips, all in front of the two old people who could not interfere however much they disapproved. You and I are proud to know where love is and there is a kind of passion in showing how I love you before those who once divided us. Yesterday we were masters of the situation under exactly the conditions which made you so miserable so many years ago – we have won our place in the sun. You were worth the passage through fire and water and now as we stand in the temple I turn to kiss you in the face of the admiring, success-adoring populace and say 'Meine Königin' [my Queen] my companion fit for a King.[94]

This public demonstration of their love thus served as a triumph of true feeling over conventional restraint.

Head sought to make a virtue of the length of time it had taken for his relationship with Ruth to achieve fulfilment. He assured Anthony Mayhew that:

> In spite of all the instability of human affairs I feel that she & I are about to enter on marriage with an unusual certainty of security. For during the seven years, throughout which she has given me the delight of her Companionship, all those differences, which must inevitably appear when two people are brought first into close contact, have passed away whilst our intimacy was still a delightful friendship. Now that it has become love we stand already closer to one another than many who have entered the married state with no such beautiful period of probation.[95]

Ruth too seemed to have mixed feelings about the extent of the courtship. Writing in 1926 to Head's niece, Hester Pinney, on the subject of long engagements, she maintained: 'if it is the Real Thing waiting does not matter, and I speak from long experience. There is a sweetness about love unfulfilled & that even the happiest marriage cannot replace'. This was true of 'women of my generation and fundamentally I expect you are not much changed'. Ruth also revealed, however, that: 'I have never owned this to anyone before – not even to Uncle Harry'.[96]

Because their courtship was extended over seven years, for most of which time they could only meet intermittently, Ruth and Henry were obliged to rely on the written word as a means of maintaining communication. They faithfully exchanged letters at least once a week. From October 1901 they also contrived an additional means of literary exchange. The two began to keep a 'Rag Book' – a series of volumes in which each would write down experiences, reflections, and extracts from works of literature. The book would then be exchanged and the other would add his or her own observations on and addendums to what had been written. Taken together, these two sources provide a remarkable window into the inner life of both parties.

The Correspondence

In the course of their seven-year courtship, Ruth and Henry regularly exchanged letters on a weekly basis. Even when Head was travelling abroad, he implored Mayhew to continue the correspondence because her letters 'made me feel as if I was bound by a silken thread to all that was pleasant at home'.[97] Most of these were carefully preserved, an indication of the significance they placed in these documents. Head's mother sensed something of the importance of the letters to the relationship when she sneered to Arthur Mayhew that Ruth wrote 'composition letters' to her son.[98] Indeed, soon after the correspondence commenced, it was apparent that the pair had invested their letters with multiple layers of meaning.

This was especially apparent when the exchange was threatened by the disapproval of Mayhew's mother. When he learned that Ruth might cease to write, Head declared:

> You have frightened me terribly with your suggestion that your letters might cease. Surely the most prudish could not object to our correspondence – for I suppose that if you cease to write I also shall have to stop writing. You have no conception how much I look forward to your weekly letter. I so hurry through my life that for days I scarcely think, apart from the cares of the moment. Then your letter comes as a breath from a life where there is time to read pleasant books and to think fair thoughts. For days I follow up the train of thinking that you have set going until at last the track is obscured and hidden out by the multifarious coming & going of my daily duties.[99]

The letters themselves and the processes involved in their production frequently figure in the correspondence. In May 1899, Head wrote: 'I need scarcely tell you what the letters are to me'.[100] He nonetheless proceeded to expatiate on the central role these artefacts had come to occupy in his life:

> On my way homeward on Tuesday evening I look to the letter that I know will be lying on my desk and can scarcely refrain from reading it in the street as I pass from my room to my lodgings. I seem to enter another land as if I suddenly passed out of the windswept dusty unlovely street through a door to which I alone have the key into a walled garden bright with beautiful things. Let me beg you humbly on my knees do not take away my key.[101]

He had the previous December described how every Saturday, when:

> the strain of the four hours teaching is over – whether I have taught well or badly – whether I have made shameful mistakes or narrowly escaped hideous pitfalls is all one for the remainder of the day is to be devoted to you – I draw my chair to the table, light my pipe and set sail for a land where I can mentally lie on my back and talk to you unhampered by an uneasy feeling that I should be up and doing. I spread out your last letter before me and turn over sheet after sheet in search of those questions that

lie hidden in your description of things that have pleased you – Questions that are such excitants to talk.[102]

Ruth's weekly missive thus served as a means of escape from the dreary and sordid routine of his daily life and the means of entrance to a private magical land that he shared with her alone.

Ruth also described the effect that the receipt of one of Henry's letters would have on her. She: 'waited with decorum til all was cleared away, til my big chair was pulled up to the fire-side and then, and then ... I had you all to myself, almost as if you had the other armchair the other side of my hearth ...' At the worst, she informed Head, 'your letters are a kind of palimpsest ... the more I read them the undermeaning, which is not stiff and hard, comes out, and at the best they are the best letters in the world to me and that I hope is what you mean them to be, is it not, Sir?'.[103] Reviewing the accumulated letters Henry had sent to her various addresses led Ruth to the realization that: 'they made up the whole story of my life they showed how your care for me had been for years the star of my wandering bark'.[104] When she received one of these missives, 'the roof opens, the house vanishes, I am living in the big world outside where men fight and work and laugh and push. I feel proud and strong'.[105] A temporary interruption to the correspondence caused by Head's illness provoked a powerful reaction from Mayhew. She was: 'horribly frightened. When there was no early morning letter today to say how you did, I felt literally sick with misery and went to school the saddest woman in Christendom'.[106]

She could not, however, forbear to voice the fear that: 'I really believe [you] like my letters better than you like me ...'.[107] This more equivocal response to Head's letters to her was to recur. In contrast to Head's unqualified lyricism, Mayhew confessed that his letters sometimes conveyed an almost eerie affect; one that she received in the spring of 1900 made: 'something of the chill of this most uncanny April ... to come into the room ...' She explained that she had 'been thinking a great deal about our letters on my lonely walks lately', and had concluded that there was an asymmetry in the correspondence:

> if you will forgive a Jacobean metaphor, it seems to me that my letters to you are the coarse durable garment of my friendship for you, but your letters to me are the royal robe of your condescension towards me, all glorious with embroidery of curious device, but not always very cosy to wear ... The royal robe is too heavy for my unworthy shoulders ...[108]

Head firmly rejected such suggestions that Mayhew's side of the correspondence was somehow inferior in value. Nor would he concede that his dedication to the correspondence could somehow be disconnected from his attachment to her. He did acknowledge, however, that sometimes when reading Ruth's letters, 'I seem to have you close to me and to be looking straight into your eyes & down into your soul'. On the other hand, 'sometimes when I am with you, you seem miles away – as if I was groping in the dark or was speaking to you through a telephone'.[109]

Later he maintained that her letters were: 'so alive that they almost seem as if I touched you – No, that is not true, for no actual touch is so intimately living as your letters; and often when our bodies are close to one another a wall of ice seems to lie between us'.[110] Ruth was moved to reply that she found herself 'in the anomalous position of a woman who is jealous of her own letters'.[111]

The significance attached to these inscriptions was evinced by the ritual care with which the letters he had received from Ruth were treated when, in June 1899, Head moved into his new flat in Great Marylebone Street. This was a milestone because: 'for the first time in my life I occupy a small plot of my own …'. He had carefully carried Ruth's letters from Wimpole Street, 'and began to put them away'. Instead of completing the task, 'I began to read and read through from May 1897 until the present time – Do you know I have 106 exclusive of certain manuscripts & cards?'.[112] Ruth's response was typically self-deprecatory – ' I hardly know what I felt most mad or glad or sad when I heard the astounding news that my letters, my horrible brazen old letters had moved into your fair clear new flat … I thought they would have gone to light your pipes long ago …' But she thought it 'wonderful' that Henry had chosen to read her missives 'on such a night'. If she could 'for one moment think them worthy of such honour, I should be to-night the proudest of women …'.[113]

Writing in October 1902, after re-reading some of the correspondence, Ruth marvelled at the 'wonderful quality of crystal sincerity' in Head's letters. She assured him that: 'They are my most precious possessions, those hundreds of letters'. She planned to continue to read them throughout the remainder of her life. Mayhew had, moreover, stipulated in her will that upon her death they were to be returned to Head because: 'it seems to me the only way of securing to you what you will I believe one day feel to be a precious heritage'.[114] The 'heritage' in question might be construed as entirely personal. But Ruth and Henry were avid readers and critics of the published correspondence of such figures as Robert Browning. There are hints that they suspected that their own letters might one day also have a similar public currency.

Through the letters, Ruth and Henry generated representations of themselves and of one another. These operations were sometimes performed in a self-conscious way. In November 1900 Ruth confessed to: 'a strange feeling, as though I were really writing to the man built up by my imagination only, as you once said I was in danger of doing …'.[115] This was part of a wider process of mutual representation that was perhaps the chief function that the correspondence performed in the course of their courtship. The letters hence served as an instrument for the dynamic production of what Lydia Ginzburg calls 'literary personality'.[116]

Through this medium, Ruth and Henry engaged in a form of interactive self-fashioning in the course of which aspects of the personality of each was subject to scrutiny, revision, and confirmation. Through writing these letters, they were, in short, composing their selves.

The process was not free from danger or pain. Head, in particular, came to grow wary when he wrote to Mayhew of unintentionally treading 'on another tentacle or [touching] another tender spot'. Such transgressions betrayed a lack of clinical tact. He wished that: 'I knew all the hyperaesthetic spots of your soul with the accuracy with which I could recognize those of your body were you ill & in pain'.[117]

Mayhew's letters contained representations of her past and future, as well as of her current, self. Her recollections of the young Ruth were tinged with nostalgia and regret. On the occasion of a trip in March 1901 by Henry to Venice, she assured him that: 'You may meet my ghost there, very prim and just 17. My hair is up and I look everywhere and love everything ... My poor happy little ghost, please smile at her, I don't mind how contemptuously'. She doubted, however, that Henry would find this girl of much interest: should Head encounter the teenage Mayhew, 'we should certainly have no conversation'.[118]

In January of the same year, goaded by her mother, Ruth conjured 'mental pictures of myself at 55 in two tiny rooms in Long Well Street at home with a tiny fire, a little candle, a patched dress and the Daily Mail'. She sought to offset this bleak vision by offering Head a word picture of her current flirtations with chinoiserie; she told him of: 'a charming flame coloured mandarin's coat embroidered in pale flowers, which I bought at Liberty's and wear every evening to my own great satisfaction'.[119] But it was just such 'extravagance' that her mother had warned would lead Ruth to penury.

In public, Ruth might have appeared an accomplished and successful woman. A thread of doubt and deprecation, however, runs through the presentations of self that Mayhew conveyed to Head. A casual remark by Henry might lead her to conclude that he found her 'far inferior in intelligence and ripeness of judgment' to his sister, Hester.[120] Elsewhere, Mayhew expressed the concern that Head considered her: 'the odious creature I sometimes fear I am becoming'.[121] She worried that these fits of depression – or her 'Katers', as she called them – would persuade Head to end their 'most-unequal friendship'.[122] Thus in October 1898, she confessed:

> I haven't written before, although I have had a perilous number of things to say, because I have been swimming about in a perfect sea of sordid worries, and I can never write to you when my wings are all draggled with salt water ... However, you know through what a miserable microscope I habitually view my own small troubles ...[123]

Her move to Brighton had heightened Ruth's melancholy. Although she was now relieved of her mother's tyranny, there were many aspects of Oxford life, such as evening service in the College chapel, that she missed. Occasionally old friends like Dolly Irving, wife of the actor Sir Henry Irving, would come to visit Mayhew in her new accommodation. But even this 'most pathetically lovely creature' could bring only temporary relief. The anxieties attendant upon being head of a school were often the immediate antecedent of a 'Kater'. She told Henry how one Friday she endured: 'a horrid visit from a fiendish parent who abused the school and the mistresses and because it was the end of a hard week and because I was a fool, I let her see it hurt me and when at last the door shut upon her, I felt very shaky as to the legs and very much vexed at my own want of callousness'.[124] Moreover, Mayhew recurred in her letters to a fear that her school would fail and she would lose her post, leaving her: 'that most pitiful of all objects, a middle aged failure looking for new work'.[125]

Underlying these particular anxieties, however, was a more fundamental sense of disillusionment. The hopes and aspirations of the youthful Ruth as she toured Europe seemed at odds with the shortcomings of the woman she had become – or whom she was, at least, obliged to play. 'At eighteen', she told Head, 'in Germany, one <u>was</u> what one seemed, but in middle life one isn't, and it is only the masquerade aspect that makes life endurable'.[126] In February 1900, Mayhew claimed to be reconciled to the fact that she would never be at peace: 'in very truth I shall always be storm-tossed, husbandless, militant'.[127]

Head adopted a variety of strategies in response to Ruth's exercises in self-deprecation. Sometimes, he merely assumed humorous, reassuring stance. Thus replying to Mayhew's suggestion that he thought her 'inferior in intelligence and ripeness of judgment' to his sister, he claimed the suggestion 'amused' him: 'for I only intended to imply that Hester answered to the stimulae better than I expected. As to knowledge and ripeness of judgment you know as well as I do that she is but a child to you'.[128] The stimulus in question was a performance of Edmond Rostand's *Cyrano de Bergerac* to which he had taken Hester, who, apparently, was to be viewed merely as an experimental subject not as Ruth's superior.

In other instances, Head adopted a quasi-clinical pose when dealing with Mayhew's 'Katers'. In January 1901, Ruth wrote that since their last meeting she had been lost in a 'murky valley of humiliations'. Even a letter from Henry 'could not shake off the misery of that morning hour'. In particular, she felt that she had 'degraded' herself in Head's eyes by her evident eagerness for a meeting with him. His reply combined an analysis of the psychological processes that supposedly underlay such anxieties with an affirmation of the solidity of their relationship:

> I hate to think of you as travelling in the valley of Humiliation more especially when you have reached that vile spot by following the will-o-the-wisp of false fears. Humiliation should only be normally reached by some bodily condition over which we have no control and we should say I have a stomachache and am suffering from humiliation. You, however, say 'It is shameful to think how eager I showed myself' and 'it seems as though I were for always degraded in your eye'. After all these years I am vexed that you know me so little for why should you not be as eager as I am for an evening with you? The bond of our friendship is founded on the stipulation of equal rights – I am openly aggressively rejoiced when I knew I am about to spend time in your company and when I think that pleasure is within my reach I ask for it shamelessly and eagerly. You on the other hand always look upon any wish to hold intercourse with me as something to be furtively experienced. The natural result follows this perverted aspect on your part.

Ruth's sense of humiliation was therefore 'entirely unnecessary and your supposition preposterous'.[129]

When Mayhew suggested that her regrets at her childless state were in some way reprehensible, Head sternly informed her: 'For a woman to desire children is the highest moral craving implanted in her by nature'. Far from such desires being detrimental to her mental well-being, they endowed Ruth with 'increased humanity' when compared with 'the atrophied womanhood of the usual Schoolmistress'. It was, indeed, 'morally wrong for you to be without such desires and it is only foolish social custom that looks upon them as reprehensible'.[130]

Sometimes Head supplemented such psychological approaches with the still more robust clinical judgment that the source of Ruth's complaints was 'purely physical'. In March 1901, when her feelings of being 'on edge' became especially acute, Mayhew, 'bearing your "pure physical" in mind, did really visit a dentist', who assured her that her teeth were 'more sensitive' than normal. She assumed that Head would laugh at this diagnosis; but she also derived amusement from the episode: 'Is it not too bad? To have not only fewer skins than other folk emotionally but to have the bone of one's teeth of an inferior quality too'.[131]

In her letters, Ruth thus exposed a troubled and vulnerable self to Head's gaze. His responses were always reassuring and remedial, sometimes verging on the analyses and reassurances a neurologist might deliver to a nervous patient. Ruth, however, found a more homely metaphorical profession (perhaps with undertones of William Morris) for her correspondent: 'My mind is your garden. You are the gardener'.[132]

From the outset of the correspondence, Mayhew also reflected representations of Head back to him. The terms in which she described him were often hyperbolic in nature ascribing a magical, superhuman, aspect to Head; this was often contrasted with a dismissive assessment of her own worth. Thus in an early letter, Ruth addressed him as: 'prophet, wizard, most wonderful of men, how did you know everything about my poor mean little soul and its petty miseries?'.¹³³ Elsewhere she described herself as a mere blackbird while Henry was the 'King Phoenix' whose personality: 'is like the sun at noon in a wide open marketplace'.¹³⁴ Given the disparity between the two – with Henry 'always smiling, lusty and debonair', while she was 'full of sighs and groans' – Ruth professed that with each letter she expected to be informed: "Bitte hören wir jetzt gleich auf' [Let's stop right away]'.¹³⁵

Mayhew chose to construe the fact that Head nonetheless continued the correspondence as further evidence of his compassion. Head was akin to an Old Testament prophet – with an 'Elijah-like mission to Humanity'.¹³⁶ This dedication to the benefit of his fellow man was most evident in his persona as a physician, and Ruth sometimes suggested that there was indeed something of the clinical in his relationship with her. In April 1900, Mayhew recalled that at a recent meeting she had been 'in a state of very real distress and the troubled weeks before had made me conscious of what Mr Raleigh calls you remember a "vast fatigue"'. Ruth had brought: 'my sore, my aching sore to you and you, the wise chirurgeon, brought the healing plaister and also gave me oil and wine, so that I have gone on my way singing …'.¹³⁷ Head's curative influence on her was: 'a miracle, strength flows out from you to me. Isn't it natural that I should overflow with gratitude for a gift so godlike?'¹³⁸

Mayhew appeared determined to insist on imputing this 'godlike' aspect to her correspondent even in the face of his protests. When Head asserted that he and Ruth were equals, her response was:

> Your attitude to me [?] always seem the archangelical and your last letter is almost aggressively the Epistle of the Great Archangel. Generally I can come to meet you in the appropriate virgin-like attitude, but not this time. I am reassured, but I am not in the least comforted. You fall back, I see, on the old fiction of our equal rights. These are you know simply non-existent. Neither physically, nor mentally, nor socially am I your equal.¹³⁹

Behind the apparent show of humility and submission there lurked an implicit critique of a culture in which it was indeed difficult for a woman to be the social equal of a man.

Corresponding with such a supernal entity was not altogether a comfortable experience. Mayhew could light-heartedly describe some of Head's letters as: 'so harsh, so dictatorial, so awe-inspiring they are well calculated to frighten a timid female like me out of all compliancy …'. Nonetheless, coupled with her protestations of awe and gratitude, there were occasional expressions of genuine disquiet. Ruth seemed sometimes unsure whether she had truly grasped the nature of the being with which she was dealing. After several years of interchange, she could still address Henry as an 'Unknown Correspondent', and declare that: 'I have never seen, nor shall ever see your face'.¹⁴⁰ Her letters to Head could elicit the 'oddest feeling' in Ruth; she found herself asking: 'Am I writing to a real person, or an imaginary one, or to a Something that is only the creation of my brain?'[141] In 1901, she could still feel that she had: 'wrin three letters all about myself to a quite strange man whose only comment on those letters was to the effect that my theory about such and such a pointed was entirely wrong'.[142] She was, however, 'proud' when on certain occasions – such as when Head shared with her his mixed emotions on being appointed FRS – he 'let me see a little bit of the face beneath the mask …'[143]

Head did indeed often use his letters to Mayhew as an opportunity to drop the mask he felt obliged habitually to wear. The correspondence provided him with a medium in which to express views about his professional and personal identity that were otherwise strictly censored. Thus in September 1899, Head gave her an unvarnished account of his reaction to the death of Theodore Beck,

> the first of my contemporaries to go. He started so brilliantly but stumbled at the first hurdle; picked himself together again and has gone out of the race simply from want of physical power. For it all seem very like one of those races where the winners do not go any faster than their companions who, one by one, drop out leaving the prize of success to those who finish – and sometimes I feel tired and a little sad about it all.[144]

Head also used the letters to articulate the frustration that the demands of 'routine practice' and teaching left him with so little time to pursue his scientific researches. Mayhew was sometimes aggravated that he was not more forthcoming about the details of his professional activities. But, for the most part, Henry sought to demarcate what he depicted as the sordid realities of his quotidian routine from the special, indeed sacred, space that he and Ruth had established for themselves. He represented himself as:

> but the man with the muckrake sorting the rubbish of life and attempting to make it do duty again. I do not expect to find the noble metals and joy the more over my task when by a rare chance something other than rubbish turns up in this sorting. You on the other hand are the temple guardian to whom the discovery that any one of the sacred vessels is not in alloyed gold comes as a living horror and almost like personal blow. But on holy days I cleanse myself and put on my best clothes to come and worship in the temple. On such days I cannot forget that but yesterday or even a few hours ago I was turning a muckheap; yet this only adds to the pleasure of the cleanly quietude and holiness of the sacred.[145]

Through his letters, Henry might confess to 'Katers' of his own. Upon returning to London from a holiday with Mayhew in June 1900, Head wrote of a growing sense of despair. A 'horrible feeling of insecurity' began to grip him as the pleasant memories of Ruth's companionship were displaced by the prospect of a return to metropolitan life: 'Now that the dirt & disease of this vile town is beginning to close round me again my joy seems a memory so remote as almost to be imaginary ...'. In this mood, his sense of self and purpose seemed to dissolve; he felt that: 'I had toiled and given up everything and gained nothing; that I knew nothing and had planned greater things than I could ever carry out'.[146]

Later in the same month the existential chasm seemed to yawn still wider. During a break in the correspondence Head re-read some of Mayhew's previous letters. This brought on: 'a desolate sense that the first half of my life was closed & my lonely old age had begun'. He told Ruth of a dream in which he had fought violently against a crowd who were trying to separate him from her: 'One person who seemed to be more officious than the rest in holding me back I fell upon and killed ...'. He awoke with 'a sense of the most awful desolation I have ever experienced' – a feeling that lasted until he received Ruth's next letter.[147]

When in February 1902 Head described himself to Mayhew as being 'emotionally atrophied', it was her turn to seek to apply a healing balm by contrasting his own representation of self with another of her fabrication. At first, she had found Head's confession of this 'monstrous hideous you' horrible. But then: 'it came to me quite clearly that the you of which you speak in your letters is the you yourself know'. Ruth, however, knew another 'you, who has given me ... divine moments and who will give me more'.[148]

Later in the correspondence Head was to emphasize the therapeutic, indeed transformative, effect of his meetings with Ruth. In some ways, this made his customary metropolitan life more difficult to bear: 'The noise of the streets, the blatant hoardings and all he unloveliness of town life was never so apparent to me as on my return last Monday'. But when walking,

> hand in hand with you on the hill the rushing fierceness of my life seemed to pass away. I was no longer a man devoted to the diseased and disordered aspects of life, a fighter who has to pluck the heart out of each moment lest knowledge should escape never to be recaptured, but just a little child walking gaily in the rain down a green hill with my hand in your dear hand. The world seemed young and fresh again and I was cleaner and simpler ... Indeed I have 'a new heart' – and that new heart of your making.

This 'new heart' was suffused with 'a kind of proud humility' that contrasted with Head's 'usual dominant self-confidence'.[149]

Head also used the letters to convey something of the unspectacular, yet satisfying, aspects of his routines. He told Ruth that his Sundays in particular were often spent in 'almost Carthusian' silence:

> Silent breakfast, silent journey to a silent Laboratory, silent dinner and the Silence of night. Very peaceful but full of that quiet activity over little things which make a handicraft so restful. I often used to think that during the preparation in silence of paper boxes, wax blocks and the other accompaniments of Laboratory work that one would think of all sorts of fine things. But it is not so – all thought centres around

the paper boxes and the wax blocks – the day passes happily and the great thoughts are not thought.

He added, however, that he would prefer to spend even such days 'stretched in a long chair' in Ruth's study.[150]

Like his correspondent, Head would in his letters present demeaning versions of himself, which were often contrasted with more flattering representations of Ruth. Thus her innocent remark: 'You cannot imagine how much pleasure it gave me to walk with you a little in the suburbs of your mind'[151] supplied Head with a metaphor with which to lament the dreary topography of his psyche:

> You say you walked in the 'Suburbs of my mind' – I fear this is too true for it is all suburbs – long grubby regular streets peopled by uniformly respectable middle-class folk and when they come out of doors the world says how conscientious how careful they are! Your mind on the other hand reminds me of some country village with its irregular houses and unexpected streets where 'young lovers meet, old wives a-sunning sit'.[152]

He also claimed that, while he benefited through his interactions with Mayhew, he left her 'empty & miserable and this vampire aspect troubles me greatly ...'.[153]

Elsewhere Henry portrayed himself as a drudge to science – the servant of the 'most exacting of mistresses'. He therefore claimed to find Mayhew's suggestion that she was writing to an 'imaginary person' a source of comfort: 'for nothing is so soothing to a mill horse, blind & tethered to a shaft, as to be treated as if he were a wild creature freely careering over the grass'. His commitment to science notwithstanding, Head professed to yearn sometimes for a release from this servitude to a more serene life where: 'people are never in a hurry and can sit down to work at their soul's pleasure or salvation without let & hindrance for days months or years'.[154] Certain locations heightened this sense of alienation from the life he had chosen. Head claimed to have 'no desire for riches or for the impediments of an establishment'. Certain rooms, however, 'fill me with envy – Rooms redolent of quiet and of books – rooms that look out on a sweet garden – rooms warmed by the winter sun – All that I shall never possess and all that contrasts so vividly with my noisy second class hotel and untidy workroom ...'.[155]

This piteous creature bore little resemblance to the Wizard and Archangel found in Ruth's letters. She responded as best she could, stressing that Head's scientific work was beyond her comprehension. If Henry was indeed a beast of burden, she aspired to no more than to be: '[a] fly on your cart wheel, but the enlightened fly may be proud that her wheel is so firm and strong and runs so well in the long race'.[156]

But her own letters show a shared estrangement from the mundane. After a lyrical account of a walk with the Master of Gardens at Corpus Christi College,

Ruth concluded: 'He always makes me feel as if I was walking with a real mystic in a quietly beautiful world different from ours ...'[157] Her interactions with Head provided something of the same sense of transcendence of the commonplace.

In November 1898, Mayhew sent Henry a 'woe-begone fable'. The fable told of a:

> little girl, who was always hungry at home, because she could never eat her fill. And it happened one day, that, as she walked on the outskirts of a great wood, she came on a large house, and the door stood open. Looking in, she saw a table spread, with lighted candles and flowers, and food, such as she had loved. And a voice from within invited her to partake. Now she desired with all her heart to obey. But at home she had three aunts, wise women, full of years and experience; and they had often warned her against the food in the house near the forest – 'It was poisoned food', they told her. 'Whoever should eat of it, should surely fall grievously sick'.
>
> Remembering their words, she turned away sadly from the fair house and the dainty food, and went back sadly to her father's house.
>
> But when she was come again to her own room, she thought how hungry she was, and how she would now never be able to eat her fill, and she fell aweeping, and cried till the stars came out into the sky.[158]

The allegorical import of this story was transparent. Ruth was the 'little girl' who experienced chronic spiritual hunger at home. Head's was the voice that invited her into a mansion where she could partake of a range of delights than she had always craved. The three 'wise' women – who presumably represented the Mayhew and Head women opposed to the relationship – prevented her from accepting the invitation, leaving Ruth bereft and still hungry. She apologized for sending Head such a 'churlish receipt'; it had been composed 'in that kind of grim mood, when every thing I had deliberately missed was more present to me than all that I had enjoyed'.[159]

Although Henry found the fable 'charming', it also made him 'very sad'. He responded by rewriting the narrative from the perspective of 'the giver of the feast in the woodland', who was:

> but a poor wight who, condemned all his life to concoct herbs for the sick, found that during the concoction of a few poor shreds came to the surface that seemed to him pleasant to the palate. In his younger days he offered these to the sick but the sick rejected them with ignominy for they cried 'give us of your broth; strain out all that will upset us and set our weak heads in a whirl'. Lonely and longing for someone to accept that which he was scorned for offering, he spread a table in the sight of the daily path of a little girl feeling that she at least would deign to look kindly on his despised viands – But he had calculated without the great world outside – of whose opinions his lonely life kept him in ignorance. 'His food is poisoned' said the world outside 'for do we not see him collect the most deadly herbs and how can wholesome food come out of so much Evil?' Moreover the Sick will no more trust him if they see that his food be good for the healthy & for the feeble'. And so they persuaded the

little girl 'not to try the Experiment for Experiments are always dangerous'. Then as he watched & waited he saw her look longingly but pass by and he sadly returned to his distillation and decoction; and when any solid particle appeared in his broth he, removing it with care, cast it with a heavy heart onto the dirt heap.[160]

The allegorical texture of Head's version of the fable is more complex than the relatively straightforward message Ruth had tried to convey. The clinical side of Head's identity is depicted in a strangely negative way: he was 'condemned all his life to concoct herbs for the sick'. These concoctions serve as a metaphor for the products of Head's intellectual exertions. When he tried to offer the finer fruits of his mind to those he encountered in his daily business, they were rejected 'with ignominy'. The vulgar were clearly incapable of appreciating the best that Head had to offer. Feeling lonely and in need of companionship, he had set out these 'despised viands' for the benefit of one who *could* see their worth. The 'great world', had, however, intervened and condemned his offerings as poisonous. The fable embodies a censure of the innate conservatism of these critics who feared all innovation on the grounds that 'Experiments are always dangerous'. But there is also the imputation of an incompatibility between the squalid aspects of Head's medical calling and the higher, spiritual goals to which he aspired. Mayhew was perhaps the only one capable of transcending this antinomy; even she, however, might be deterred by the prejudices of the ignorant and vulgar.

It is noteworthy that in 1898 Ruth and Henry found it necessary to write of their relationship by way of the indirect device of a 'muse'. But reflections upon their shared meetings provided a further way to reflect on the significance that each had come to play in the life of the other. The occasion of these meetings was usually artistic or literary in nature – a visit to a gallery, a play, or the shared reading of a piece of prose or poetry. But in the letters that followed these events, Ruth and Henry would expose the deeper levels of meaning each encounter had for them. After one such encounter in January 1900, Ruth exulted: 'How good you were, how blest was I – *you really must let me write it*'.[161] To think of this meeting – in which the conversation had apparently been mostly about German poetry – was 'to keep a happy secret quite close my heart like the bird I love to talk about ...'. This metaphorical 'bird' had, moreover, contrary to the usual course of things, succeeded in frightening away Ruth's ever-lurking 'Kater'. The joy of the moment was only alloyed by the fact that Head was obliged to leave. After his departure, Ruth consoled herself by re-reading one of his recent letters.

For his part, Head would take the time from frantic preparations for a trip abroad to provide Mayhew with his own impressions of their last encounter:

> Is there any use in my attempting to say how much I enjoyed Sunday. First the quiet morning in the bright air with the talk I love and get so seldom. Then the ride, that

wood and last the hill. What more need be said of the exuberance of that joyous state than that it made a sunset happy and the moon a subject of jest? Then the still evening by a fire in the soft light of a lamp with books upon the sofa between us. You have again woven a coloured thread into the fiery web of my life.[162]

Head was especially intent in seeking through his letters to preserve and convey the memories he had taken away from these meetings. He was interested more generally in how his recollections were preserved and revived in unexpected ways. While wandering around a house he had known as a child, 'I suddenly came on one [room] with bed, furniture and belongings that seem to awake memories of my childhood. I had that uncanny feeling of something burnt in on that memory in every small detail so long ago that one is in doubt whether it is not a dream or one of those freaks of recent memory'.[163]

Henry's memories of his encounters with Ruth also included much circumstantial detail; but at their heart were the affective traces the meeting had left. He took the first opportunity at the end of a 'terribly busy week' in November 1899 to preserve one such set of recollections of a weekend he and Ruth had spent together:

> First the pleasant quiet Saturday Evening so different from the noise and [?] of the Saturdays when we have met at Buckingham. Then the glorious walk in the Sea spray on Sunday afternoon followed by the delight of Sunday evening beneath the rosy lamp with books and sober talk. All this has left a memory behind that is in its own way perhaps the pleasantest of all memories of our meetings. For the memory of this tiny holiday is coloured by a sense of security and peace.[164]

Such recollections could be striking in their intensity. While leafing through a book, Head would have lapsed into reverie:

> had I not been rapidly arrested by a revival or representation of pleasure so acute as to be almost a pain. I seemed to feel the contrast of a hot sun and cold white marble, to hear your voice reading each well balanced sentence and bursting into exclamation of admiration or elucidation, and over the whole memory picture noted a well formed Cock chaffinch an integral part of that exquisite Spring morning. The original pleasure was vehement and rational – the revival on the other hand was an emotion so violent as to be almost painful. Do you ever have this?[165]

The question was an invitation for Ruth to compare her own mental processes with his. She was, however, somewhat disconcerted that Henry had found this memory to be almost 'painful' in nature. In a later letter Head therefore expatiated on his precise meaning. 'When', he explained, 'I said that a certain recollection was almost a pain I meant that the pleasure returned in so pure a form that it was almost painful ...'[166] He was determined to give precise verbal form to his feelings.

Anything that distracted Henry from the generation of these cherished impressions was a cause of irritation. The 'braying of the orchestra' during an interlude when he might otherwise have been able to enjoy Ruth's conversation fell into this category. But Henry resented any unwarranted intrusion upon his orderly mental processes. He even complained that: 'This [Boer] war is the most terrible obsession'. Head had begun to turn his thoughts to the composition of a poem for Mayhew's further edification when: 'the cries of the newspaper vendors, "Great British Victory" "Defeat of the Boers" again started that terribly instant train of thought which has robbed me of all quiet thought for a fortnight'.[167] In other contexts, Head was forthright in the expression of his patriotic views. When it came to his interactions with Mayhew, however, the great events of the outside world were merely extraneous noise.

The letters served also to share something of what Ruth and Henry had experienced during the intervals between their meetings. Each would, in particular, dutifully describe their trips abroad for the benefit of the other. Head's account of a visit to Munich – during which he had bumped into 'Bertie Russell' and 'the great [Bernard] Berenson'[168] (see Chapter 5, p. 203) – moved Ruth to exclaim:

> It was very good of you to tell me so much, all I wanted keenly to know. The whole thing, the opera, the pictures, the statues, the suppers, the Prince, the Baron, the English author, the American amateur from Italy, reads like a delightful chapter from some new 'Novel without a hero' ... I feel so proud at being admitted as it were into the circle. It is as though I had 'dreamed true' ... and had, with you as a guide, been invisibly present at the scenes you describe, and had lived through little bits of the pageants of your [days].[169]

Anecdotes afforded scope for ironic comment on the foibles and mores of their contemporaries. Thus in March 1900, Henry gave an amused account of a visit to a 'model couple'. The husband:

> enjoys doing all the proper things in the proper way. For instance he formally sharpens the carving knife on a steel before a meal begins; he repairs the washer in the hot water tap – regulates the gas fittings. She is a sensible commonplace woman with no marked natural taste but much common sense.[170]

Mostly the tone of such observations on the customs of others was thus one of amused and somewhat superior detachment. Head, however, was sometimes led to reflect upon the vacuity of some of the social practices in which he was himself obliged to participate. Thus the supposedly august business of answering consultations 'becomes weary work and the answer becomes more and more platitudinous as time goes on and as the letters pile up'.[171]

Overtly, Head was gregarious and took an active part in the medical and scientific community. In his letters to Mayhew, however, he allowed himself to reflect on the tedium and vacuity of much of this social intercourse. Thus he expounded on the monotonous, repetitive nature of conversation at Royal Society dinners:

> The next day at lunch Lord Lister led the Conversation and we talked of the Soudan, the College for Soudanese, Sir H. Kitchener ... And then I felt as in a nightmare and kept saying to myself the next move in the game is 'Public dinners'. Sure enough we next discussed the dinner at the Mansion house, after dinner speeches & the whole weary round of the night before. Thus it is clear that any man in a public position who dines out often puts on the same monotonous decent Conversation in uniformity with his dress suit.[172]

Such vapid ritual represented the antipode to the authentic, spontaneous conversation with Ruth to which he attached such value.

This intense and extensive correspondence flourished during the period of Ruth and Henry's courtship. On 27 April 1904, shortly before they were to marry, Henry composed a coda that stressed the mutual benefit that had flowed from this exchange: 'And so closes a series of letters into which for your dear sake I have put my best thoughts and the sweetest moments of seven long years. All the beauty of my life had been made by you as you say your strength has come from me'.[173]

After they married, the two still occasionally exchanged letters; there was, however, a noticeable change in the tone and content of these later missives. The agonistic and often painful character of the period when they were divided by distance, family antagonism, and perhaps mutual apprehension, was replaced by a confident celebration of their newfound intimacy and understanding.

If anything, marriage strengthened the sense that they jointly possessed an exclusive realm of feeling and understanding from which others were excluded. This belief helped Head prevail over the travails of his mundane life. After 'a miserable day', during which he had been obliged to put on a show of false cheerfulness to the wife of a colleague who was also a patient, he 'went to bed rebellious and tired'. Head compared his mental state to that of a prize-fighter 'after a round in which he has been heavily punished'. However, he drew strength from what he shared with Ruth: 'we will make at any rate the walls of our citadel impermeable however much the outposts of our affection may fall before the battery of life'.[174]

Engagement and marriage also relieved them of the need to speak of their feelings for each other through fables and allusions. Head was able openly to compare his passion for Ruth to his ardour for his only other 'mistress'. At a professional dinner held in Edinburgh in 1909, Head found himself seated next to Byron Bramwell (1847–1931) who remarked on how well he was looking:

> I answered 'I have got the secret of Cortical Sensibility, and that acts as a tonic'. But I did not tell him that there was another more secret elixir of youth working in my

body. For it is more good fortune than any man has a right to to be in love, madly in love and at the same time to have made a discovery that men have been wanting for 50 years. Was this my letter mad? When I am alone and particularly when I have your darling presence still warm upon me I am sure I am sometimes mad. I see you so vividly I long to sit down and draw picture after picture of you as Rembrandt did of Saskia – then when I have you again in my arms all images are swallowed up in the joy of sense.[175]

Such lyricism pervades these later letters. Henry was at last able openly to address Ruth as 'Dearest of Women', and to assert: 'You have spoilt me for the company of those who do not care for beauty but have heightened my love for talking with them who occupy themselves with thought of the mind. Life with you is such a beautiful thing ...'[176] Ruth was the 'flame of my life;' without her, Head's life was: 'blank prose'.[177]

For her part, Ruth also felt transformed by her marriage. A visit to Oxford in May 1905 made her realize: 'how happy my secure life is with you all the more for this peep back into the troubled life of my young years'.[178] Henry was: 'to me the Sun and the Moon and all the Stars'.[179] In these later letters, Ruth also sometimes ascribed to Head the more homely status of the child she had never had. Thus she rejoiced when his mother transferred some childhood photographs of Harry into his wife's keeping. The pictures she already owned represented to Ruth 'the man whom I trust and lean on and reverence and adore'. But in the older photograph that Hester Head had kept in a carved frame: 'you are the child Harry I find again who leans his head on my breast now and whom I can care for and tend and cherish with tenderness and wonderful joy'.[180]

Head's happiness was, however, sometimes tinged with sadness at the inevitable finitude of his union with Ruth. While on holiday in France in the summer of 1910, he 'wept copiously' at the wedding of the pharmacist of the village where he was staying. Characteristically, he felt the need to analyse his own psychological processes. Weddings, he explained to Ruth,

> always makes me cry I believe for the following reason. So long as one is engaged on work, the feeling is always present that however little may survive ultimately the thing made became part of the force of the universe as the mote that floats with the sunlight is after all a part however light of the material universe. But the happiness we have together is mortal – is the sport of chance. Of necessity you must exist in order that I may have the happiness and your force it is discharged forever. Nothing will make it part of universal happiness; it belongs solely to you and me.[181]

The Rag Books

On 18 October 1901, Ruth and Henry began to keep a 'Rag Book'. They would exchange these volumes at their periodic meetings, while the entries were written while they were apart. The 'Rag Book' thus served as a further literary medium for them to maintain communication during the course of the courtship. Once they had married, Henry ceased to make contributions. Ruth Head, on the other hand, continued to make occasional entries.

The 'Rag Books' were from the outset viewed as an essentially *private* medium. On the title page of the first volume Mayhew wrote: 'Ceci n'est pas pour le public, c'est de l'intime, et c'est de âme, c'est pour un'. ['This is not for the public, it is intimate, it is of the soul, it is for one (person).']¹⁸² The 'Rag Books' were thus to serve as a means of exploring the inner world, including such aspect of consciousness that might otherwise remain unarticulated. There is, however, an ambiguity about the identity of the 'one' for whom they were intended. The process of inscription was, in the first instance, one that enabled self-analysis. But the fact that each entry would be scrutinized by another, who would in turn offer his or her own revelations, introduced a dialectical element into the processes of composition.

The 'Rag Books' thus afforded Ruth and Henry a further means of projecting aspects of their own self-imagining onto the consciousness of their partner in this enterprise. On a number of occasions, Head sought to express this function of the 'Rag Books' through literary allusion. Thus on the title page of one volume, he quoted from Henry James's *On the Wings of a Dove*: '"They could think whatever they liked about whatever they would – or in other words they could say it. Saying it for each other, for each other alone added to the taste"'.¹⁸³ Later, he found import in an extract from George Eliot's novel, *Felix Holt*: 'These words are charged with a meaning dependent entirely on the secret consciousness of each'.¹⁸⁴

Quotations – sometimes extensive – from novels and poems that had attracted Ruth's or Henry's attention figure prominently in the 'Rag Books'. Head might indicate for whom a particular extract was deemed to be especially apposite. Thus a series of passages from George Gissing's, *The Private Papers of Henry Rycroft* (1902) were labelled: 'Für mich', 'Für dich', 'Für uns'. ('For me', 'for you', 'for us'.)¹⁸⁵

These extracts were sometimes accompanied by descriptions of reading habits. In the course of a discussion of the peculiarities of the 'advanced' French novels that she so admired, for example, Ruth confessed to possessing: 'what French novelists call "une petite joie intime" in introducing my books to each other. My Horace Walpole Letters are naturally deeply interested in my Letters de Madame de Sévigné. All this evening I have been reading them in parallel lines, pencilling in references and wishing I could descend to the Shades to show the great Horace the beauties of my Edition ... A new introduction was held between Walpole and Browning ...'.[186] But she would also devote herself for periods to the exclusive reading of a single author. Thus June 1902 was: 'all Flaubert for me; beginning with the afterglow of Madame Bovary and continuing with the fascination of an odd volume of his letters and the beginning of L'Education Sentimentale'.[187]

Ruth's admiration for Flaubert was, however, surpassed by Henry (see Chapter 5, p. 216). Head was especially enthused by *Madame Bovary*; he reflected: 'When I think of "Madame Bovary" I have an impression that the book is of immense length. I see upon my table the small volume in its worn boards and wonder that so small a compass can contain so much that I have learnt from life'.[188] The 'Rag Books' contain a good deal of analysis and criticism of the works with which Ruth and Henry engaged. In this instance, however, what signified was the peculiar effect of *Madame Bovary* upon Head's psyche – an effect he was anxious to communicate to Mayhew.

In the 'Rag Books', the two often also wrote at length of events in their everyday lives that seemed imbued with special meaning. Ruth was wont to reflect upon aspects of femininity. One reason that she found Henry James 'wonderful' was that: 'He knows more about us than any man writer who has ever invented a woman. Most authors paint us from the outside as they see us. He paints us from the inside as we know ourselves to be'.[189] But Ruth also recounted episodes from her working life that seemed to cast light on the essence of woman. One day in November 1901:

> there came to my rooms a dozen of my big girls in age varying between 17 and 19. I showed them photographs of the Italian masters, many Madonnas and many many Bambinos. And from the girls, the serious hard-working really rather learned young women, it was one cry – the child, the beautiful Baby – everything was noticed with admiration and love ... And it was all so natural, so spontaneous I could have cried to think these are the girls who, says the Man of Science, are to recognize from the first that the joys and sorrows of marriage are not for them and are, deliberately, to shape their lives to live alone. Ah! God only knows how little the Man of Science knows.[190]

The passage gains poignancy from Mayhew's own conviction that she was fated to remain a spinster denied the joys of marriage and motherhood. Perhaps in his capacity as 'the Man of Science', Head annotated these remarks with a reference to a discussion of female education in François de Curel's *La fille sauvage* (1902).

The alleged peculiarities of female psychology were indeed one of the few issues covered in the 'Rag Books' over which open disagreement arose. In response to some comments by Ruth on Flaubert's letters, Henry was moved to remark: 'When a woman confides some personal experience to a man she is happy so long as he listens quietly. Signs of any thing more than polite interest upon his face frighten her. She fears she may have told him the Truth'. Her response was sharp:

> This is horribly unjust: I hope it is a quotation, not an independent judgment. The aim of an educated woman, when she talks, is to record sincerely what she has observed, enjoyed or in any way made her own. She does not wish to be conceited nor egotistical, but she wishes her words to reflect truly her thoughts. Insincerity is an indelicacy to her. Why do you make her out such a fiend? You out flaubert Flaubert. He only means, what is alas true that between human beings, absolute confidence is impossible and every story told will be flavoured by the personality of the narrators, however crystal clear her mind.[191]

Head's exposition of his own personality was more overtly masculine inasmuch as his contributions to the 'Rag Book' tended to be expressed in a more 'philosophical' idiom. He might provide a purely impressionistic account of the landscapes he experienced on a visit to Italy, but would accompany these with more overtly intellectual exercizes. Thus a conversation about Renaissance art in Florence with the art critic Berenson led Head to pose the question: 'Why do the majority of painters now hold pessimistic views with regard to the future of Art whereas in the 15th Century men believed that art was the one certain line of progress? Is it not that in the 15th Century all that was young went into some form of Art. Now all the youth of the world goes into Natural Science'.[192]

As well as such attempts at epigrammatic insight, Head would sometimes engage in more extended analyses in which he drew parallels between past and contemporary societies. For instance, he might expatiate on the parallels between ancient chariot races and the modern cricket match.[193] The psychology of the twentieth-century crowd had, in his view, barely changed from that evident in New Testament times.[194] Head might also embark upon an extended discussion of how in a 'civilized community people above the poverty line are roughly divided into three classes by their attitude to life'.[195] A chance remark in a book he was reading led him to draw upon his professional experience in order to refute the claim that an individual who showed courage in one context would necessarily be brave in another setting.[196]

But the 'Rag Books' also afforded the opportunity to record and share various of his own psychological quirks that seemed to demand no further explanation or analysis. In August 1902, Head felt moved to report:

> Tunes have a curious way of returning to me unrecognized at certain Seasons.
> For many years I have found myself humming the Frühlingslied from the Walküre on some morning when spring first invades the air. My feet go to a tune I do not recognize. Suddenly words are there and I know why the music seemed made to fit my mood. Analogous external conditions will also bring back tunes to me.[197]

Head's daily life and how it affected his moods provided further material. The 'für mich' extracts from *The Private Papers of Henry Rycroft* that Head entered into the 'Rag Book' revolved around the more obnoxious aspects of city life. With these sentiments Head could empathize. When waking in his London flat, 'One of the first sounds I hear is the ill tempered bicker up and down the lift. The voice of my quiet respectful housekeeper becomes that of a termagant dealing with a sodden husband'.[198]

The handwritten entries in the 'Rag Books' were accompanied by newspaper cuttings, programmes from performances Ruth and Henry had attended, and other miscellaneous documents. One of these – a print of a cat stalking through an avenue of trees – inspired Ruth to write a poem:

> The Wild Cat walks the Woods all night
> He's never the Woman's friend
> He never shares the smallest bite
> He's never going to mend.
> But if Harry says, he best alone
> I know it's only pretend
> For sometimes he shares the Woman's bone
> And he is the Firstest Friend.[199]

This humorous stab at verse allowed Ruth to record something of her own conception of the 'Harry' with whom she composed the 'Rag Books'. He still remained something of an enigmatic figure – a solitary wild cat who, by his nature, could not be 'the Woman's friend'. Yet there is also the hint that, thanks to their extended pursuit of mutual understanding, this feral creature might yet be domesticated to become a true companion.

Our Happy Double Life

After their marriage Ruth and Henry set up house at Montagu Square in London. Here they entertained friends and acquaintances drawn from literary and artistic circles as well as from the worlds of medicine and science. Robert Nichols, who had been a frequent visitor, in 1931 fondly recalled 'the never-quite sedate quakerness of Montagu Square'. In the same letter, Nichols also remembered that on one of these visits he had discerned a 'slight cleavage displayed in Ruthie's disciplined personality'. What, he asked, 'did the past mentor of the young think? What the wise & careful companion of the distinguished Fellow of the Royal Society? Somewhere deep down in her [was] an eighteenth century granddame …'.[200] Nichols was later to assert that the married Ruth was: 'au fond a cosmic Bolchevist of the deepest lye'.[201] This was a far cry from the fragile creature depicted in her letters to Head in the 1890s.

Ruth gave up her career as a schoolmistress and devoted herself to her new domestic role. As well as managing the budget and supervising the servants, Ruth also served as an unofficial receptionist for Head's practice. Maintaining the finances of the Head household sometimes proved a challenging task, especially when during the First World War Henry abandoned private practice in order better to serve his country – a decision that evoked 'violent emotions'[202] in his wife. Ruth's letters to Janet Ashbee afforded her the opportunity to complain about the hardships of those dark days. In October 1915, for instance, she lamented: 'H. has a bad cold, it is pouring with rain, I have been writing lots of cheques and <u>all</u> the patients are gratis!'.[203] Ruth used her correspondence with Ashbee to voice certain frustrations with her husband (whom she privately called 'The Oracle'); Head's lack of interest in or understanding of household finance was particularly vexatious.

Despite such frustrations, the predominant view of Head that emerges from Ruth's letters to Nichols and to Ashbee is suffused with affection and admiration. When she described Henry in an exhausted state or laid low by ill health, Ruth's affection took on a maternal aspect. In January 1920, after having two teeth extracted, Henry 'returned home a Wreck, hands cold, quite shell-shocked. The fire was bright in his dressing-room, so I whipped off his clothes, popped him into bed and left him to sleep'.[204] After he succumbed to Parkinsonism, Henry's dependence on Ruth became permanent. Her devotion to the invalid was total.

The duties of housewife, unofficial practice manager, and occasional nurse did not entirely consume Ruth's energies. The outbreak of the First World War made the purely domestic realm seem limited and unsatisfying. Ruth wrote to Janet Ashbee in June 1915, that her time was spent: 'in all sorts of domestic jobs, making cushion covers and the like … I sit at home a great deal and people come and see me and tell me how busy they are and pleases them and interests me and so the days go'.[205] Finally, 'tired of living in clotted idleness', she began to undertake voluntary work preparing surgical dressings at the Hospital Supply Depot in Cavendish Square. It was: 'the most fearful monotonous occupation possible, only we are too thick on the ground, 2000 females on the books and no sanitary inspector would pass our work-rooms'.[206] Her days at the Depot left Ruth exhausted. Nonetheless, she persevered with this drudgery to the end of the War; indeed, it was there in November 1918 that she heard news of the Armistice: 'I can tell you the tears ran down my cheeks! Guns to cease after all these years!'.[207]

But Ruth also found time to develop her literary interests. She edited compilations of the writings of Thomas Hardy and of her particular literary idol, Henry James.[208] When trying to obtain permission of the former, Ruth assured him of how 'well I know what joy extracts from great works do really give to sick people, to the weary, to travellers, to those who sleep badly, to all those who from one reason or another cannot for the moment have many books at their disposal'.[209]

Henry supported her in these efforts; Ruth informed Hardy that: 'my husband is my unfailing helper in all my work'.[210] While on holiday in Sussex in the summer of 1921, the two jointly selected what extracts from Hardy's prose and poems were to be included in the volume. As Ruth told the author, the process was more a joy than a labour:

> I wish you could see my husband and me close together on a white bench in the garden, well screened by an enormous lavender-bush, the well-fingered green volume of your Collected Poems between us, he reading to me, and I to him by turns, now laughing, now wellnigh crying with pleasure and with pain, as we discuss and mark and finally forget everything in simple enjoyment.[211]

On perusing the finished compilation, Hardy was moved to remark: 'In your arrangement of the selections you trace associations of ideas where it never would have occurred to myself, the instances seeming quite distinct; which does credit to the vividness of your imagination'.[212]

Moreover, she published two novels of her own; a third remained in manuscript. Ruth Mayhew had shown early aspirations as an author. In May 1898 she ventured to send Head 'a little reminiscence' of a morning she had spent at the Foundling Hospital. She invited his criticism and begged: 'if you ever think it could be polished up enough, could you let me have a godfatherly sort of note to introduce it to the Westminster withal?' With the self-deprecation characteristic of her early correspondence with Head she however added: 'But if you are busy, to the flames please anyway, because you know I would bury any offspring of my pen fathoms deep in oblivion before I would have you waste one moment of your time or one thought of your brain, on me'.[213] Later she felt the need to apologize to him 'for the futile bit of work I sent you'.[214]

Ruth Head, on the other hand, had more confidence in the offspring of her pen, although a certain diffidence remained. She described her first novel – *A History of Departed Things* – as her 'poor babe', and told Ashbee that she regretted the choice of title. Ruth claimed that the work had been composed: 'for Harry's ear only and [I] was extraordinarily surprised when he said it was worth publishing'.[215] Ruth's early efforts to find a publisher were disappointing. Chatto & Windus 'had been ecstatic about my "History of Departed Things" but declined publishing it as so few people would appreciate its delicacy'.[216] Kegan Paul were, however, less concerned about this alleged excessive delicacy and published the book in 1918.

Ruth's second novel – *Compensations* – appeared six years later. Robert Nichols came across the book during his stay in Hollywood and was, in particular, taken by its depiction of female psychology. Ruth thanked him for reading her 'horrid little book' so attentively, and added: 'For my part, I think the most interesting thing about fiction is the revelation it makes of the utter and entire and insuperable difference between men and women'.[217] Nichols had also remarked that the utterances of one of the male characters in *Compensations*: '(dare I say it?) seems not so remote in its wisdom from that of a certain celebrated neurologist'.[218] An autobiographical element is, however, still more evident in Ruth's first novel.

A History of Departed Things is written in epistolary form. It tells the story of two sisters, Bettina and Charlotte, who had grown up the children of an Oxford professor. The novel's depiction of Oxford is equivocal. The culture of the place is marred by intense snobbery: 'For no hierarchy had ever more closely defined class-barriers than that home of democratic learning'.[219] On the other hand, the Oxford of Bettina and Charlotte's youth is also intellectually and socially stimulating. With its well-stocked libraries Oxford is, moreover, a place where Bettina can satisfy her voracious appetite for literature. She is to retain this passion for reading in later life, telling her sister: 'I think, on the whole, what I read makes more impression on me than what I experience; it becomes more part of myself'.[220]

She only fully appreciates these amenities after an ill-judged marriage to a young doctor named Francis who sets up home with his bride in a remote Devonshire village – which Bettina refers to as a 'real toad-in-the-hole' place far from 'the full stream of life'.[221] She soon finds the marriage stifling. Francis is kind and a conscientious general practitioner; but he is too limited intellectually to be a fitting partner. Within nine months, Bettina is complaining to her sister: 'not once have I heard anybody talk about any of the real things we used to hear discussed all day long ...'. She longs to visit Oxford and begs Charlotte to 'pack me up all the books you can find'.[222] Even her beloved books cannot, however, wholly fill the spiritual void in Bettina's existence: 'I lie here and read "Aurora Leigh" till tears of self-pity trickle down my nose, a sickening state of decrepitude'.[223]

Bettina is rescued from this sad state when while in Oxford she meets 'an old friend of her husband's, fresh from a German laboratory, eager for fun, quick and untiring as herself'.[224] The friend is a London consultant and medical scientist named Lucas Beck. 'Lucas' and 'Beck' were the surnames of Head's maternal grandparents. It is therefore plausible to take 'Lucas Beck' to be a fairly transparent cipher for 'Henry Head'. Indeed Head evidently used 'Lucas Beck' as a kind of *nom de plume*; the name appears on the title page to a collection of his poems that he had printed for private circulation.[225] This identification is strengthened when epithets such as 'Quakerly sober'[226] are applied to Beck.

Ruth Head's depiction of aspects of Beck's character parallels the representations of Henry found in their correspondence, and hints at perceptions that were never fully articulated in the letters. To Beck:

> each new woman he encountered was a new problem, to be handled with care, turned inside-out mentally, and placed subsequently in one or another of his specimen drawers. Yet he was the pleasantest of companions, too tactful to give pain, and so kind-hearted that his rage for discovery was turned often to good account by offering him excuses for generous help. I don't know if Bettina suspected him of a microscopic interest in the workings of her wayward brain.[227]

Moreover, in Lucas Beck, Bettina finds a man who shares her passion for literature and the other arts. She begs him to:

> Send me more books, and send them now, dear Lucas. Never mind my bald commentaries; you have better things to do than read my school-girlish criticisms, I can only tell you that I am a new creature since you began to lend me literature, and I more than suspect you invented the system as a tonic, far more successful, by the way, than the quinine Francis sends in from the surgery. Let it be a great packet this time, please; and, please, let there be some Ibsen in it, to keep me from crying myself sick because I cannot see the 'Doll's House'; and, oh, if you have a set of Wagner's operas – the words, I mean – I should so love to read them. You remember how we walked in St John's Gardens that lovely afternoon last summer, and you told me all about the Nibelungen Lied and taught me what a <u>motif</u> was? It was like a draught of clear water when one had been very, very thirsty.[228]

Ibsen had been one of the shared passions that Ruth and Henry had pursued during their courtship. It is tempting to surmise that in this passage Ruth was alluding to an incident during that period of their lives when Henry had first shared with her his enthusiasm for the music of Richard Wagner.

Bettina's expressions of gratitude to Lucas also echo and expand on those of Ruth to Henry. His 'hand has been ever since my sure guide every step of the way'. He had found:

> such a raw, inexperienced girl, tossing on the open sea without compass or pilot, and you listened, and talked, and helped me to begin to express myself in words: then you wrote to me, and I went on growing, it seems to me, in understanding, till I learned to find myself equally in writing: or am learning, am I not? You have made everything seem more worth while to me, and what, before, I was only feeling vaguely after in the dark has grown clearer and more shapely – it has been, in a way, what conversion is to the followers of Wesley. I feel so much stronger for your help, and you have made my wilderness blossom like a rose.[229]

But as the novel progresses, a darker side to Lucas Beck's personality emerges. His commitment to science comes to verge upon obsession. In particular, he resents the fact that his most promising student, Ambrose Parry,[230] chooses to marry rather than concentrate all his energies on research. Beck tells Bettina:

> I have sent him [Ambrose] out armed with all the Lydgate and Rosamund episode to read and ponder upon. That's a fine antidote to sentimentality. George Eliot should have a statue erected to her in the College of Physicians for the service she has done young doctors in her 'Middlemarch'. I don't care <u>who</u> they marry before thirty – it's bound to be the wrong woman.[231]

When Ambrose continues to insist on marrying, Lucas describes him as a 'young fool'.[232]

The fictional narrative echoed an event recorded in the Head–Mayhew correspondence. In May 1903 Head was outraged when a promising chemist whom he had 'educated and supported' at the Pathological Institute of the London Hospital resigned his post to pursue private practice. He had done so in order 'to get married'. Head was obliged to concede that his protégé was entitled to prefer personal happiness to a career in science: 'Nevertheless I cannot help crying out when my ideals are rendered difficult and work suddenly stopped in consequence of the desire of a man with an income up to his needs to increase it by a few hundred pounds'.[233] As she read this letter, Ruth confessed that: 'all my sympathy is on the other side ...'. Indeed, she found Henry: 'a pretty sort of monster' for resenting his assistant's decision. The incident brought home to her the intensity of the 'Moloch flame which is your Work'.[234] She reproduced a similar fanatical devotion to science when she came to write the character of Lucas Beck.

In time, Lucas comes to blame Bettina for luring Ambrose away from his work; he sends her a 'cruel, detestable letter'. Bettina laments that: 'The chill of your horrible words penetrates me as I write and takes the warmth from the sunshine'.[235] By the end of the novel, Beck is estranged from both Bettina and Ambrose. He dies a lonely and embittered man. In his will, he establishes a research fellowship that is to be restricted to bachelors. Bettina comments that Beck's intention was: 'that in years to come lots and lots of young Ambroses will arise, free and unfettered, enabled by his generosity to follow the stony paths of pure research and flee the snares of wife or mistress. Horrible, I call it!'.[236]

A History of Departed Things is of course a work of fiction. While 'Lucas Beck' bears more than a passing resemblance to 'Henry Head', there are also obvious differences between the two. Lucas is perhaps best regarded as an exaggerated depiction of certain aspects of Ruth's conception of her husband's personality. The fact that the book was at first written for 'for Harry's ear only' suggests that he was meant to be entertained, rather than mortified, by this facsimile of himself. Ruth, however, took as the epigraph to one chapter of the novel, Robert Louis Stevenson's dictum: 'Science carries us into zones of speculation, where there is no habitable city for the mind of man'.[237] There is at least a hint that she used the character of Lucas Beck to show the spiritual price that a complete dedication to Henry's 'other mistress' could exact.

Henry Head's devotion to science did lead him to postpone marriage at considerable personal cost. In later life he was wont to expound upon the 'the devastating consequences of long engagements'.[238] Head was also, it is true, frustrated when one of his assistants decided to give up research in favour of medical practice in order to support a wife. Unlike Lucas Beck, however, Head did not allow his commitment to the pursuit of knowledge to become a fixation that displaced all other passions. Ruth – the ' light of my secret life'[239] – had become as integral to Head's sense of self as the rigours of investigation and the ecstasy of discovery. As he told her in September 1922:

> You have woven yourself into the texture of my life and are indeed the complement of my widest and highest joys. The scientific work will I hope be preserved in print. But the common joys will be lost without the combined delight and knowledge of two, so intimately bound together are they with our happy double life.[240]

5 THE CULTIVATION OF FEELING

The Land of that Inner Life

In August 1900, Henry Head's sister, Hester, married Reginald Pinney (1863–1943), an officer in the Royal Fusiliers. Her brother's reaction to the match was remarkable. In an angry letter to Ruth Mayhew he declared:

> I was sorry angry & ashamed at the vulgarity and commonness of the whole business. I do not mind her [Hester's] marriage ... I do not mind her desertion of the things she had been taught to enjoy, for that too was inevitable – But her excitement made me sorry, her frivolity angered me and her acceptance of the blatant publicity made me ashamed exactly as if she had danced on the stage of the Swiss Gardens.[1]

In a later letter to Ruth Mayhew, Head chose to describe his sister as 'Major Pinney's wife'. He insisted that he had no interest in listening to this woman's increasingly strident and foolish opinions. For: 'it was Hester I loved & Hester is dead'.[2]

Mayhew – usually so sympathetic to her correspondent – confessed that she was appalled by the vehemence of these remarks. When Head sneered at 'the gross ineptitude of [Hester's] Anglo Indian life',[3] Ruth admitted that: 'Your last letter made me miserable'. Head's mother, who was also aware of the extraordinary nature of Henry's reaction to his sister's marriage, had alleged that her son was: 'jealous of her husband and you loved Hester more as a lover than a brother'. Mayhew found this suggestion of some form of incestuous attraction: 'monstrous and you may think how it hurts me. I cannot believe it ...' Nonetheless, she was left bewildered. 'Why', she demanded, 'are you so cruelly bitter about it all?'[4]

This incident deserves study because it reveals important aspects of Head's worldview. He categorized the world and the people that inhabited it according to an interrelated set of binary opposites. He attached immense importance to certain values and attributes, and greatly esteemed those individuals who manifested these traits. Those who were indifferent or even inimical to these ideals were worthy only of contempt and derision. By marrying Pinney, Hester had, in her brother's view, committed the worst transgression imaginable: she had allowed herself to fall from the first of these categories to the other. Thereby, she was guilty of: 'the true "sin against the Holy Ghost" of which we hear so much but understand so little'.[5]

Before her marriage, Hester had shared some of Henry's intellectual and artistic interests. She had, for instance, sometimes accompanied him to the theatre – occasioning a certain jealousy in Ruth Mayhew. Ruth feared that, compared to Hester, Head would find her: 'inferior in intelligence and ripeness of judgment'.[6] Although Head was quick to dismiss such fears, it is clear that he did regard his sister as a worthy spiritual companion.

But when Hester transformed into 'Mrs Pinney', she seemed deliberately to eschew beauty and refinement. She had, Henry raged, 'cast her mind into the mire and can no longer either write, spell or draw ...'.[7] Head claimed that his sister had willingly assumed this new, infinitely less admirable, persona in order to match the vulgarity of her new husband. Reginald Pinney was a man who possessed: 'no humour, no sentiment no culture ...'. And it was 'to this being Hester is going to approximate herself'. An example of this 'spiritual death' that Head found especially painful was the new Hester's reaction to one of her brother's poems. Hester: 'cried when she read the Pastoral; but she was afraid to tell him [i.e., Pinney] she was affected by it, could not get him to read more than the first stanza & he refused to let her read it to him'.[8]

Head's feelings about his sister and her husband were to moderate over time from utter contempt to mild disdain. He took a particular malicious pleasure when Siegfried Sassoon caricatured Pinney in his *Memoirs of a Fox-Hunting Man*.

Mrs – later Lady – Pinney had joined the ranks of the vulgar and philistine. Throughout his correspondence with Mayhew, Head applied a quasi-clinical eye to diagnosing these defects in the character of those he encountered. In making these judgments he operated with a peculiar form of dualism. Head postulated that human beings were composed of two radically opposed elements – body and soul. In lower types the two co-existed amicably enough. But: 'as refinement increases, the intrusion of the body into the regions of the soul is disturbing and offensive. In a lower stage of development both body & soul can comfortably dwell together under one roof like the Irishman & his pig without offence to either tenant of the room'.[9]

As he went about his business, Head made a point of seeking to discriminate between those he encountered in order to determine what point on this scale of refinement they had attained. His almost physical revulsion at the working-class denizens of the East End that he encountered in the course of his hospital work has already been noted. But he was even more sensitive to evaluating the attainments of those who aspired to some degree of intellectual attainment.

These judgments sometimes smacked of unabashed snobbery. Thus in March 1900 he noted that his partner at a dinner-party: 'belonged to that class of person who ruin conversation not by cutting it short but by debasing it with their ineffable stupidity – the sort of person who supposing Watteau were the subject of conversation would point out that "Long also paints dainty pictures"'.[10] When staying in the 'ugly & cheerless rooms' of a Cambridge undergraduate, Head amused himself by subjecting the cultural pretensions of the usual occupant to a mordant and gratuitous analysis:

> On the mantelpiece stood his sister, as she was presented, looking exactly like every other girl who ever made her bob to the queen, Bishop of London and Welldon, the late headmaster of Harrow, in one frame, flanked on either side by Sophie Fulgarney [i.e., the actress, Irene Vanbrugh, in the role of SF] covered with a scrawled autograph – Among these offensive household gods stood your dear letter making the social, religious and artistic impulses of this youth seem increasingly crude & repulsive.[11]

A rival's failure to reach the required standard in scientific achievement likewise evoked in Head: 'the sense of aloofness that the artist experiences amongst those who are ignorant of art and its methods'.[12]

Head often seemed to derive satisfaction from ruthlessly dissecting the aspirations to taste and accomplishment of those who ventured within the scope of his gaze. One of his mother's friends was: 'very much dressed the grand lady'. Head, however: 'did not think she was much good for she seemed to have little of that quick perception of atmosphere that is so fascinating in the really fine women of her Class'.[13] He also liked to compare men and women of the same 'Class' in order to establish their relative merits. Thus the wife of a Cambridge physiologist was dismissed on the grounds that: 'she does not care for science and although she persistently talks books she sways with the crowd & has little power of individual appreciation'.[14]

In contrast, Head found much to admire in Evelyn Whitehead, because: 'Her whole mind seems fixed on enjoyment of life as if she had suddenly awoke to the knowledge that she had not tasted the full pleasure of living – Birds, flowers, colours & scents have begun to enter a life whose pleasures in the past had been mainly intellectual'.[15] This combination of intellectual power and sensual discernment possessed special significance within Head's system of values. The absence of such attributes occasioned pity or contempt. As he wrote in the 'Rag Book': 'I dislike the man who does not recognise and enjoy his momentary passage through some beautiful phase of life. He is like those futile persons who do not know what flowers adorn the rooms in which they live'.[16]

Head exulted when he occasionally encountered such a kindred spirit in an otherwise banal society. Thus he rejoiced in the company of a colleague, 'a jew [who] has all that astonishing love for beautiful things that is so common in his race'. One Sunday, after having to endure the company of an 'extremely philistine friend and his wife and family', the two were finally able to speak freely about their shared artistic interests. Head and his companion: 'walked up the Waterloo Road arm in arm and stood on the bridge watching the lights ... and his lips still moved as he repeated to himself verses that pleased him'.[17]

But of all Head's friends, none approximated to his ideal combination of intellectual and aesthetic attributes more closely than Ruth Mayhew. In December 1900, Head praised Ruth's 'delicate soul', insisting: 'Beauty comes to you so naturally as a component of life that you do not understand with how rare a sense you have been dowered and how ardently others, whom you envy, long for what is to you as simple and as vital as the breath of your nostrils'.[18] Indeed, whenever Head saw: 'a beautiful thing feel a true emotion or think a clear thought ... my mind treasures them for you or refers them to you as a standard'.[19]

In Ruth, Head felt that he had finally discovered the perfect partner: one with whom he could share all the subtleties and intricate delights of the inner life. These were rare innate attributes that could not be acquired by mere education and upbringing. 'How uninteresting women are after you!', he declared. Other women of Ruth's class:

> have had no experiences, many of them have felt but they are not honest – Every thought is dressed up by those secondrate modistes Education & Co. & I like their thought as little as I like their frocks. You dearest of women have a mind as beautifully frocked as your body – the colours harmonise, the lines are charming...[20]

Ruth and Henry's relationship was necessarily an exclusive one. They belonged to: 'that class of human beings who by some curious inward force are compelled to seek our happiness apart from that of our race'.[21] The letters they exchanged and the Ragbooks they jointly composed formed the record of the intensely private domain the two had created. Within this realm the exploration of the finer feelings took precedence over all other considerations. The pair indeed aspired to *create* a special world for themselves from which the squalor of mundane life would be banished. In this regard, Head regarded himself as among those 'superior men' whose deep thought and refined sensibility could enhance received reality.[22]

Head described this personal realm as: 'that land of that inner life which alone is always beautiful & clean & cool'.[23] The cultivation of that inner life rivalled the pursuit of science in his system of values. But that land also served as a refuge from the pressures, vulgarity, and tedium attendant upon so much of his daily life. The time he, 'spent amongst things that are unpleasant and troublesome is rendered possible only by looking at things as they are and by the suppression of all feelings of disgust or beauty emotional pains and pleasures ...'.[24] In short, it was a space in which he could refine an alternative to his professional and scientific self – one which was deemed somehow more authentic and wholesome than the various guises Head was obliged to wear in the common world.

Head employed a number of metaphors to express this contrast. He sometimes spoke of the secret world he shared with Mayhew as a 'garden' that: 'you & I have tilled for many years – Each year the flowers grow more thickly and more beautifully from the fact that they are descended from earlier blossoms'.[25] Head developed the figure further in a letter written in February 1902, using the fable form in which he and Ruth were wont to represent their relationship:

> Once upon a time a woman and a man lived far from one another in a bleak and inhospitable country. But now & again they would meet and together would journey in search of the pleasant valleys and quiet gardens that lay hidden away far from the place of their daily toil. Wonderful were such journeys for the man and the woman rarely failed to find some quiet spot bright with the babble of water or the song of birds – Here the woman would build an arbour covered within and without with sweet-smelling flowers ...[26]

The reference to the 'inhospitable country' from which Head longed to escape expressed a deep sense of alienation from much of the professional, metropolitan life that consumed so much of his time. On occasion he made this disillusion and despair explicit. Thus in November 1902 he lamented:

> I am crawling away to you, dear Ruth, from one of the most uncomfortable weeks I have lived through for a long time. People I trusted fouled me. I am engaged in a violent controversy on a Hospital matter with a member of the Committee and the weather has been of the vilest November quality. Every morning I go down to my nine o'clock lecture in a train, late with 12 to 15 people in a carriage. Even my research has been of little comfort for my assistant is in for his final examination at the University of London and it will be five weeks in all before he is again available and no new patients have turned up on the clinical side.

Finding a letter from Mayhew waiting for him when he returned home, however, transformed Head's mood. Her reference to a piece by Walter Pater: 'took me to the Winckelmann essay – I drew my chair up to the fire and a kind of warm peace seemed to lap me round. The long sensuous sentences appealed to all that is wanting in these hurried smoky November days'.[27]

A letter from Ruth that Head received while on a trip to Germany in 1898 provoked still greater lyricism. He read it while at the opera: 'Throughout the pauses of Götterdämmerung your description of your church going on Sunday morning glowed like a Peter de Hooch interior hung in a room of Rubens'.[28]

Better even than Ruth's letters were the visits that Head sometimes paid her. 'When a man and a woman are associated together', he wrote after one such meeting, 'they say that the days they spent one with another contain the beauty of the world'. He conceded that this was: 'the one common knot between the high & low, the wise and the foolish'. But Head maintained that he and Mayhew had raised the beauty inherent in the relation between the sexes to a new height: 'we have had the good fortune to adorn the texture with threads of exquisite Colour: and of all the many lovely threads was not last Sunday one of the most beautiful?'.[29] Henry and Ruth's 'association' was thus an aesthetic achievement, itself a work of art.

When he returned to London from one of these weekends with Ruth, he had only his memories for comfort:

> On Monday morning I travelled to London Bridge in the hot summer weather with the feeling of the beautiful blue day and the exquisite joys of the Sun upon me. The long down, silent except for the calling of the cows & the bleat of the sheep, the scent of the grass the clear blue sky over the sea and you close to me laughing & talking formed a picture so delightful & so vivid that the journey passed insensibly – Then came the worries of the day – patients to be seen notes to be taken, difficulties to be faced – and it all seemed a beautiful unreal dream.[30]

Head encapsulated this sense of a divided reality when in June 1903 he declared that:

> I often feel as if life was for me like a cathedral with two towers. The one looks to the north the other to the South; but from either tower the movement of the world seems an idle show. High up on the one I am lapped in lyrical joy and breathed upon by the soft wind from the South. In the other it is bleak & cold & lonely but still I am happy to be there.[31]

The architectural figure also expressed a sense of detachment from and superiority to the world he inhabited – 'from either tower the movement of the world seems an idle show'. But it also articulated a sense of an existence polarized between a bleak, cold, and lonely everyday world and an alternate realm suffused with lyricism and joy.

The Southern Prospect

Although Head's primary passion was for literature – and particularly for poetry – his artistic interests were wide-ranging. Instrumental, orchestral, and operatic music all appealed to him. He was an avid theatregoer. He took a keen interest in the pictorial arts. Robert Nichols stressed the breadth of Head's tastes:

> He loved art and artists in all media. I have heard him in one afternoon describe ... the peculiarities (due to incipient lunacy) that gave a particular nuance to the performance of Dan Leno, contrast the conducting by Richter and by Mahler of Tristan, comment on Coleridge, tell a story about Henley (whom he had known personally), and finish up with an excursion into the mind and personality of Piranesi.[32]

While enjoying all these art forms at a purely sensual level, Head also took a keen intellectual interest in understanding the nature of their appeal to the mind. Thereby Head hoped to gain a better insight both into his own mental processes and into the psychology of aesthetic appreciation more generally.

His responses to musical stimuli were often extreme – especially when the experience was seemingly adventitious. Thus:

> On many a Sunday during the last two Summers when I have seated myself at my table the sun streaming in upon the paper joy has come to me from the wonderful playing of someone in Welbeck Street whose room looks onto the back. I can hear the grand piano as if I were in the next room. The execution is wonderful. The player's taste is catholic and he or she is gifted with a rollicking humour. Beethoven or Liszt played with a technique only given to one who has treated music as a profession will be suddenly followed by the theme of a popular tune with astonishing variations. Many a time I have jumped up & laughed aloud at the dexterity and astonishing incongruity.[33]

Although Head found metropolitan life obnoxious in many ways, he was aware of its compensations – notably the wealth of musical performances on offer in London. Thus on Good Friday 1900, after seeing an 'interesting patient' in the morning, he took advantage of the respite from his teaching responsibilities, to attend an afternoon concert at the Queen's Hall. It was, he told Mayhew: 'a royal feast of good things'. The programme included Tchaikovsky's 'Pathétique' Symphony. Head carefully noted the effect the music had upon him:

> The first movement always fills me with the feeling that I am riding to death under a leader I do not trust for a cause in which I do not believe – It affects me more than any instrumental music I have ever heard .... If I were a King I would order it to be played to me whenever I was peculiarly conscious of my power & might as the triumphing general had one with him in his Chariot to remind him that he was mortal.[34]

But it was the Prelude to Wagner's *Parsifal* that had the strongest impact. The music invoked a remarkable stream of memories of events and of the moods that accompanied them:

> with the sacramental theme I was back again at Bayreuth in the good days 13 years ago with life before me and my first work finished but as yet unpublished – A time of joy when one felt like a woman about to be a mother that the eyes of many friends & rivals are fixed upon her fruition & who knows from the movements of her child that with God's help she will bring forth a lusty infant.[35]

Head had taken time from his scientific pursuits during his early visits to Germany to make the pilgrimage to Bayreuth to hear his first *Ring* cycle. Many more were to follow. For instance, while on a visit to Munich in October 1898, Head attended a performance of: 'all four nights of the Ring uncut and according to the best Wagner tradition'. He was not an uncritical Wagnerite; hearing the operas uncut made him notice that: 'The whole is full of crudities and inconsequent occurrences & explanations'. Nonetheless, 'in spite of all this the thing is colossal'. He now felt that he had at last attended: 'a satisfactory performance from start to finish' of Wagner's masterwork.[36]

A devotion to Wagner's music became central to the enthusiasm for German culture he displayed in later life. Indeed this devotion sometimes verged on the religious. A performance of *Tristan und Isolde* that he attended in Milan led Head to reflect:

> I have always envied Catholics in that wherever they go they find an office they know from childhood being performed as part of the life of any foreign country in which they may be homeless strangers. But Wagner now gives me exactly this envied feeling – For whether I am in France or Italy I can now go & hear some opera every note of which is associated in the closest way with both my real & adopted home ....[37]

Head's trips to Europe also provided opportunities for him to pursue his interests in the pictorial and plastic arts. He showed none of the indifference to the visual that Ann Banfield has discerned as characteristic of the Cambridge culture from which he had emerged.[38] During an extended stay in Italy in 1892 he kept methodical notes of the works of art he viewed while touring churches and museums in Rome and Florence. His comments were sometimes trenchant. Thus he found Albrecht Dürer's painting 'Christ among the Doctors' – 'disgustingly ugly'.[39] The journals Head kept during this stay reveal more than a casual touristic attitude to the paintings, sculptures and buildings he viewed. Head showed a scholarly interest in the style, provenance, and technical detail of these works. He was already well versed in art history and in the archaeology of the ancient sites that he visited.

The honeymoon that he and Ruth took in the spring of 1904 was largely occupied with tours of Parisian galleries. The newly-weds were, in particular, drawn to the exhibition of Flemish 'Primitive' paintings and illuminated manuscripts at the Louvre and Bibliothèque Nationale. They also took a keen interest in the impressionist paintings on display. A serious intellectual approach to the appreciation of art was again evinced – this time by attendance at a lecture on the history of French art. Ruth recorded that: 'The room was stuffy and the audience mostly unwashen [*sic*]. But the lecture was full of quiet vivacity and told us a great deal we knew already so gaily that we listened with eagerness and were rewarded by hearing also something new'.[40]

While on a trip to Bavaria in 1898, Head took a particular interest in the work of the Munich Secessionists and toured the galleries sometimes in the company of his brother Alban. These tours were undertaken in a solemn, almost reverent, fashion. The brothers would afterwards engage in intense discussions of what they had seen over dinner – 'for we of course work separately and are silent in the Gallery'.⁴¹

Head reported that on one such visit, while:

> I was looking at Titian's Charles V and stepping back 3 paces to see it better collided with a little man upon whose heels I had trodden – We swung round to face one another & it was Bertie Russell. His wife was with him and she took me off into the next room & introduced me to Mrs [Mary] Costello[e] and the great [Bernard] Berenson ...

Head joined the group for lunch and took the opportunity to engage in another form of connoisseurship – evaluating the conversational traits and attainments of his companions:

> The whole of that set have a dangerous way of running off onto inept and untrained observation of their feelings and call this pastime 'Psychology'. Russell, the trained and astute metaphysician scoffs or is silent, and the conversation becomes a monologue – Yesterday at lunch however they succeeded in all talking their best. By a little management such diverse elements as Berenson, delicate in sensation as a woman, Russell [... word torn out] cold & hard as a piece of steel, Daniel a metaphysician with an artist's aims & Mrs Costello[e], a mentally-adoring American woman, all talked well and to the point and nobody held forth in monologue.⁴²

Head already knew Bertrand Russell through their shared acquaintance with Alfred North Whitehead. He was, however, particularly pleased to make the acquaintance of Berenson, whose works on Renaissance art he admired. Later the two were to meet again at Berenson's house in Settignano near Florence. Over dinner, they 'talked about the drawings of the Florentine masters and especially of Leonardo da Vinci. He believes that even the greatest masters had not developed the faculty of continuous application. They did not know the meaning of work in our sense of the word ....'⁴³

While touring the Pinakothek with Alban, Head had another thought-provoking encounter, this time with 'a professional pianist who was there for the first time although he had been in Munich nearly a year'. Henry and Alban took it upon themselves to conduct the pianist around the gallery with which they evidently felt familiar. They showed him various sculptures that they found particularly admirable, but were puzzled by the response: 'No, they were ugly and nothing would convince him that we found them beautiful – "They give me no Emotion" "I cannot sit before them and feel I am better"'.⁴⁴

Head tried to change the subject by talking of his own love of music – a topic with which he assumed the pianist would sympathize. Head told him in particular of how: 'an organ beneath my window made me dream wonderful dreams and that one of my pleasantest associations with my home was the possibility of playing classical music on the piano by means of the "pianola" ....'. The pianist 'shuddered but did not understand'. Artists, Head concluded, 'sind beschränkte Leute [are limited people]'.⁴⁵

This incident characteristically instigated a chain of reflection. It set Henry and Alban 'talking of the reasons that people admire a picture and we invented absurd reasons as we walked round the Secessionist Galleries together'. Head concluded that an episode from his own experience 'seemed to bear the greatest verisimilitude'. Henry had seen many paintings by the German artist, Franz von Lenbach (1836–1904), 'but never cared for him'. A visit to Lenbach's studio had, however, altered the way in which Head viewed his works. Now that he had:

> seen where he works he interests me more than any modern Artist. And of all his pictures the one I like best is that of Marion [Lenbach's daughter] and Duse. For I saw Marion playing in the studio gardens and you know my associations with Duse.⁴⁶

Eleanora Duse (1858–1924) was an actress who held for Head a special fascination (see below). She had shortly before Head's visit to Munich been portrayed by Lenbach in the company of the artist's daughter, Marion. This association, along with the opportunity to see Lenbach at work, had evidently transformed Head's response to the paintings. The general point that Head seemed to derive from this and his previously noted experience in Munich was that reactions to works of art did not merely vary between different individuals; they were also contingent upon the associations at work in the mind of the same person at different times. The implication would seem to be that works of art possessed no intrinsic or absolute merit. Their beauty was dependent upon the conditioning of the eye of the beholder.

Head's observations in the 'Rag Books' sometimes toyed with the notion that the canons of perception – and therefore of beauty – might themselves be historically contingent. He noted that:

> In all early art hands & feet are characteristic stigmata for the painter or Sculptor because they do not normally lie in the focus of attention. In looking fixedly at an object the eyes are focussed so that certain parts lie in the centre of vision other parts at the periphery of the visual field. So it is with attention – Not only are our eyes focussed towards a certain point but our attention is similarly focussed and non-essential parts escape that strict attention devoted to those which are more essential – With the progress of civilization the field of attention became wider until the artist lost himself in a mass of detail. Conscious contraction of this wider field is now a necessary preliminary to production.⁴⁷

These observations were preceded by a lengthy discussion of the life and career of the German sculptor, Tilman Riemenschneider (1460–1531), some of whose work Head had also viewed in Munich. Head's particular concern was to distinguish the various stages of Riemenschneider's output, and to account for what he saw as a decline in quality of the sculptor's work in the latter stages of his life.

Head claimed that the point at which Riemenschneider's sculptures allegedly declined:

> corresponds with his entry into politics, with his election to the town council. He was seduced from his art by the immense & immediate importance of the political atmosphere in which he lived. Later there seems little doubt he was willing to permit the altars & statues he had constructed with so much art to be destroyed by the political party to which he belonged owing to the association they now had for him with the tyranny he abhorred and the idolatry he distrusted.

Thus in his latter years, Riemenschneider became 'a good citizen & played his part on the side of progress that has ultimately given the world the Reformation & the German Empire – The price he paid was his art'.[48]

Head's explanation of the alleged decline of Riemenschneider's work is open to dispute. These comments do, however, manifest an impulse to seek to place the processes of artistic creativity in an historical context. The chief interest of this passage is, however, what it reveals of Head's own hierarchy of values. Political engagement and religious devotion are depicted as deleterious distractions from the far more important ideals to which the artist should aspire. 'Today', Head insisted, 'we know little of Riemenschneider's political aims we care less for his religious views but we search every corner for fragments of his art'.[49]

Head gave his most extensive discussion of the aesthetics of the visual arts in letters written to Ruth Mayhew in October 1898. During his recent stay in Munich, Head had been much preoccupied with painting and sculpture: 'Poetry has gone to the winds for all our thoughts have been expended on colour and line'. Head's discussions with Alban of what they viewed had led him: 'to a much clearer understanding of the painter's aim and the value of intention in judgment of the result produced'. Moreover, although he initially applied this understanding to the plastic arts, Head had come to see ways of applying 'a somewhat similar test to poetry with, I think, good results.'[50]

Mayhew asked Head to clarify this notion of artistic 'intention' as a criterion by which to judge the worth of a painting or poem. Head began by insisting that the aim of art was not to attempt to simulate 'reality'. This was a vulgar fallacy of:

> the layman who was under the impression that the greatest compliment he could pay a painter was to say 'how real!', 'how like life'. Now the painter who paints a great picture paints a picture – Line & colour are used not for the sake of reality' for such is unattainable but to produce a thing pleasing to the senses. Thus no great painters ideal is 'reality' –

Nor, Head insisted, did art have any *moral* import: 'no great painter paints a picture to point a moral only. In so far as a picture preaches it is not a picture but a sermon'. Thus, in Head s view: 'a painter's intention must be pictorial and not (1) scientific or (2) moral'.[51]

Although Head cited no sources for this theory of art, they bore obvious similarities to those of the late nineteenth-century Aesthetic Movement[52] – and, in particular, to those of the painter James McNeill Whistler (1834–1903). In the 'Rag Book', Head recounted a visit he paid in 1901 to Whistler's lodgings in North Audley Street:

> He had a bad cold and was cowering over the fire in his room wrapped in a dressing gown & blanket looking quite old and feeble. He, usually so absurdly brisk and jaunty, seemed shrivelled up into an old man – Fog had prevented his work & to crown his misfortunes his French servant was knocked down by an Omnibus outside the Studio. She had discharged herself from the French Hospital whither she was carried & taken up her quarters in the same hotel as her master, to his considerable discomfort – 'I hope she is now better' evoked his characteristic retort 'Thanks we are recovering from our first sympathy'.[53]

This anecdote is immediately followed by an account of the effect the sights of the city had upon Head. For the most part this reiterated Head's customary disgust with the sights, sounds, and smells of the metropolis. 'The streets of London', he declared, 'never produce a more depressing effect or appear more dreary than when illuminated in winter by an unrelieved row of arc-lamps. The cold brilliance of the lights emphasises the grime of the houses and the slimy mud of the footpaths and the heavy shadows cast on every passing figure give the impression that all the world is in mourning'. At the same time, however, he appreciated a certain beauty in the tableau: 'towering over the golden gas lamps and the red & white lanterns of the road traders[?] an arc-lamp gains something of the grace & beauty of a lily in a bed of jonquilles & tulips'.[54] The juxtaposition of the two passages is suggestive: in his word-picture of the streets of London Head seems to be gesturing towards Whistler's view on the pre-eminence of pattern and colour in painting.

In the remarks on the theory of aesthetics that he penned in 1898, Head certainly posited a relationship between the criteria that should govern pictorial art and literary production. He maintained, in particular, that the canons of criticism he had outlined in respect of the visual arts: 'can be advantageously applied to poetry'.[55] Just like the painter, the poet could not be expected to depict the world in a realistic fashion: 'No one could describe the appendages of the crayfish in poetry with the necessary accuracy to satisfy a S. Kensington examiner'. Nor should the poem aspire to producing a moral message – 'poetry cannot be directly moral and that is why hymns are bad poetry ... The Poet's aim is a poem & not a morality'. In his own poetry Head claimed he had deliberately attempted to achieve pictorial effects with mixed results. Although he could: 'get fairly definite "line effects" they mostly lack colour owing to the want of singleness of vision. Where I have obtained colour it has been simply because the thing has been so vividly "seen" that the true poet's intention has been reached accidentally'.[56]

In later years Head took an interest in others' attempts to explore the psychology of the visual arts. In 1923 he reported that: 'Roger Fry gave us a paper at the British Psychological Society on "What pictorial art demands of psycho-analysis"'. Head considered this: 'one of the ablest, and at the same time most wrong-headed addresses, I have ever heard'. Fry had contended that the appreciation of pictorial art depended chiefly on the recognition of relations in space: 'That is to say, it is an intellectual exercise, which has no direct emotional origin'. Head was intrigued by Fry's contention that the satisfaction the viewer derived from art: 'depends upon the perfection with which the problem of relations is solved, and not in any sense on the significance of the picture or on the emotion it otherwise excites'. But he regretted the fact Fry 'went off into "Wish Fulfilment" with "Fantasy Building", and so spoiled his whole argument in this direction'.[57]

Head was also an enthusiastic patron of the theatre. In the 'Rag Books' Ruth Mayhew scrupulously recorded the performances that she and Henry had attended. Their taste in plays was wide-ranging and included the classics. In November 1897, Mayhew, for instance, remarked: 'We have even the most charming trilogy of Shakespeare Plays together haven't we?'[58] Through their acquaintance with Charles and Janet Ashbee, Ruth and Henry were also among an invited audience for a performance of Ben Jonson's comedy, 'The New Inn' presented in January 1902 by members of the Guild of Handicraft. Head took a close interest in details of the staging, noting in the 'Rag Book':

> A raised portion of the Studio overhung by a gallery, that forms part of a passage on the first floor, formed the stage – Exit from the ... ? of the inn upstairs was given by the steps to the gallery to other parts of the inn by a door to the stage left: but exit or entry to or from any part outside the Inn was through the audience in the hall. Thus many exits became beautiful processions and the actors sang some catch or short refrain as they passed up one lane between the chairs and down the other.

He was less impressed with the acting, noting that: 'The only actor of any merit was Ashbee as "Lovel". During the earlier parts he was over emphatic as the diffident lover but the great speech on Love and that in the last act on Valour were magnificently given'.[59]

But Head's greatest passion was for modern innovative drama that seemed to challenge artistic and social conventions. The ultimate criterion by which he judged pieces of dramatic art and the actors that he saw performing in them was, however, ultimately personal. Head showed an intense interest in the psychological effects that theatrical, as well as musical, experiences had upon him.

A play in which Head showed a particular interest was Edmond Rostand's *Cyrano de Bergerac*. He saw the play in London soon after its Paris premiere with both his sister Hester and later with Ruth Mayhew. In September 1898, while touring lunatic asylums in Berlin, Head found time to attend a performance in German translation at the *Deutches Theater*. In this instance, he felt that the experience had provided him with important lessons about the differences between the French, English and German national characters. Indeed, the performance:

> taught me more of what was essential to the German Spirit than years of life in the country. For I take it that the essence of Cyrano consists in a recognition that 'most foolish persons say the most foolish things with an air of the greatest solemnity whilst others treat of very important matters in a light and playful manner' (Morelli) The sage in France plays the fool and only the fool appears wise. In fact Cyrano is the glorification of the domino of the soul in the masquerade of life. This we can understand in England and we but differ from the French in that the Frenchman clothes himself as Harlequin whereas our dominos are white and expressionless, our faces starched & blank. In Germany however such an aspect is unknown. A man <u>is</u> what he seems – If he is not he is a frivolous and empty person who does not count in the pageant of life.[60]

This was one of the few instances in which Head, prior to 1914, expressed negative sentiments about German culture. He felt that the Berlin production had entirely missed the essence of the play. *Cyrano*: 'was turned into a kind of Wallenstein's Lager[61] with a hero who was badly used & an object for pity – Now one who demands pity is perilously near to earning contempt and [Josef] Kainz as Cyrano was contemptible'. Indeed, the whole experience: 'was lamentable and but for the sharp lesson in national character which I had learnt I should have felt the evening wasted'. He even found the German translation of the triolets vexatious; as he left the theatre, Head found himself:

> saying aloud to clear my ear from so foul a tone
> 'Nous sommes les Cadets de Gascogne
> De Carbon de Castel jaloux &c'.
> until my back stiffened and my body swayed with the glorious swagger of the lines.[62]

Later the same year Head reflected on the widespread critical attention that the success of such plays as *Cyrano* had excited. 'Everybody', he noted, 'is talking now of the revival of the Romantic Drama'. Head, however, maintained that a profound aesthetic shift had occurred in the course of the decades that separated Hugo from Rostand, arguing that such plays as: '"La Princesse lointaine" & "Cyrano" differ profoundly from anything we have ever had in this century'. For instance, heroes such as D'Artagnan 'sought material gain in his most chivalrous deeds'. In contrast, 'the whole essence of Cyrano & La Princesse seems to me to be in the idea that enormous physical trials can be undertaken for the sake of a dream – an idea – without desire of any physical reward'. Head saw such 'Quixotism' as a natural response to an era 'when the whole of daily life is regulated towards material ends ....'. He counted himself as one of an: 'ardent band of followers who spend laborious days fighting for their material wants but only really living in dreams hidden from the world'.[63]

Around 1903, Head and Ruth Mayhew became members of the 'Pharos Club', a society dedicated to the staging of new – and, in some instances, somewhat controversial – plays. The Prospectus stated that on the Sunday following a performance: 'there will be a discussion on the production at the Pharos Club'.[64] The plays Head and Mayhew attended at this venue included Hermann Sudermann's *Es lebet das Leben!* and Laurence Housman's *Bethlehem*. The latter could only be staged in private theatre clubs because the prevailing rules of censorship forbade the depiction of biblical characters on stage.

Head moreover attended a number of productions put on by The Stage Society – a body also founded in part to avoid the Lord Chamberlain's censorship – at the Imperial Theatre, Westminster. In February 1903 he saw Harley Granville Barker (1877–1946) play the lead role in William Somerset Maugham's *A Man of Honour*. The Head's were to become personal friends of Barker and his wife (see Chapter 6, p. 264). Head was, however, unimpressed with the play itself, remarking: '"A man of Honour" is the work of the author of "Mrs Craddock" which shows the same flashes of insight and the same constructional blindness as the play'.[65]

Head saw a performance of Ibsen's last play, *When we Dead Awaken*, at the same venue. 'Of plot', he noted, 'there is little or none. But the play is full of terrible things said with equal force from the side of the artist who is anxious that nothing shall disturb his production and by the woman who cares for him and detests his art as such'. After the play Head, his friend Everard Feilding, and some other members of the audience retired 'to a little supper party at a little house in the neighbourhood of Curzon Street'. Head found this gathering at least as fascinating as the play itself, recording:

> We were very gay; but over the younger members of the party brooded a feeling of discomfort. Something unpleasant not understood. Fitfully they drifted back from time to time to the play. What did it mean? Was she really mad? It was really awfully funny. They were easily headed off and we persuaded them to take it as a colossal joke.

Later he and Feilding withdrew and: 'quietly discussed the reason that such a play as we had seen that night could never move even a selected English audience'. They were especially concerned to establish why 'did it only produce a feeling of discomfort, of unrest like the memory of a visit to a dentist to be banished as soon as possible'. The conclusion of this discussion appears to have been that there was an essentially philistine strain to even educated English society. In countries such as France, in contrast: 'art & science become matters of daily interest and such plays as the "Nouvelle Idole" & those of Ibsen find an audience'. Not so in England where: 'the Artist & man of Science are first of all gentlemen'.[66] Head evidently felt that he, and a few select others such as Feilding, possessed special qualifications that enabled them to perceive and analyse the shortcomings of their compatriots. The incident also suggests that Head showed no eagerness to be regarded as a 'gentleman'; indeed, for him that epithet seems to connote a coarse sensibility and lack of discernment.

Head had seen François de Curel's play *La Nouvelle Idole* performed at the Royalty Theatre in Soho in March 1902. It had been an exhilarating experience that prompted him to speculate about the likely future course of dramatic literature. In the 'Rag Book' he enthused: 'This play seems to me to be the germinal shoot of a new type of theatrical representation'. Head argued that previous kinds of drama had dealt with one of three kinds of issue:

> (1) Action in the conflict of a will against obstacles material or moral (Hamlet)
> (2) Pictures that evoke feeling and result in complex emotions (Maeterlinck)
> (3) Everyday types dramatically treated so that they become inevitable symbols of forces at work in Society (Ibsen).

To whichever of these types it belonged, 'if the play is a great one the resultant however moral, emotional or sociological must be dramatic & must appeal plastically to the human mind'.[67]

Curel, however, 'has discovered the dramatic value of ideas'. The scientist protagonist of *La Nouvelle Idole* was portrayed as directed by the ideals that governed his life. A 'commonplace writer' might, in order to create dramatic conflict, depict 'the scientist impeded by material difficulties, by the destruction of manuscripts or by the consequences of a wife's extravagances'. But Curel rose above such banalities to show the scientist opposed in the pursuit of his ideals by opposing intellectual forces, notably: 'the anti-scientific force of a woman's emotion'. Thus, 'whereas the old drama exposed its hero to chains and physical torture from without the new drama binds him by the ideas of the multitude and the torture of logical doubt'. Head concluded that: 'Never before have the ideas of the Scientific worker been exposed more concisely or with greater exactitude – At the same time the defects of his temperament are not spared'.[68]

This revolutionary contribution to drama was, however, betrayed by the quality of the performance the play received in London. Curel's 'great lines' were entrusted to:

> a band of actors trained in the conventions of the English Stage. Albert Donnat the Scientist was played like the betrayed husband in a military drama, Louise his wife

was a shrew who did not understand one line of her part, Antoinette the young girl was the conventional 'saintly ingénue' of the melodramatic stage always ready to fall on her knees and fold her hands to heaven with or without excuse.[69]

Head appended a cutting from the *Daily Chronicle* to demonstrate how little English critics also understood drama of this stature.

In his play-going Head thus sought out works that seemed to embody modern conceptions of the form and purpose of drama. He was not content with the output of the commercial playhouses, but patronized private theatre clubs that provided a venue for new plays that might fall foul of the censor. He showed, moreover, a bias for continental European authors, making no secret of his disdain for English tastes and standards.

Head's remarks on the significance of Curel's work demonstrate a conviction that art in general, and perhaps drama in particular, could reflect and even shape developments in society. Curel was 'modern' in that he appreciated something of the novel significance of science to western civilization, and sought to create new dramatic forms that could adequately represent the role of the scientist in contemporary society. But for Head the theatre was also an intensely personal experience that offered him exceptional opportunities to pursue the introspective study of his own psychic processes.

Thus in December 1898 Head was moved to write to Mayhew of the striking effect 'Mrs Patrick Campbell's performance as Lady Macbeth' had upon him. 'No written words', he maintained,

> can quite give the effect of 'To bed to bed. There's knocking at the gate. Come Come Come Come give me your hand'. The four 'Comes' were a long moan. Each was said rapidly with a moaning sound at the beginning and end. For the first time in my life I got a sense of horror in this scene – And it haunts me now – it was all so like a sick child.

He brought something of his clinical experience to the analysis of the phenomenon. The actress had: 'grasped the idea that that horror of delirium lies in the lunatical absence of word modulation. For in delirium the phrase is a modulated cry – the words are monotonous'. Head regretted that Ruth had not been present to share the experience – 'It is so difficult to convey in writing any idea of the effect it gave me'.[70]

It was the acting of Eleanora Duse, however, that had the greatest effect upon Head. In the 'Rag Book' he recorded that: 'I have now seen Eleonora Duse in many parts – I can remember "La Locandiera", Santazza (in "Cavalleria Rusticana") Magda, Paula Tanqueray, Francesca, "Dame aux Camellias" and lastly as Hedda Gabler'. Head was aware that it was fashionable for English critics to belittle Duse's talents as an actress. But: 'I know, as a psychologist skilled in self analysis, that she produces in me an effect I cannot otherwise obtain from acting'. Much as hearing the Prelude to *Parsifal* had revived a stream of vivid memories, so seeing Duse in *Hedda Gabler* took him back to his early reading of Ibsen while working in Prague. Duse gave him the same 'extraordinary feeling of emotional reality' he had derived from his first experience of Ibsen's writing. This was, moreover, an exceptional occurrence in Head's psychic processes: 'for in ordinary life this is a state that is foreign to me. With me the sense of reality is intellectual. I say "I know that to be true because –". Rarely in my life, after times of great emotion, I seem to have become convinced of something not by reason but emotionally as a revivalist may become "convinced of sin".'[71]

Head left the theatre:

> not only feeling that I had seen the only possible Hedda but that this Hedda was a play that embodied the very tricks of speech of a woman I had known, of a woman who had made me uncomfortable but had filled me with vivid interest as a scientific observer when she had passed through my life – Had the original I saw upon the stage been one of my patients? How did she come into my life? These were the questions that constantly cropped up in my mind.

When this exceptional emotional state had abated, Head realized that no such 'prototype' had existed among his real-life acquaintances. Yet under the extraordinary influence that the actress was able to exert, 'the most highly trained critical faculty becomes useless; Duse's appeal is to a mental state in which reason is in abeyance – as irrational as love'.[72]

Head's comments on the painting, music, and theatre reveal a conviction that the different branches of art were subtly interconnected. The experience of seeing Duse on stage produced a comparable mental effect to reading Ibsen's words for the first time. Moreover, analogous principles could be applied to the production of telling poetic as well as pictorial images.

But despite the breadth of Head's artistic interests, and his evident eagerness to experience every kind of beauty in all its forms, literature enjoyed a special prominence for him. This was reflected in the breadth of his reading. Head was not, however, content to be merely a passive receptor of literary experience; he aspired to be a writer of some merit himself. This ambition was most apparent in his efforts at poetic composition. But, although he never wrote a novel[73] or short story, some of his prose writing sought albeit tacitly to have some literary merit.

Reading

Head had since his school days been a voracious reader. As an adult, despite the many professional calls on him, he found time to continue to consume an astonishing amount of literature written in at least three languages. Indeed, according to Ruth Mayhew: 'your book bill exceeded even your shirt-bill and Heaven knew how long that was'.[74] This was no mere recreational activity. Head engaged intellectually and emotionally with the works of the authors he most esteemed. Moreover, he drew no sharp distinction between his clinical and scientific concerns and these literary interests. On the contrary, he maintained that the kind of psychological insights and analytic skills that were of use to him in the consulting room could also be applied in reading works of fiction. Indeed, these skills might even be honed to a higher degree of perfection through reading authors who were themselves accomplished 'psychologists'.

Head was especially alert to the emotional effects that certain works could invoke in him. These results could be extreme. He recalled that the first time he had read Ibsen's *The Wild Duck*: 'was on a summer evening in a little Beergarden where I frequently supped in Prag'. He had gone out at about 7p.m. with the newly translated book in his pocket to peruse it over his customary Vienna schnitzel and glass of beer:

> When I rose to go home I thought I was drunk. But the two round mats told me that I had only drunk two glasses of beer in more than three hours. Moreover when I rose next morning I was still drunk. The daily world around me seemed unreal. I felt as if I had gone through some colossal personal experience of which the memory could only end with death.[75]

Head had: 'experienced this condition more than once and twice' when reading other books.

He invested the large library of books he had accumulated with great sentimental value; indeed, he personified these possessions. Head wrote in August 1899 of how he was trying to: 'get my books onto my new shelves'. This endeavour was, however, continually thwarted: 'I start with much energy and then some old friend, some forgotten marker or scarcely remembered note catch my eye and all progress ceases–– I think I shall call in my housekeeper to keep me under her stern eye I shall not dare to dawdle with my friends'.[76]

The range of Head's taste in prose was extensive. As well as novels, he enjoyed biography and collections of letters. He took an interest in history, archaeology and ethnography. He was engrossed by the subversive doctrines of James George Frazer's *The Golden Bough* (1890), buying a new edition as soon as it appeared. He was also familiar with contemporary German examples of the new demystifying approach to the study of Christianity. When touring Rome in 1892, Head sought to put this reading to use. He studied early Christian iconography with care in the hope of discovering: 'a further link to the "Baum Cultus" and Bacchic worship'.[77]

Head's view of Roman history and that of the early Church was also coloured by the work of his favourite historian, Edward Gibbon, to whose work he attached extraordinary significance. In March 1901 Head recorded:

> I have today read the great Chapter XV & XVI [of Gibbon's *Decline and Fall of the Roman Empire*] and have an early restless desire to show some act of physical reverence to so stupendous an attempt at pure Truth – I feel as if I had been listening to a great modern scientist, who is what they are not, Eloquent ...[78]

The chapters in question dealt with the rise of the Christian religion and the official Roman reaction to the phenomenon.

Further reflection served only to enhance this admiration for Gibbon's genius. Head was:

> rapidly beginning to believe that the germ of most of the questions one constantly asks oneself about corporate human actions or modern institutions is to be found in Gibbon – For instance, you know I have been long interested in certain German walled towns & have proposed to go a short walking tour through them with the fortunate youth. I have found their origin in Gibbon. Then the origins of the French 'Préfet' and how he came by his name & power often puzzled me – See Gibbon – So with a hundred & one things ...[79]

Head's tastes also extended to reading the correspondence of eminent literary and historical figures. In 1899 his friend, John Waldegrave: 'sent me Dorothy Osborne's letters saying "I know how fond you are of letters"'. Head found Osborne's letters:

> fascinating and so [?] modern. I am beginning to believe that in every age some woman is born the unconscious heir of all the great letter writers of the past. Men on the other hand seem to stumble along each in his own little cell with all the limitations of a beginner in the art. Even the best men's letters seem to me to be prose (illuminated here and there by wit) but those of a woman are lyrics full of temperamental melody.[80]

The chief interest of such writings to him was the psychological insights that they offered. When reading the letters of Robert and Elizabeth Barrett Browning, for instance, Head tried to assume the pose of 'a psychologist, one's business in life is to note mental facts and to express no emotion whatever phenomena may appear'.[81] Love letters were particularly fascinating because they obliged the writer to seek literary forms for the articulation of the most intimate of feelings. When successful, these effusions were of especial aesthetic value. 'In a letter', Head declared, 'as in a lyric, temperamental expression if scientifically true must as surely be beautiful'. While the Browning letters were of some psychological interest, Head found Osborne's letters to her lover more pleasing; the charm of her letters: 'lay in the brilliance with which she expresses her warm feeling in measured phrases. She never became incoherent or uninteresting although there is no reason to suppose that her love was less intense than that of E. B. B'.[82]

Reading love letters led Head to reflect on the contrast between truth in science and in art:

When dealing with natural knowledge things that are true need not be beautiful and when I say 'what a beautiful case' I mean that it exemplifies some truth though it may be in itself foul. But directly we consider either letters or lyrics which depend on emotional colouring of life for their truth, beauty and truth become one.

Thus a passage in a letter by Jonathan Swift, 'in which he imagines Stella riding on horseback ... is beautiful because of the colours imparted to the daily life of his friend and this because of the rightness of its tone'.[83]

Head's taste in writers of fiction was catholic. He was sufficiently fascinated by the work of William Makepeace Thackeray, for example, to spend part of a holiday in September 1903 cycling around Cambridgeshire in an attempt to discover 'the scene of the early chapters of Pendennis'. He concluded that: 'Peterboro which Thackeray probably knew well would do excellently for the Cathedral town, the name for which might have been taken from a small neighbouring town'.[84]

Head's bias was, however, towards novelists who sought to describe and explicate the mental processes of their characters. He was thus drawn to writers who were also accomplished 'psychological' observers. The accuracy or otherwise of such observations became a criterion by which to judge the value of a novel. Thus in the 'Rag Book' Head noted that: '"The house on the Sands" by Charles Marriott is a failure. But some of the characteristics of the two women are well observed'. For example, Marriott succeeded in accurately depicting what went through the mind of one of his female characters when she received a letter announcing the imminent arrival of her lover.[85]

Sometimes a mediocre novelist would betray psychological truths inadvertently. This was true of Elizabeth von Arnim's novels. Head considered *The Solitary Summer* a better book than *Elizabeth and her German Garden* because: 'as long as she is gossiping about her garden and telling stories about her children and the man of wrath, she is charming; but as soon as she describes her commerce with another woman she at once sinks to a lower level'. This trait drew attention to the 'curious paradox that a woman may look upon her husband and her children with sufficient objectivity to talk of them wittily and well but is incapable of looking at another woman without emotion destroying the clearness of her vision'. Head had observed the same phenomenon in the women of his own acquaintance. Ruth Mayhew's 'description of your talks with my mother exactly bear out the same idea. She can be clear and original in her observation but when she talks to or of you and Hester her conversation at once drops to that of any other woman'[86]

A great psychological novelist could, however, capture aspects of the human mind better than any other. Such a writer could create a representative character – a 'complete figure' – that could 'reveal our state better than any individual outpourings'. The 'Rag Books' were in large part to be composed of extracts from works of fiction chosen by Head and Mayhew that exemplified such 'complete figures'. This exercise would, Head maintained, provide him and Mayhew with insights into their own characters. Indeed, 'the Rag Book becomes a double diary without the crudity of attempt at mutual analysis'.[87]

Head regarded Joseph Conrad as a master in the exploration of character and motivation. According to Robert Nichols, when they first met in Palace Green Hospital, Head at the outset: 'asked me if I liked Conrad'.[88] After Head's death, Nichols tried to recreate one of the conversations patient and doctor had conducted on the subject of Conrad's writings. The discussion had evidently focused upon the question of: 'why in Conrad's Lord Jim, did Jim leave the bridge of the refugee ship Patna in that disgraceful fashion?' Head tried to steer Nichols towards the conclusion that Conrad had provided the reader with 'a perfectly sound & particularly psychological reason' for Jim's decision, even 'tho' never, as I remember, openly declared'. The reason was thus left implicit, Head suggested, 'Because Jim would be ashamed to declare it. Yet it exists & can be inferred'. Head eventually revealed that the 'reason' that finally overcame Jim's reluctance to join 'the disgraceful & all too busy group' that was preparing to abandon ship was the death of one of their number from a heart attack. This created a sudden vacancy in the lifeboat: 'This is the last straw--& Lord Jim jumps'.[89]

Other of the discussions with Head that Nichols recorded dealt with the explication of the character and motivation of actual acquaintances – such as the seemingly bizarre behaviour of the novelist 'X' (see Chapter 2, p. 92). It is clear, however, that the same canons of psychological scrutiny and explanation applied when dealing with fictional characters. For Head, Conrad was thus a true 'psychologist' whose insights into human personality possessed a real scientific validity.

Gustave Flaubert was another writer to whom Head ascribed this kind of quasi-clinical insight.[90] In particular, *Madame Bovary* was a novel for which he felt a special fascination. 'Emma Bovary', he declared, 'is the portrait of an egoistic hedonist painted by a supreme constructive artist'.[91] In the 'Rag Book', Head observed: 'When I think of 'Madame Bovary' I have an impression that the book is of immense length. I see upon my table the small volume in its worn boards and wonder that so small a compass can contain so much that I have learnt from life'. He was lost in admiration for Flaubert's skill as a writer, noting that: 'No metaphor in this book fails to add to the vividness of the picture'. Head carefully copied into the 'Rag Book' several examples from the novel illustrating this gift. But what most impressed Head was the expertise with which Flaubert knew how to employ his literary technology to achieve striking psychological effects and insights.

For instance, Head noted that:

> An event happens of moment in our lives; but we are left with an incomplete picture even though we may wish to impress the smallest accompaniments upon our memory. Attention is too much occupied with the thing that is happening to make an inventory of the furniture. In the same way when realising some scene presented by an author a long description obscures the sense of movement. And yet that we may fully appreciate the action it is necessary that the scenery should be adequately set.[92]

In *Madame Bovary* Flaubert had, however, demonstrated that it was possible to solve this literary-cum-psychological dilemma.

As an example Head cited the depiction of the character of the 'Imbecile with lupus of the face'. Flaubert gives a full description of the hideous appearance and existence of this wretch, and adds the detail that: 'As he ran after the carriages he sang a little song about the birds, the Sun & the green leaves …'. In Flaubert's description, the harsh sound of the imbecile's voice is combined with the clatter of his clogs and the thud of his staff. This dismal medley of sounds: 'penetrated to the depths of [Emma's] soul like a whirlwind into an abyss and swept her away into regions of a boundless melancholy'.[93]

Later in the novel, when Emma is dying, 'this creature ravaged by a foul disease, idiotic, obscene anointed with the dregs of illicit love is her last impression of that world of which she has desired so much'. A complete restatement of the character of the imbecile would, however, have retarded the progress of the narrative. Flaubert therefore achieved the desired effect by merely making the words of the imbecile's song, together with the noise of his clogs and staff, the last sounds that Emma hears at the moment of death. That, Head marvelled,

> is all – The details that complete the hideous vision are given incidentally on the two previous occasions upon which the creature appears in the book. They are given in such a way that they lie stored in memory ready to be evoked in their full horror on the supreme occasion by forty words of prose and eight lines of a little song.[94]

Head shared his admiration for Flaubert with Ruth Mayhew; they read *Madame Bovary* together on at least two of their meetings. Mayhew was also an avid reader of Flaubert's correspondence. But her own special passion was for the work of Henry James – Head remarked, 'As Flaubert to H. so HJ to R'.[95] Although he had read some of James's work before he met Mayhew, it was she who persuaded Head to become more fully acquainted with the work of this author. As he wrote in October 1903, 'Your account of the "Ambassadors" makes me itch to read it. But for you I should never have read any H. James more recent than the Daisy Miller set. What a number we have now read together …'.[96]

Most of the extracts from James in the 'Rag Book' are by Mayhew, who maintained that James 'knows more about us than any man writer who has ever invented a woman'. Head however, did extract one passage from *The Wings of the Dove* because it contained: 'another characteristic, well observed'.[97] Part of the appeal of James's work to Head may have lain in a shared sense of the fluidity of moral categories and of the manifold ways in which the phenomena of mental life could be interpreted.[98]

Head's emotional response to some of James's stories was extreme. When Mayhew leant him her copy *The Aspern Papers*, Head claimed that: 'You had I suspect little conception of how it would affect me'. On reading James's account of the narrator's doomed attempt to gain access to the dead poet's letters, Head:

> almost cried and felt as if I had suffered an analogous loss. For Henry James has described with consummate accuracy what all investigators feel when a research is baulked – I have been made miserable for days by the loss of an opportunity of versifying or placing the coping stone on a series of observations and I was made equally miserable by living through this loss of opportunity even though fictitious.[99]

The impact of the story thus derived from James's success in representing the psychology of the archetypal 'investigator', a category central to Head's understanding of his own nature.

The reading of prose thus provided Head with a range of emotional and intellectual satisfaction. However, he attached still greater significance to poetry. He told Mayhew in July 1899 that:

> Some even say 'I have worked hard and now I am going away to get drunk' but I frequently say 'I am going to have a holiday and read poetry'. For poetry will send me into a state utterly unlike my cold grey everyday self and provide an exultation like that of wine in which deeds and words are foreign to my sober state.[100]

Poetry thus had the power to reach the deepest recesses of the mind and to release profound emotions that would otherwise remain latent.

Head took a keen interest in the efforts of Robert Graves and other to apply psychological concepts in an attempt to elucidate these effects. In *Poetic Unreason* (1925) Graves had argued that the subconscious mind played a crucial role in the creative process. Head told him that he had read the book: 'with the greatest pleasure. My only difficulty in writing to you about it is, that you have opened up so many points that I should like to discuss with you'. Head agreed with Graves 'with regard to non-logical thought and its importance in the life of the mind'. There was: 'no question that this level of consciousness is extraordinarily creative'. Head cited a poem by Walter De La Mare where a long-forgotten association provided the poet with a word that was inaccessible to his conscious mind as an example of the creative potency of 'non-logical thought'. This was true of scientific as well as artistic creativity: 'You mention that there are days in which it is impossible to write poetry; there are equally weeks or months in which scientific activity seems ploughing the sand'.[101]

Head, moreover, believed that the psychological processes involved in the creation and appreciation of poetry were amenable to a form of empirical investigation. He was especially interested in the 'extraordinary variety of the appeal [a verse] makes to different people'. Thus he had experimented with the effect of one of Graves's examples – the rhyme 'How Many Miles To Babylon' – on a number of subjects. Ruth Head, for instance: 'almost exactly anticipated your friend's statements as to his feeling about it. She had innumerable feelings of significance and grandeur; she at once saw the point of "Three-score miles and ten", and "Candle Light" gave her the sense of something intimate and secure, and a scent of incense'. When, however, Head 'tried the verse' on the writer Colburn Mayne, 'she, on the other hand, like myself obtained a sense of mystery and beauty only, and wanted to say the words over to herself for the pleasure they gave her'. Head suggested that Graves use this rhyme as a test on a larger number of individuals 'who could be trusted to introspect truthfully'.[102]

In a postscript, Head consoled Graves about a 'stupid review' *Poetic Unreason* had received. 'These critics', he averred, 'can never understand that the appreciation of beauty is relative and not absolute. What we are out for is to discover these relations'.[103]

Writing to Robert Nichols, Head confided that he found Graves's book 'curiously unequal'. Graves had, however, 'undoubtedly got hold of two valuable lines of thought'. Firstly,

> there can be no doubt that the creative uprush, particularly of poetry, comes out of a realm of non-logical thought that is profoundly susceptible to the same laws as dreams and other sub-conscious activities. Secondly, granting that poetry has to do with this level of consciousness, it is obvious that poetic appreciation must vary profoundly according to the particular psychological make-up of the reader. Several individuals may obtain a thrill from the same verse or line of poetry, although introspective analysis shows that this appreciation is founded upon often diametrically opposed perceptions, images or ideas.

Head expressed the: 'hope that Graves will continue to make observations on himself and his friends with regard to how and why they come to enjoy or dislike certain poems'.[104]

Among the reminiscences Nichols compiled after Head's death was an account of a discussion of two poems by John Keats that took place at Head's house in Montagu Square around 1919. This seems to show him adopting a similar analytic approach to the one he commended to Graves: Head's principal concern was with how the various psychic constitutions of different individuals affected their response to the same verse.

Nichols presented his memories of the discussion in dialogue form. He did not claim that these dialogues reproduced Head's 'exact phraseology, still less the cadence of his speech'. However, 'the <u>sense</u> of his utterances I can guarantee to be present'.[105] The dialogue begins with Nichols reading aloud – presumably at Head's bidding – the 'Ode to a Nightingale'. Head stops him at the Fifth Stanza:[106] 'at "The murmurous haunt of flies on summer eves". Tell me what happens to you when you read the fifth stanza of the <u>Ode to the Nightingale</u>'. Nichols's reply is that he 'sees' the flowers mentioned in the verse. Head proceeds to correct him by showing that Keats explicitly states: 'I cannot see what flowers are at my feet'. His allusions are to the *scent* of the flowers. The reference to the 'white hawthorn' Head dismisses as being: 'chiefly emotional as the "pastoral" in "pastoral eglantine" most certainly is'. Head continues that in this 'emotional appreciation' of the object, of which Keats is a master, visual stimuli play little part. Nichols had, Head concluded, misconstrued the poem because of his psychological constitution. 'You take him as you do, Robert', he argued, 'because being predominantly visualist you instantly & unconsciously translate what you receive from him into visual terms'.[107]

The discussion then moves on to the 'Ode on a Grecian Urn'. Again Head maintains that Nichols misconstrued Keats's intentions when in response to the first line of the poem – 'Thou still unravished bride of quietness' – he saw 'a Grecian bride robed in a pale yellow peplos with a purple ... [?] – pattern border to it'. Nichols only saw this figure: 'because your apprehension instantly translates the given word into its visual counterpart – even, as here, before you have finished the phrase'. In poetry, 'as in music, there are phrases which are emotive as well as symbolical in grammatical wholes'. This led Head to muse upon the possibility of: 'a study of lyric poetry which might proceed upon a basis of the "Gestalt" of the lyric and seek to establish in what manner this Gestalt was declared throughout the poem & how the Gestalt of particular phrases contributed to the Gestalt of the whole'. He lamented the fact that literary critics were 'not better acquainted with some of the possibilities the modern psychological science is revealing ...' Head complained that if these critics did explore the possibilities of any psychology, then it would be that of Freud. This, however, would lead to 'a horrible mess'. Freud had, in Head's view:

> made a great discovery, a discovery of capital importance in the complex. Unfortunately he later lost it in the lavatory, whither I am afraid the literary critics, when they have tumbled to Freud, are like to follow him & where, such is the perversity of human nature, they are like for some years to remain.[108]

These dialogues need to be treated with some caution: they represent Nichols's recollections of conversations that had occurred many years in the past. But they can be taken as giving the gist of how Head sought to combine his literary and psychological interests. The reliability of these dialogues is strengthened by the fact that in the 'Rag Book' Head does, in the course of an analysis of William Butler Yeats's *Ideas of Good and Evil* (1903), expound views on the psychology of poetic apperception akin to the one ascribed to him by Nichols. Head remarked on Yeats's claim that: 'All sounds, all colours, all forms because of long association evoke indefinable and yet precise emotions'. He added the observation: 'This is singularly true and is the essence of all true lyrical or temperamental poetry. The word scarlet occurring in the first line of the verse will give the mind-picture or colour that must be resolved and cannot be blotted out'.[109]

The 'Rag Books' also preserve some of Head's reflections on Keats. These passages deal, however, with the poet's correspondence with Fanny Browne rather that with his verse. Head approached these letters very much with a clinical eye, arguing that: 'The fact that they were written during the active advance of tuberculosis of the lungs makes it impossible to accept them as revealing Keats' true nature'. The letters were, however, 'overlaid with manifestations of protopathic activity which gives them a slight non-literary interest'.[110] Head thus casually deployed the concept of a more primitive form of feeling that he had developed in his researches with W. H. R. Rivers to elucidate the pathological forms of sensibility that he detected in Keats's letters.

Elsewhere Head remarked that: 'In 1818–19 the Edinburgh Quarterly Review demonstrated for ever the dangers of dogmatic shallowness – For in one year the Review ridiculed Keats, the poet and denied that the white matter of the brain was made up of fibres'. These were errors of judgment of comparable magnitude since: 'Recognition of the latter fact has been to physiology what the appearance of Keats has been to literature'.[111] There was thus, in Head's view, a correlation of taste in questions of science and of poetry.

As well as Keats, Head also appreciated the work of the other English Romantic poets. Wordsworth, Shelley, and Coleridge figured among his favourites. Of later English writers, Robert Bridges was also much admired – perhaps in part because he too was a physician-poet, and a personal acquaintance.[112] In 1897, Head proposed that he and Ruth Mayhew have 'a "Bridges" morning'.[113] Head was also familiar with Bridges's book on *Milton's Prosody* (1893).[114] Bridge's verse always gave Head a feeling of: 'richness and restraint'.[115] George Meredith's poems, on the other hand, produced 'an extraordinary feeling of youth'.[116]

The poems of W. B. Yeats, as well as his critical writings, also made an impression on Head. He remarked that: 'In "Ideas of Good & Evil" (W.B Yeats) are many true things concerning the elements of poetry'.[117] Head was sufficiently intrigued to seek a meeting. Ruth recalled that:

> Alban Head once took us to pay a mid-night call on Yeats after an after-supper party! It was when he lived the back of a Bloomsbury slum ... in an attic lighted by enormous altar-candles and filled that also with the scent of incense.
>
> He sat in bard-like isolation with queer people worshipping around his feet, but not a word fell from his lips that we could 'carry away' as the pious used to say of sermons in my young days.[118]

However, Head felt a special affinity for the poems of William Ernest Henley (1849–1903). Head was acquainted with Henley and recorded a number of encounters with him in the 'Rag Book'. Thus in November 1901, shortly after Henley returned to London, Head sought him out in his new lodgings in Prince of Wales Road. He found Henley in the company of the author and journalist, Arthur George Morrison (1863–1945).[119] Head and Mayhew devoted part of a weekend in that month to a reading of Henley's recently published collection of poems, *Hawthorn and Lavender*.

Head was especially fascinated by the imagery of the final stanza of the tenth poem in the collection:

> And I long, as I laugh and listen,
> For the angel-hour that shall bring
> My part, pre-ordained and appointed,
> In the miracle of Spring.

Knowing Henley's 'fondness for the Bible it seemed to me that the "angel-hour" was intended to call up the angel that troubled the water at the pool of Siloans? My view was strengthened by the immense gain in poetic value throughout the whole verse by such allusion'. Head put this suggestion to Henley, who:

> was very struck by this reading but denied he had intended any other significance to be attached to 'angel' than that of a messenger. 'I am very simple & not complex in what I write – If I had thought of it I should have written a whole poem around it'.[120]

The episode again exemplifies Head's efforts to discover the subconscious springs of poetic inspiration.

Henley was, however, among the first to build a bridge between poetry and pictorial art by introducing impressionism into English verse.[121] This style of writing, which sought to translate immediate sensory impressions into literary form, was one that was of particular interest to Head. Moreover, he was not content to explore its possibilities merely as a passive reader. He sought to investigate the possibilities of impressionism through his own writings in both verse and prose.

Writing

'Impressionism', although implicit in his literary practice, is not a term that Head employed in his more reflective discussions of poetry. But in a letter to Mayhew of 1900 he did remark: 'It is curious how in Germany France & England all the younger poets are phenomenalists – Each strives to paint a word picture that will bring its concomitant emotion without actual expression'. He added that he had attempted to achieve similar effects in some of his own poems – trying to evoke 'a strong emotion by the vivid contrasts in the picture it evokes – an emotion which is in no way expressed in words'.[122]

'Phenomenalism', connoted a receptiveness to sensory experience – free from any presuppositions about what might underlie those sensations – on the part of an observer who then sought to give these impressions an adequate linguistic representation. As such, it was a stance that might be adopted as much by a scientist as by an artist.[123] What distinguished the artistic from the scientific response to the phenomenal world was, as Head stressed, attention to the emotional aspect of the subject's response. Thus when sending Mayhew one of his poems, he explained:

> I have carefully cut out all explanation and attempted to make it a pure Impression – So many such things in English explain what they mean and thus to my mind destroy the pictorial emotional value by interposing the painter between the spectator and the canvas.[124]

Head's letters to Mayhew and the 'Rag Books' contain several such essays on impressionistic writing. Many of these were attempts to provide word-pictures of land and seascapes. These pictures could be either of rural or urban settings. Thus while on a visit to Venice, he wrote:

> The night was pitch dark but the canal was illuminated by hundreds of morning lights. I was placed into a Gondola with my luggage and was borne away into the darkness of a network of silent canals. A slight rain was falling so that each lantern and lamp emitted ray light that fell on the side of the houses giving them the look of scenery in a theatre. As we turned into each narrow canal the vista of houses with ends set irregularly to the canal seems flat .... And then the silence that seemed almost terrible after the long accustomed clatter of the train and the clang of street car bells in Milan. I sprang out of this little cabin that obstructed my view and sat at the side in spite of the rain. You know what long streak of light the lamps turned across the roadway on a wet night in London. Tonight in Venice the lights threw the same long streaks but they turned and twisted like snakes on the rippling water. Occasionally a figure would pass over a bridge but the dim light of the lamp was sufficient only to show the passer by as darkened shadow against the partly illuminated houses.[125]

Such passages arose from an impulse to record a particularly striking set of sensory impressions and the affective response they invoked. Sometimes the painterly inspiration for such passages was made explicit. Thus seeking in August 1899 to convey the effect of the London sky, Head wrote: 'At sunset the whole air seems pearl grey and the western sky is barred with Salmon pink or splotched with old gold. Exactly as if a child had dipped a paint brush in gold size and made random strokes here and there'.[126]

Although the stimuli that occasioned such descriptions were predominantly visual, the response of the other senses could also be recorded. Thus, while Head's account of an exceptional autumn day experienced while walking back from his laboratory in Claybury noted the colours and forms of his surroundings, it also dwelt upon 'the aromatic smell of fallen leaves'[127] and the birdsong that helped to make up the gestalt.

Writing of this kind served to crystallize exceptional sensory experiences and to hone Head's literary skills. On occasion they also appear to have served a therapeutic purpose. Thus, while still reeling from his sister's metamorphosis into 'Major Pinney's wife', Head sought solace in a walk in the park:

> the air had that touch of keenness that cried, Autumn has come – There was a scent of wood smoke in the streets from the early fires, that scent which only comes when the first fires are lighted. On my return the dark roofs & minarets of Regents Park stood out on the orange ~~golden western sky~~ sunset and crimson streamers strayed across the uniform dove coloured sky. It was such an evening as one only gets in early autumn.

Moreover, the salutary effect of these impressions was augmented by the literary and other associations that they invoked: 'My thoughts have gone back to Heine as they always do in the autumn and I long for another day such as we had two years ago'.[128]

Head appears to have been especially receptive to impressions so vivid as to merit such lyrical depiction when on vacation. As he wrote while cycling through the Cambridgeshire fens in 1903, 'On these holidays I rejoice in my senses, I see the sun upon the grass, I smell the leaves and hear the wind in the trees – Beauty comes to me again with an intensity increased by the hideous surroundings of my daily life'.[129] While in Milan in March 1901, he told Mayhew that: 'Impressions crowd upon me, beautiful things come before me, some odd word or cry strikes my ear and I think of you and long to communicate them to you at once ....'.[130] Sometimes this crowding of impressions could be almost overwhelming.

Extreme physical exertion could heighten Head's sensitivity to a point where the normal rules of syntax could no longer contain his response to the stimuli acting upon him. While cycling through Germany in the spring of 1902, a kind of delirium overtook him, which could only be represented in a rhapsodic form. Again the experience invoked a literary association – but a more unexpected one than his beloved Heine:

> Then I became drunk with –
> My respiration & Inspiration the beating of my heart &
> passing of blood and air through my lungs
> The sniff of green leaves and of dry leaves, & of hay
> in the barn
> The play of shine and shade on the trees as the supple boughs wag
> The delight alone in the rush of the roads or along the fields & hillsides
> The feeling of health the full noon thrill, the
> song of me rising from bed & meeting the Sun.[131]

The unascribed quotation was from Walt Whitman's 'Song of Myself'. Whitman was not a poet in whom Head otherwise seemed to take much interest. In this exceptional instance, however, his form of free verse, together with his celebration of sensuality, appeared the only idiom in which Head could do justice to what he was feeling. While in this state,

> all but pure scientific knowledge seemed repulsive – All poetry with one exception became nonsense. Wordsworth sickened me by his harping on the human aspects of nature while within me I was – Stuff'd with the stuff that is coarse and stuffed with the
> stuff that is fine.
> I was the Earth in me & not me in the Earth. I felt and the scent of my clothes became the scent of the fields.
> Healthy free the world before me
> The long brown path before me leading
> wherever I choose
> Henceforth I whimper no more postpone no more
> need nothing.
> Done with indoor complaints, libraries,
> querulous criticism.
> Strong and content I travel the open road.[132]

Head's own words again flow seamlessly into those of Whitman.

Passages such as these may be viewed as Head's attempts to provide literary forms that did justice to the richness of his inner life. From that perspective, they assume a scientific aspect: they were attempts to describe a portion of his phenomenal life that was not amenable to the dry prose of a psychological paper. But what was of the most pressing interest was not the sensible phenomenon as such, but the aesthetic response it evoked. In other words, through writing sensations of a particularly striking kind, Head was exploring the nature of beauty.

His aesthetics presupposed that beauty was not an abstract or absolute category, but an aspect of the organism's concrete interactions with its environment and therefore contingent upon the modes of apperception that a particular sensory mechanism could sustain. Head thus sometimes noted, in a somewhat detached manner, certain peculiarities of his own senses and sought to relate these experiences to aspects of the pictorial arts. Thus, while viewing the landscape near Settignano, he observed: 'A cypress between me and the setting sun is not round but flat exactly as if it had been cut out of a sheet of black tin. But as soon as I advance so that I see the light on the tree – as soon as it no longer only blocks out the light – it becomes a solid object'. He went on to speculate that this effect might account for the fact that: 'The early Florentines always painted trees between the Spectator and a low sun not daring to attempt trees otherwise than in outline'.[133] Another striking impression gained on the same trip reminded Head of: 'an effect only attempted as far as I know by Turner & perhaps by Claude. This might be called an effect of "solar perspective"'.[134]

Because human beings possessed a shared physiology, certain consistencies and consonances in the ways in which they experienced reality were to be expected. However, since each organism was also to an extent unique so the particular perception of beauty was the exclusive gift of each observer – dependent upon 'the incalculable angle at which experience may strike'[135] each individual. What distinguished the artist was his capacity to give a permanent form to the perpetual flux of his impressions.[136] By seeking to translate some of his most striking impressions into words, Head was thus in more ways than one seeking to write himself.

The inner life thus consisted in part in the mind's register of stimuli emanating from natural phenomena. Head also moved, however, in a social environment that produced another set of impressions that also called for literary representation. Head regarded himself as an acute observer both of individual psychology and of inter-personal dynamics. In March 1900 he sent Ruth Mayhew an arch account of an encounter with a 'model couple'. The husband:

> enjoys doing all the proper things in the proper way. For instance he formally sharpens the carving knife on a steel before a meal begins; he repairs the washer in the hot water tap – regulates the gas fittings. She is a sensible commonplace woman with no marked natural taste but much common sense.

Head evinced an obvious condescension for these people whom he clearly regarded as less sophisticated than himself and Ruth. Their personalities and relationship were of sufficient interest, however, to set Head speculating on the possibility of writing 'a tragedy on the following lines'. Given:

> a couple such as this and to them enter a man who wrecks this happiness, not by any intervention of sex, but by casting them the apple of intellectual discord. Let the woman for instance gain a certain sense of the beauty of poetry or painting & imagine she can write or paint. The husband on the other hand should remain wedded to some scientific pursuit and be only capable of seeing his wife's creative sterility but quite unable to understand her perfectly correct artistic appreciations – Don't you think this would make a fine tragedy [?][137]

In this case Head saw an actual encounter as the inspiration for the composition of a 'tragedy'. It is unclear whether this projected tragedy would take the form of a novel, short story or stage play. But the letter also reveals an experimental approach to the study of human relations. Rather than merely observing the extant interactions between the model couple, Head suggests authorial interventions to vary the relations between them in order to explore more fully the nature of their relationship. Literature, in this view, can serve as a kind of psychological laboratory.

Such experimentation was dependent upon the imaginative capacities of the psychologist. Head was critical when novelists made ill-informed use of psychological theories in their works. Thus he regarded Paul Bourget's use of the notion of dual personality in *La Duchesse Bleue* (1897) as unhelpful. Reflecting on this book however led him to reveal:

> I have always felt (and in my humble way have always experienced) that an artist can lead an entirely imaginary life. Since I was a boy I have always known characters that I have never met – It is true that they were founded and shaped on persons I had known – But once formed they acted coherently and moved in paths different from anything I had ever experienced. It is as if one had dressed dolls to take some part on a marionette show of one's own devising – But when that show began the dolls became alive and carried out a play of which their author could only be a surprised & amused spectator.[138]

Elsewhere, he seemed to blur the boundary between the real and the fictional in a playful fashion. Thus after Ruth Mayhew had regaled him with an amusing anecdote about an encounter with two tourists she had met while touring Germany, Head claimed that: 'If I had been with you I should have thought & thought until I had invented a long story about them and then have gone away and called it "Experience".'[139]

This ability to extrapolate from experience of human character might seem to have made Head exceptionally qualified to be a writer of fiction. Unlike Ruth he never attempted to write a novel. From his youth, however, Head on occasion adopted an authorial voice akin to that of a novelist in some of his letters and journals – for instance when seeking to give adequate expression to the predicament of Fraulein Weiss during his vacation in the Böhnewald. In the accounts of journeys he took in later life, Head combined impressionistic accounts of landscape with psychological analysis of the individuals he encountered. Ultimately, however, these letters constituted exercises in self-analysis. Head's final aim was always to document how these experiences of foreign parts cast light on aspects of his own personality.

On a number of occasions Head holidayed in Bavaria, staying part of the time at Schloss Neubeuern as a guest of Baron Jan Wendelstadt. Head seems to have met Wendelstadt through his brother Alban. Wendelstadt was something of a patron of the arts, with a special interest in painting – indeed, a 'lame portrait painter' was in residence at the castle during Head's one of stays there. While staying at Schloss Neubeurern in October 1902 Head wrote lengthy letters to Ruth Mayhew in which he tried to share his impressions of his surroundings and of the company he kept. Although these reflections were meant to seem spontaneous, it is apparent that Head took considerable care in the composition of these documents, sometimes crossing out and rewriting phrases in order to improve a sentence. The product of these efforts was an account of his social and physical surroundings with a marked bias toward exploring the psychology, mores, and interpersonal dynamics of his fellow guests.

These wry observations began with an account of Henry and Alban arriving at the railway station at Raubling to be met by: 'a small brown waggonette waiting for us with a middle aged coachman on the box of the most profound gravity'. An English Coachman, Head remarked, 'always looks as if the world began & ended with his horses; but his German imitator carries impassiveness to a fine art'. Upon arrival at the castle Head noted that it had undergone considerable renovations since his last visit in keeping with the Baron's aspirations to be known as a patron of the arts. Thus, 'The old brown dining room has gone & has been replaced by white wood work with a dark frieze on which Bacchic and Bacchaulic figures are portrayed according to the newest Munich mode'. Head did not forebear to demonstrate his superior taste in these matters – 'the white dining room is beautiful with its fluted panels but the frieze is bad ...'.[140]

The castle was 'full of guests'.[141] Head provided character sketches of some of these. For example the Frau von Arters – 'separated from an intolerable husband' – was a long-time resident who spent her time shooting or viewing the building work. She was 'a large woman utterly unlike any German I have ever seen'; von Arters did, however, remind him of one of his English acquaintances. There was: 'in her the same directness the same direct outlook on life and the same intolerance of emotion of any kind'.[142]

Frau von Arters was accompanied by her daughter – 'a small positive, mocking, wisp of a girl ready for any mischief'. Head was in particular fascinated by the interactions between this girl and his brother Alban. Fraulein von Arters's:

> relations with Alban are very odd. She is musical, but scorns all Art and if praised for her playing which is good will probably break into the wildest music hall song or the music of the latest operetta. She knows exactly on which side her bread is buttered and, in her own limited sphere, has a terrible knowledge of life. And yet the impossibility of fitting Alban into any type she has ever met or heard of attracts her like a moth to a candle. She quarrels with him and he goes his way entirely unmoved. She whistles him back again only to insult him again. He talks to her as if she was something routinely outside the serious aspect of life, lectures her, laughs at her. They remind me of a great puppy and a kitten when they are together – and after all the puppy is less sensitive.[143]

More generally, it was the character of the female inhabitants of the castle that Head found most fascinating. He recorded that the Baron 'is entirely unchanged. He wears the same jäger coat & hat and his laugh has lost nothing of its infectious character'. But Head spent far more time in descrying the subtle changes since his last visit he discerned in the Baroness. In contrast to her husband, 'externally', she was: 'more the great lady. But under the exquisite clothes the jewels and the bold handling of social difficulties lies a soft humorous woman disliking effort and fond of physical warmth & mental intimacy'. Among her foibles, Head had remarked was to:

> place her chair so that she could hear Frl. von Arters & Alban bickering ~~in a corner~~ together. She would say nothing for a long time but suddenly would find some expression that exactly described the situation of these two odd creatures. Her sympathies were always on his side and the more he was teazed & harried the more she stood up for him.

Head's gaze had also cast light on her relationship with her husband. In public, the Baroness was 'the lady of the Castle'. On more than one occasion, however, 'when they thought themselves alone I surprised two lovers in the shadow of the great library'.[144]

Head thus haunted the castle in much the same way the narrator inhabits the pages of a novel. When writing of the care the inhabitants of the castle took of a relative who was 'defective in intellect', Head described how: 'I sat in the half darkness after tea on a wet evening and listened to a humorous account of a visit paid by her two old maids to their charge ....'[145] He had the same ironic detachment complemented by an intimacy and empathy that enabled him to detect aspects of his characters' personality of which they might themselves be unaware. He ascribed these insights to the fact that he was a trained 'psychologist'. But for Head, the best novelists were also astute psychological observers.

In this narrative, Head also gave due attention to the aesthetic effects of his physical surroundings. 'Although a good deal of the simplicity has passed away', he reported, 'the superb eye for effect of both the Baron & his wife prevent the luxurious surroundings from being anything but beautiful'. Thus he tried to convey the striking impressions he derived when the company dined:

> Every evening the colour scheme was changed & with the change in scheme every thing on the table, excepting the silver candlesticks, was changed. One day it was mauve with tiny asters, another evening the table was a-blaze with crimson single Dahlias. The dinners were short but the food superb and in the middle of one side of the oblong table sat the Baroness pencil & paper by her side that the cook might hear at once how his efforts had fared. She herself had been forbidden most of the good things provided for us. This abstinence on her part gave me always the sense that we were a ~~set of~~ lower order of creation presided over & catered for by a Circe who cared little for such sensual enjoyments.[146]

More generally, Head wrote of his stay in Neubeuern as an idyllic experience. He woke each morning to be pampered by an exquisitely courteous footman anxious to ensure that the 'Herr' enjoyed every comfort. Head described himself reading Mayhew's letters to him while luxuriating in an unaccustomed late lie-in. When he eventually rose, he would: 'go onto the Terrace to see the mist filling the valley like cotton wool. The mountains were bright in the morning Sun and far away up the valley lay the snow hills with their crests looking as if they had been powdered with one of the great silver sugar-dredgers that stood on the dining table'.¹⁴⁷ Germany still retained for Head its status as a fairytale land.

Afternoons were spent cycling or playing tennis according to rules that 'I believe existed in England somewhere in the late Seventies [which] were laid down for us by Frl. von Arters'. Head had not played tennis for years, but: 'renewed my youth, served the old 'shooters' of ten years ago, placed & became excited over the result of the game exactly as in the days that seem so long ago, in the early Eighties'. After these pleasant exertions that recalled his less stressful days student days,

> To enter my room wet & tired from tennis or bicycling was the greatest joy of the day. Early in the afternoon my great green stove had been lighted and allowed to burn itself out. The room was filled with a soft warmth sweet with the scent of mountain wood. On my writing table stood a great lamp that filled the whole room with a pink light. The curtains were drawn, my books were spread all over the table and the welcome of the warm lighted room seemed almost personal.¹⁴⁸

Head was anxious to convey the olfactory and tactile, as well as the visual and auditory impressions, that combined to make his experience of the Schloss such an altogether delightful gestalt. In the evenings, 'A fire of logs crackled gaily in the open hearth and a huge green "Ofen" radiated heat throughout the great room'. Those of the guests, 'who came down early enough discovered that the vivid scent of burning wood was intensified by the manipulations of a footman who violently agitated a pair of wood ashes in the middle of the room a few minutes before we were expected'. When the Baroness finally appeared, the doors of the dining room were thrown open and the guests filed: 'into the white dining room lighted by many candles in silver lustres'. After dinner, the remainder of the evening was spent: 'with music songs or games'.¹⁴⁹

So passed Head's days at Neubeuern – 'and you asked yourself what have I done? There seemed time for nothing the days were so full and yet at their close there was nothing to show but a sense of pleasant easy living'.¹⁵⁰ Head thus tried to convey the sense that this vacation constituted an approach to the ideal kind of existence that he so often craved. It was in particular the extreme opposite to the stress and harassment of his everyday professional life in London.

When the time came to leave the Schloss, the idyll continued. The Baron gave Head permission to shoot one of the chamois that grazed on the estate and despatched his 'Förster' to assist his guest. Although the pair found the herd easily enough, the chamois caught their scent and fled. Head, however, admitted:

> I was not greatly disappointed. The long lying in the mountain grass that had with the true Autumn smell and the feeling of the hot sun in the middle of the back was enough apart from the intense interest of watching these shy creatures at so close a

range. Bernaner made another cast leaving me on one of the tops where I frankly slept – the most delicious sleep in the sweetest smelling bed I have ever lain in. In about an hour he came back saying it was no use so down we went to his Gasthaus at Nundorf to Mittagsessen somewhere about 3.30 p.m.[151]

This 'beautiful day on the hills' put Head off the idea of returning to Munich. Instead he cycled into Austria along the River Inn, admiring the scenery and taking especial note of 'a perfect instance of the Renaissance Castle' that he passed.[152] As he rode into the town of Schwaz, the surrounding hills were:

> covered by thick mist and long whisps hung about the higher parts of the road. But the river was a great joy – It raced away from me steel grey or milk blue swirling around obstacles forming mighty backwaters a strong male thing that in spite of all obstructions knew where it wanted to go and got there.

Even the signs of industrial activity in the landscape filled Head with exhilaration: 'Flour was ground, planks were sawn & at one place an immense electric supply station was run entirely from the river'.[153]

It was only when he entered the town and found his way to an inn that Head's mood of euphoria began to dissipate. He found: 'the hotel full of grumbling, weather bound English complaining that they had been cheated into coming here and that the waiters stole the wine. Ignorant for the most part many of them nasty tempered, I shall be glad to be rid of them tomorrow when I get to Munich'. This encounter with his countrymen therefore served as an unpleasant reminder of the mundane existence that he had for a while escaped. Head took some satisfaction, however, from the fact that: 'the three days in the open air have made my skin feel alive again and my legs have lost their London flabbiness'.[154] His body as well as his spirit had thus been renewed by this temporary escape from the toil and tedium of the metropolis.

The letters that Head wrote on this and on some of his other foreign trips – such as those he took to Montenegro and Morocco – served overtly to allow Ruth Mayhew to share his experiences. But these writings also allowed him to take on something of the role of a novelist seeking to descry the physical and psychological phenomena that passed through his purview. The final point of reference of these journeys was Head own conception of self and how this was revealed and perhaps modified in the course of the journey he recounted. Perhaps the most extreme instance of a radical shift in his sense of personality came at the end of his travels through the eastern Adriatic in the spring of 1901, which he described as an 'adventure'. He had, he declared:

> travelled the path of hundreds of tourists but we returned as they say those return who have drunk of the roman fountain, sick with desire once more to go back before we die – For we had been to the back of beyond, we had heard the East calling but it

was the free East where Everyman is a Soldier and a mountaineer and all the women beautiful & unveiled –
Shall I ever go back?[155]

These letters were as much lyrical as analytical. Prose, however, was too limited a medium to express the full range of Head's thoughts and feelings. For this, he was obliged to resort to verse.

The Poems

It is not clear when Head began to write poetry. But by the later 1890s poems by him were circulating among family and friends. His mother and his sister, Hester, were especially avid readers of these offerings. One year Hester Head senior asked her son for a particular sonnet as a birthday present. Somewhat reluctantly, for he was still unsatisfied with the poem, Harry complied:

Mass on Easter day

Across the awakening hill the dawn mists fly
To hollows where as yet no sun has been
The Woodlands, redolent of budding green,
Re-echo with the Stock dove's amorous cry.

For down the vale a tiny church doth lie
Where altar lights glow dimly through a screen
Of odorous incense and a bell unseen
Proclaims Christ's sacrifice for man is nigh.

Bowed, like a Virgin Lily ripe for seed,
You kneel beside me and intently pray
For grace to tread ~~that dim inhuman way~~ that dimly lighted way
Of death in life – But surely God hath need
Of all you doom to premature decay
'Midst sterile Comforts of a Sexless creed.[156]

The poem faithfully reflects Head's views of Christianity. Why his pious mother felt drawn toward it is, however, more puzzling.

But it was in Ruth Mayhew that Head found an ideal reader who combined the attributes of empathy and critical insight that he craved. In July 1899 she informed him that for her poetry: 'is a religion and an hourly companion and the thought of it runs in and out of the most sordid employments of the most difficult days'.[157] Head nonetheless was sometimes apologetic about how large his poetry loomed in his conversations with Mayhew. 'Whenever we meet', he confessed, 'I seem to either talk about the verses I am going to produce or lament the weakness of those that are done'.[158]

Head and Mayhew's shared devotion to this religion was to form an important thread in their early letters. He would send poems to her 'hot from the bakehouse' of his imagination. Moreover, on occasion Head specified that these offerings were for Mayhew's eyes alone, enjoining her: 'not show it to anybody and do not write to mother about it. She can have it in good time'.[159] In response, Ruth dubbed Head: 'Herr Bäcker [baker]', and assured him of 'how delighted I am with the latest production of your spacious bake-house'.[160]

During an unhappy stay in the Netherlands in the spring of 1897, Ruth took much comfort in the sonnets that Head sent her. A tapestry Head had seen had inspired one of these poems, and Mayhew observed: 'I suppose you saw the tapestry ages ago and the germ of the idea has just lain in your brain ripening and the sonnet is the perfect fruit'.[161] In response to a later offering, she wrote that: 'I do hope the new sonnet will prove a healthy vigorous child. I am much looking forward to seeing it and I feel much honoured to think that you allow me to see it in its swaddling bands'.[162]

Head did not, however, send her his verses simply for Mayhew's edification. He conceded that she had sufficient taste, judgment, and erudition to make suggestions on how they might be improved. Indeed in 1897 he urged her to be 'harsh and critical' about two lyrics he had forwarded.[163] Mayhew replied in a deferential tone. Nonetheless she did venture to question the justness of one line: 'shrivelled petals slowly fall'. When Ruth saw: 'the cherry blossom petals, for example, with my inward eye, they lie like smooth velvet on the grass and only shrivel and turn brown after rain'.[164] With time, Mayhew was to grow more bold in her suggestions. It is evident from the extant manuscripts of Head's poems that in some cases her input to the final text was considerable.

Moreover, with time, Mayhew ceased merely to be the recipient of poems that Head had composed and became central to the inspiration of much of his later verse. In August 1899 Head professed that he had steeped himself in the 'descriptions' found in Ruth's letters – 'In many places you will recognize your own words and by even such phrase you will own that the verse has gained in actuality'.[165] Increasingly, not only her letters, but also his memories of their interactions and in particular the places where they met provided Head with stimuli to write poetry.

Mayhew acknowledged this in a letter she wrote to her brother, Arthur, in 1919:

> We are both now 'bathed in bliss complete', but it makes the old days none the less delightful in the memory. You have seen all the surroundings that inspired the poems quite surprisingly and they do bring feelings back too, don't they? the tearing sheet noise of the sea on my pebbled shore in Brighton and my brown room with its trois cent, quatre vingts chandelles en cire blanche.[166]

Head's acknowledgment of the debt his art owed to Ruth was unequivocal. When a selection of his verse was published, the dedication to the volume read:

TO HER
without whose touch the strings would have been mute[167]

Sometimes when composing poetry Head used Mayhew as a kind of experimental subject. He sought to explore the subtleties of her psychological processes and then to incorporate these mental states into a poem. In May 1898 he asked her to: 'give me light on two matters, but if the first is in anyway painful pay no attention to my request'. The first of these matters was 'a certain Condition of mind in extreme Sorrow' that Ruth had mentioned at one of their meetings. This condition he had 'crystallized for future use' through the lines: 'With eyes part weeping & with forehead bowed/To the still earth, my sorrow outward went'. Mayhew had also 'said something about feeling as if all your mind was outside & that you felt as if you were looking on at its working'. Head asked her to: 'amplify this?'[168]

He also took a keen interest in his own mental states when composing poetry. Thus when in November 1899 Mayhew enquired about the 'genesis of the last poem', Head supplied a reconstruction of the psychic processes involved. The poem 'began to come' to him:

> as I rode back through in the darkness with the Chopin still humming in my head. As to the simile of the orchard do you not remember as we walked back on the Sunday I first visited you how we discussed the likeness of the rows of lamps to this and that? I said I thought they looked like fruit trees hung with golden light and you rather scorned the idea. I was at first a little timid in using it but I gradually became certain that the simile was a good one.[169]

Head was especially preoccupied with the question of how the right metaphor could serve to give just verbal expression to a particular sensory impression. While wandering through the woods near Settignano Head took careful note of the fact that:

> The sound of the wind in the trees varies with the nature of the tree. Thus with elms & oaks it resembles the distant sea – Pines produce a sound like a railway train at a distance, olives rustle like a silk shirt. The cypresses make a curious sound somewhat like that of the wind in the bare masts of ships in harbour.[170]

Such fine distinctions in auditory apperception had implications for the range of metaphoric resources available to a poet seeking to represent these phenomena.

One of Head's poems contained the lines: 'With a hungry roar grey water leapt the seawall, ground upon the shingle like a tearing sheet'. These had initially been written:

without any allusion to sound in the first line. The second line then made nonsense for the ear could not shake of the association of 'shingle' and 'like a tearing sheet'. But directly some allusion to sound was put with the first line the thing came right. This is rather a neat technical point that I was pleased to learn and that I shall now look for elsewhere.[171]

Sound literary technique thus rested upon a proper understanding of the psychological processes underlying the appreciation of poetry.

Whatever their subject-matter, all Head's poems may be regarded as 'lyrical', inasmuch as they were chiefly concerned with the emotional response evoked by a particular set of impressions. It is possible to relate some of these works directly or indirectly to aspects of the author's life and experience.

Some, for instance, seem to reflect aspects of Head's scientific identity. 'Premonition' is both a lament at how the relentless passage of time threatens to overtake his labours and ambition and an assertion of his ability to raise an enduring achievement:

> No longer when the cold and sterile moon
> Paces her virgin watch across the sky
> Calling the hours of life's long afternoon
> Must I from out the deep in answer cry.
> With rhythmic tide she swept the foreshore land
> Whereon I often see my heart's desire
> Leaving is a barren strip of watery sand
> A mirror for the moon's chaste silvery fire
>
> For I have built a barrier 'gainst the Sea
> No more the moon swept tide my fruit devours
> The seed in set and in security
> I watch the silent passage of life's hours
> May that Sea wall till harvest time abide
> Steadfast against the ever restless tide.[172]

Head originally entitled another poem 'The Yew Tree'. Later, however, he changed this to 'The Seeker after Truth':

> Hid from sight of pasture lands,
> Behind the Church, a yew tree stands,
> Banished from the cheerful fields,
> For the deadly crop [published version: 'fruit'] it yields.
>
> Year by year it waxes tall,
> Hemmed within the churchyard wall;
> For the Church must ever keep
> Poisoned fruit from silly sheep
>
> Without a knot its branches grow,

> Each to form a yeoman's bow,
> Evergreen and never old,
> For they spring from Churchyard mould.
>
> A solitary from the throng,
> I fashion weapons for the strong;
> But every thought within my head
> Has its roots among the dead.[173]

The later title provides the key to the imagery of the poem. The yew tree represents the scientist – the 'Seeker after Truth', who through his labours fashions 'weapons' for the intellectually strong. The Church figures as an obscurantist force that seeks to keep 'poisoned fruit from silly sheep'. The poet remarks upon the irony that, as a medical scientist who relied greatly on autopsy material to make his discoveries, 'every thought within my head/Has its roots among the dead'.

Other poems reflect the resentment Head sometimes expressed for the sacrifices that his devotion to science and medicine had involved. This is especially striking in one entitled: 'The Man. The Healer'.

> The crowning seal of passion never fell
> Upon these lips: for all my happiness
> I sought with those whom misery and distress
> Drove to my dwelling by the healing well.
>
> And never woman sought my lonely cell
> But to some foul distemper would confess
> Or cast on my compassion the redress
> Of wrongs she had not courage to repel.
>
> Now half my life is spent with lavish hand
> My soul lies calm in spiritual death
> And under Love's reanimating breath
> My heart no longer glows – a burnt-out brand
> Each whirling gust of passion scattereth
> A few gray ashes on a dead sea strand.[174]

Other poems refer directly to aspects of hospital life. 'Ballade of the Sister' is written from the point of view of a patient and expresses gratitude for the good offices of the ward sister during his stay:

> Of all the weary weeks of pain
> There in the Hospital I lay
> My sister, you will ever remain
> A happy memory – Day by day
> My mother's sorrow you'd allay
> Or for some privilege would sue ....[175]

In 'Spring Tragedy' Head also attempts to enter into the mind of a patient. In this instance, a woman who was incarcerated in Devonshire County Asylum after committing infanticide.

'Scent Memory', written in May 1898, is more interesting because it seeks to represent Head's own experience while working at the London. The poem includes both generic aspects of that experience and the recollection of an especially striking incident:

> With careless heart and practised ease
> I followed the tracks of a dread disease;
> As I rapidly pass from bed to bed
> And illumine the living with light from the dead
> Foul symptoms like a beacon glow
> Over hidden/sunken? Rocks of disease below
>
> But my words come thickly, I wander ^{stammer [RM]} and grope
> Through a scented halo of heliotrope
> From a nosegay that gilds [deletion RM] ^{thrust through [RM]} an arid text
> On the sick man's pillow – My mind perplexed
> By so fierce an assault on its ordered train
> Sways, staggers forward and stops again.
>
> Then with rustle of silk my love appears
> Laughing scorn her sweet mouth wears
> She stoops to the bed, she does not speak
> But plants a kiss on that sin scarred cheek –
> And the anxious sick man wonders why
> I stare in his face with so vacant an eye.[176]

The first stanza recounts Head's mental processes as he went from bed to bed seeking to gain intellectual mastery of the cases presented to him – trying to make the 'foul symptoms' he witnessed the source of insight into the fundamental nature of the disease. But this orderly process is interrupted by an unexpected and incongruous sensory stimulus: the scent of a nosegay on a patient's pillow. Head's consciousness is wrenched away into an unexpected set of associations that culminate in the fantasy of his lover appearing at the patient's bedside to provide comfort with a kiss. The text of this poem was extensively amended by Ruth Mayhew and gives a good impression of how active she was in seeking to improve Head's verses.

A number of poems dealt directly with Head's relationship with Mayhew. 'Love the Intruder' deals with the moment when they ceased to be merely friends and acknowledged their love for one another.

> With silence or with careless talk
> Joyous our days have been

> Until upon that fatal walk
> A stranger came between
> Felt surely but unseen.
>
> Our souls were one, yet either heart
> Beat free of Love's commands
> But now we loitered far apart
> Lest he should join our hands
> In everlasting bands
>
> We chattered of each passing thing
> And dared not silence keep
> Lest he into our talk should spring
> And from our lips should peep
> What either heart hid deep.
>
> With heads close joined o'er many a book
> We'd oft sat knee to knee
> Now either shunned the other's look
> Lest eyes should mirrors be
> Where ~~each himself~~ we ourselves [emendation by RM] should see.
>
> Why did he trouble the clear source
> Of our bright friendship's stream
> And mar that sexless intercourse
> Till all our actions seem
> ~~Aimless as in~~ Unstable as [emendation by RM] in a dream?
>
> Love's sweetest hour can never bring
> Return to that calm state
> From our hearts' ashes there will spring
> The tolerance for a mate
> Or hate, eternal hate.[177]

The final stanza would appear to indicate that the transition from friendship to love was one that evoked equivocal feelings.

The 'Prologue' to the 'Seedtime and Harvest' series however presented an idealized vision of the relation between the sexes:

> In the beginning God created man
> Perfect in all things Lord of land & sea:
> It seemed in form to ape divinity.
>
> From Adam's side a rib he therefore took,
> And making woman halved that form divine:
> Who now would on God's perfect image look
> Must every grace of man or maid combine.

> The severed halves of man's once perfect soul,
> Dwelling apart, their wailing never cease
> Till they be joined in one primeval whole
> And in reunion find perfect peace.[178]

The poems that Head, with the assistance of Ruth Mayhew, produced in the 1890s and early years of the twentieth century circulated in privately printed editions among his family and select friends. Some were published in periodicals such as: *The Yale Journal*, *The English Review*, and *The Dublin Review*. Several also found their way, in modified form, into *Destroyers and Other Verses* (1919).[179] This volume included another set of verses that had been written during the First World War, and reflected aspects of Head's wartime experience.

The title poem derived from a visual impression that Head found at once stirring and poignant:

> On this primeval strip of western land,
> With purple bays and tongues of shining sand,
> Time, like an echoing tide,
> Moves drowsily in idle ebb and flow;
> The sunshine slumbers in the tangled grass
> And homely folk with simple greetings pass,
> As to their worship or their work they go.
> Man, earth, and sea
> Seem linked in elemental harmony,
> And my insurgent sorrow finds release
> In dreams of peace.
>
> But silent, gray,
> Out of the curtained haze,
> Across the bay
> Two fierce destroyers glide with bows a-foam
> And predatory gaze,
> Like cormorants that seek a submerged prey.
> An angel of destruction guards the door
> And keeps the peace of our ancestral home;
> Freedom to dream, to work, and to adore,
> These vagrant days, nights of untroubled breath,
> Are bought with death.[180]

Through his work with casualties of war, Head was all too familiar with the toll of lives that the maintenance of his country's freedom demanded. In 'Died of his Wounds' he recorded one such loss:

> Death set his mark and left a mangled thing,
> With palsied limbs no science could restore,
> To weary out the weeks or months or years,
> Amidst the tumult of a mother's tears

> Behind the sick-room door,
> Where tender skill and subtle knowledge bring
> Brief respite only from the ultimate
> Decree of fate.
>
> Then, like the flowers we planted in his room,
> Bud after bud we watched his soul unfold;
> Each delicate bloom
> Of alabaster, violet, and gold
> Struggled to light,
> Drawing its vital breath
> Within the pallid atmosphere of death.
> That valiant spirit has not passed away,
> But lives and grows
> Within us, as a penetrating ray
> Of sunshine on a crystal surface glows
> With many-hued refraction. He has fled
> Into the unknown silence of the night,
> But cannot die until human hearts are dead.[181]

It is notable that while another poet might have taken comfort in the belief that the soul of the dead soldier has passed to another life, the only hope of immortality that Head would profess was that his memory would persist in the living.

Head dedicated himself to public service during the war at considerable financial toll to himself. In 'The Price', however, he described the psychic cost of these exertions:

> Night hovers blue above the sombre square,
> The solitary amber lanterns throw
> A soft penumbra on the path below,
> And through the plumed pavilion of the trees
> A solemn breeze
>
> Bears faintly from the river midnight bells;
> While at this peaceful hour my spirit tells
> Its tale of arduous joys,
> Pain conquered, Fear resolved, or Hope regained
> Swift recognition of some law divine,
> Shy gratitude that could not be restrained,
> All these were mine,
> And so, supremely blest,
> I sink to rest.
>
> Through labyrinthine sleep I grope my way,
> Feeble of purpose, sick at heart, and sure
> Some unknown ill will lead my steps astray,
> Till, cold and gray,
> The dawn rays through my shuttered windows steal

> And with closed eyes I thank my God for light,
> For the fierce purpose of another day,
> When work and thought forbid the heart to feel.[182]

War could, however, also present spectacles that lifted Head's heart. Thus in 'Pegasus', Head rejoiced in the elation the sight of an aeroplane produced in him:

> The wind is still; from far and wide the air
> Resounds with Sabbath bells, calling to prayer,
> And from the vast, unfathomable blue
> Hums a propeller's penetrating drone.
> We stand enchanted, and our eyes pursue
> An aeroplane, that climbs the summer sky
> To drift alone
> On mountainous clouds of ever-virgin snow,
> Suspended like a black-winged dragon-fly,
> That turning gleams,
> Dove-gray and silver in the morning beams;
> Or like a dead leaf, loosened from a height,
> Spins in its perilous flight.
> We catch our breath like children at a show,
> Of martial music and heroic deeds,
> On every glittering incident intent,
> Forgetting for a time terrestrial creed
> For joy that man now rides the firmament.[183]

Head was fascinated with aviation – exploring for instance the physiological effects that flight produced on the human body. In his clinical practice, Head also took a special interest in airmen who had been traumatized by their experiences of war. But Head's enthralment by flight went beyond the purely professional and scientific. As the last lines of the poem show, he derived real joy from the amazing fact that 'man now rides the firmament'.

The war thus served to stretch the boundaries of Head's experience into realms that were at once terrible and sublime. But in the series entitled 'Sun and Showers' he reaffirmed that the centre of his universe remained unchanged:

> I have wandered round an Empire
> To the kingdom whence it grew,
> And the coast-line of my country
> Flashes white beneath the blue.
> ....
> Home and country, kingdom, Empire,
> All the universe to me
> Is a little laughing woman
> In her brown room by the sea.[184]

The last line recalled the room in Ruth Mayhew's Brighton home where the two had spent so much time exploring the delights of the inner life.

****

Among the manuscript verses that Head left was a strange, somewhat disturbing, set of lines:

> Upon the fragile fabric of my brain
> Speech weighs, like fallen leaf in fairy mesh
> Already burdened by the dewy rain –
> So slender is the flesh.[185]

The fragment is undated, but seems to be among the later writings in the collection. It seems to betoken a sense of growing cognitive confusion and an awareness of the fragility of the bodily mechanisms upon which intellectual activity depended. The later years of Head's life give these ruminations an added pathos.

6 THE TWO SOLITARIES

A Life of High Passion

During the first two decades of the twentieth century Head approached the pinnacle of his career. His private practice flourished. His efforts to establish a Pathological Institute at the London Hospital, despite local opposition, came to fruition. Most importantly, the research programme he had cultivated with the help of various collaborators since the early 1890s advanced inexorably. These activities made heavy demands upon Head's energy. But he showed that he had the inner resources to meet these challenges while still retaining the capacity to cultivate the vast range of his other interests.

Gordon Holmes detailed Head's hectic yet orderly routine. In these years, Holmes recalled:

> his private practice reached its peak and teaching and hospital work made large demands on his time, at least two mornings, or even whole days, every week were reserved for examining suitable patients and recording the observations we had made, while on one evening each week we met to discuss our recent work and plan investigations into points which required further elucidation. The most lasting impression of these years of active association is of his enthusiasm and energy, and of his vivid imagination which constantly suggested new lines of approach to every problem which arose.[1]

In addition to these activities, between 1910 and 1925 Head assumed the editorship of the neurological journal *Brain*. Holmes was of the view that: 'the high reputation it has attained in medical and scientific literature is largely due to his work'.[2] Head undertook the tedious and often thankless duties of editing a journal with his customary verve. No detail of the management of *Brain* was too small to escape his notice. One of his earliest acts was to review the exchange list that *Brain* maintained with other periodicals, sending a curt note to the publisher to point out: 'I see you have still continued to send "Brain" to Prof. Dubois Reymond in Berlin, in spite of the fact that he has long ceased to be Editor and we have received no copy of the "Physiologisches Centralblatt" for years

'...'. Moreover, even though Head had in the previous year supplied Macmillan with an up-to-date statement of journals to be included in the exchange, he was taken aback to receive a list that 'is the exact copy of the one sent for correction last year, even to the same typographical mistakes. Cannot this be brought home to the responsible person?'.[3] Despite Head's best efforts, these problems appeared to be ineradicable. Three years after his first protest, he was obliged to return to the topic:

> I hope you will not consider me overcritical but these lists have never been sent to me without one or more gross & in some cases laughable mistakes such as could only have occurred in consequence of a fundamental ignorance of foreign languages. The enclosed is no exception – you will notice that Dr Kurt Mendel's address to which presumably his copy of Brain is to be sent is 'half past four to six o'clock'...[4]

Head was also entrepreneurial in seeking new subscribers – for instance asking those who attended a course of lectures to which he contributed in the United States: 'if you would care to subscribe to "Brain". It is, as you know, the journal in which the work of English neurologists appears & occasionally we have the honour of publishing papers by leading American authorities'.[5]

The lectures in question were given in September 1912 at Fordham University in New York. Head was invited by the Dean of the Medical School, James J. Walsh, to contribute to an extramural course on nervous and mental diseases. Gordon Holmes accompanied him on his trip. While in New York, he found time to see patients at the Montefiore Home, where, according to a newspaper report, he discovered an 'interesting case' in which the patient suffered from a lesion of the optic thalamus. Head described the case in his lecture. But he also took the opportunity to propound his new views on sensation to his American audience. While he was at Fordham, Head received an honorary degree from the university. In introducing him Walsh maintained that: 'in scarce half a century of life his accomplishments have been the most noteworthy of our generation'.[6]

During his three-week stay in the United States, Head was treated as something of a minor celebrity – a status that he seemed to relish. A reporter from the *New York Times* interviewed Head while he sat in 'one of the small drawing rooms of a Fifth Avenue hotel'. Head was drawn on his views of the current state of western society in general and of the role of the modern woman in particular. 'Dr Head is', the reporter averred, 'not only an investigator in the discovery of the seats of sensation: he is a student of social problems as well'.[7]

The insights into social problems that Head offered were of a mostly light-hearted nature. He claimed that the twentieth century was a 'decadent age' – yet also an 'age of happiness'. The modern era, in Head's view, most resembled: 'second century Rome but it is to be compared, as well, to ancient Egypt, to decaying Greece, to Babylon'. It was possible, Head argued, for a society to be decadent and for its people: 'to live according to their ideas of what is best and getting the best out of life'. Indeed, widespread contentment was itself a sign of decadence. Personal happiness had been less a feature of the 'striving ages' of the past.[8]

Women, in particular, lived a 'soft' life in the present-day. They participated in the generally 'happy' state of contemporary civilization and enjoyed freedoms denied to earlier generations. But this liberty was bought at the price of an increased prevalence of 'neurasthenia'. Moreover, Head foresaw unfortunate consequences for the future of the species. 'I can see', he proclaimed, 'the race is going to die out if the evolution of women keeps on the way it has begun'. While he could only admire modern women for the 'splendid, intelligent, interested, healthy creatures' that they were, he could not but worry about the biological consequences of female emancipation. 'What is good from the viewpoint of the individual', he concluded, 'is very bad from the standpoint of nature'.[9] These remarks are probably best read as a rare example of Head seeking in a semi-serious way to take on the public role of the man of science as prophet and seer. In private he also hinted that there was an incompatibility between the freedom to which the modern woman aspired and her biological functions. It is, however, again difficult to gauge the seriousness of such remarks.

The 1912 visit to Fordham appears to have been Head's only trip to North America. On this occasion, he found time between lecturing, discovering unique cases, and pontificating on the state of western civilization, to pay a visit along with Gordon Holmes to Adolf Meyer in Baltimore. But Head travelled regularly to visit colleagues, give lectures and attend conferences in various European countries. He told Ruth Head that a lecture he had given at Würzberg in April 1904 had been a 'wonderful' experience – technical difficulties notwithstanding. Head had arrived at the lecture theatre early:

> in order to practice with the Lantern and see how my voice carried. But the man who works it could not be found. Then there was no electric current. Then it was discovered that they had forgotten partly to darken the Hall. This they proceeded to do by covering the skylights with paper. Suddenly with a crash part of one of the skylights fell in. However at last about 9.30 I got under way and I think all went well. The work was entirely new to almost everybody there and the young men were enthusiastic the old complimentary – the most satisfactory result under such circumstances.

Head felt somewhat bemused by the honour with which his hosts treated him. It 'seemed extraordinarily odd to sit with the Geheimräthe at the Presidents table. I, a simple doctor of medicine who cannot even write "Privat Dozent" before his name. Geheimrath Kraus is my own age. He was a young man with me in Prague and has now made a great career for himself.'[10]

The experience of German domesticity also roused memories. 'All the old associations', Head remarked, 'have been awakened by that life'. The food he was offered was especially evocative: 'We have had the same dishes for Mittagessen I remember nearly 30 years ago, first soup, their "Supperfleish" with horse-radish. Beetroot ... Then some baked meat such as pigs trotters with "Sauerkraut"'. Thanks to this assault upon his senses, Head was: 'carried back to my student days in Halle when I wondered why supper always began with Soup until I learnt that it was normally called "Soupe"'.[11]

In the following August Head attended the International Medical Congress in Budapest. Harvey Cushing was also among the delegates.[12] The opening ceremony was: 'A Court function at which Evening Dress and Gown were obligatory'; an Austrian Archduke presided over the proceedings. Head somewhat smugly noted that: 'Dr Pavey and I were the only English who had the proper clothes'.[13] Academic meetings were supplemented with frequent social gatherings, which Head enjoyed despite the stifling heat. One reception, hosted by the Mayor of Budapest, was: 'a hot and terrible crush'. Head's party adjourned to a nearby garden where: 'We soon became the centre of people coming and going as we sat under the most unreal looking trees in the purely theatrical garden. It was like a scene on the Drury Lane Stage with people in Evening frocks sitting at tables in a gas lighted garden'.[14]

Despite these distractions, Head's mind constantly reverted to the previous year, when he and Ruth had toured Budapest together. 'Every step I go about this town', he confided, 'I think of you ... The cafes where we sat, the streets we walked all speak to me of you. And all the time as I chatter to that Geheim Räthe I say I love it for her sake. The success, the friendliness, the intellectual freedom.'[15]

Even in wartime, the business of international scientific exchange continued. In April 1916, Head and some of his British neurological colleagues travelled to France to confer with their French counterparts. Head was already acquainted with such leading figures of French neurology as Joseph Jules Dejerine (1849– 1917) and Joseph François Félix Babinsky (1857– 1932). The latter was especially welcoming, offering Head: 'a private sitting in his hospital' – the Hôpital de la Pitié.[16]

When the formal proceedings commenced, Head noted: 'We had a most extraordinary meeting. All the neurologists in France were there'. He found the difference in the scientific culture of the French and British participants fascinating. While: 'some of the speeches were very like what you might hear at the Roy[al]. Soc[iety]. [of] Med[icine]', others 'were so different that one had to rub one's eyes to remember that we worked on the same material'. The British neurologists at the meeting appeared to be 'inarticulate individualists' when compared to their hosts who were 'always seeking for a formula'. The French, for instance, 'fought tumultuously over the word "reflexe" which had been chosen by Babinski to express the "Maladie" he was describing'. The British participants, on the other hand, were mystified by such abstruse dialectics. Head could: 'see the impatience expressed all over the face and behaviour of my three English colleagues to whom the whole thing seem gross waste of time'.[17]

Observing these difficulties of communication between men working in the same field and living in neighbouring countries led Head to wonder how such mutual incomprehension could be overcome. The war, he surmised, might actually serve the valuable purpose of helping to bring the two cultures closer together. Because of the British medical presence, 'France is now full of insular medical men who are necessarily brought into Combat with the ironical logical Frenchman'. This obliged the Anglo-Saxons to enter into regular discourse with their Gallic equivalents. Head was convinced that: 'Nothing will bring the two together unless both unite in the world of ideas'. He felt that this general principle extended beyond the realm of medical science. He discussed these ideas with the writer Valentine Thomson who was working for the War Propaganda Bureau that the British government had established shortly after the commencement of hostilities. Head was assured that: 'this is the line on which they are working'.[18]

Despite his cosmopolitan leanings, Head was firmly embedded within the community of British neurology. The links he had with colleagues working in areas of research related to his own were as much personal and sociable as purely intellectual. In September 1904 he joined a party that included Victory Horsley and Farquhar Buzzard, in a sojourn in the Cambridgeshire countryside that provided the opportunity for social intercourse and physical recreation as well as scientific discussion. Most of the neurologists present had brought their wives to the gathering. The morning of the twenty-fifth included: 'a hard day's working but some very pleasant shooting'. The previous day the 'men had exhausted themselves at [the racquet game] fives'. They nonetheless sat up late, after the ladies had retired for the night, and: 'had a really good talk on a neurological subject which interests us all – nearly all talked well and I got an insight into the trend of English neurology (which shows me we are all on the track of an entirely new conception of one part of the brain.)'.[19]

Head took evident delight in such opportunities to engage in animated debate with others who shared his intellectual interests. While in Cambridge for a conference in May 1907, a '"little talk" with the Regius Professor [Thomas Allbutt] on Saturday morning turned out to be a tremendously animated conference in which six of us talked, argued, persuaded until nearly half past one, two and a half hours talk'. These conversations spilled over beyond the formal sessions of the conference. Even after the participants had dined, there was: 'further talk in the beautiful Caius combination room until eleven. It was a debauch of talk saved by its variety and the living interest of the subjects discussed'. The following day, Head visited the home of the physiologist, Hugh Kerr Anderson (1865– 1928): 'where again I was exposed to a bombardment of criticism and had a most useful discussion'.[20]

Head's enthusiasm for scientific intercourse did not, however, consume all his energy. His passion for life and sociability also found other forms of expression. When in London he sometimes took the opportunity to engage in society. In June 1922, Ruth was obliged to wait patiently at home until her husband returned from a 'man's dinner in town'.[21] It is, however, apparent that he participated in such gatherings with a considerable measure of ironic detachment. Thus at a dinner-party he attended in June 1912, Head noted that: 'the conversation was of the sloppiest character'. He and a friend who was also present took evident pleasure in discomfiting their fellow diners: 'I think [Everard] Feilding and I startled them "blue"; for he railed and mocked the psychical experiences and I upheld the value of poetry as a statement of psychical processes and as necessary to sound feeling and thinking'. Various incidents of 'psychological' interest were also worthy of notice – for instance that: 'Lady Johnstone evidently fell a victim to Feilding's cold impishness and overwhelmed him with advances but fell back each time as from the entrance of a glacier'. Nor could he altogether cease to be the clinician, remarking that: 'On my other side sat a poor little woman covered with beautiful jewels, the hideous victim of Congenital Syphilis – forehead, nose, scars, teeth and all'.[22]

When he could, however, Head escaped the metropolis for the pleasures of the countryside. Shooting was a pastime in which he took particular pleasure. It enabled him to combine his taste for physical exertion with the opportunity to enjoy the beauties of landscape. Head's accounts of such expeditions roused him to lyricism:

> As we turned out after breakfast a gleam lit up the castle and the Copse below making the bare trees look as if they were transparent – as if they were drawn out of glass. Then the gleam fell upon the arm of the sea and it shone like a silver mirror. It was very cold – The woods were as I always love them still and bare so that every tap of a beater's stick and the patter patter of the running pheasants was carried for great distances. In the long quiet waiting only broken by the distant report of a gun life seemed very quiet & peaceful & the fret and fever of a London life very far away & very trivial.[23]

The experience of shooting produced something more than a soothing effect. When out hunting rabbits with his father in 1900, Head, ever attentive to his own psychic processes, noted that: 'As rabbit after rabbit turned head over heels and lay still in the heather there came a sense of power'.[24]

On at least two occasions, Head indulged in this sport when he spent his Christmas break at the Merionethshire home of John Rose Bradford (1863–1935). 'Here', Head declared, 'we are a party of young men entertained by a dear kind maiden aunt'. He had, indeed: 'never been in so non-feminine a house. We talk, we laugh and she "loves to see us so gay"'. But when he returned to his room and perused Ruth's letters, Henry's mood changed:

The very scene of them reminds me of a woman's room with a fire a large sofa & a sweet bed. Gay clothes lie about spread out for her to choose, the blue, the orange tawny with cloaks, and fur for she is going out. I see her run into my arms and can recapture the thrill that rushes over me when I hold the quintessential woman who has taught me that there is something more than the gaiety of assembled men.²⁵

Because Ruth was unable to join him on these trips, Head sought by means of word pictures to communicate something of what was clearly an exhilarating experience. 'Every morning', he wrote, 'we see the sun rise but today it was a glory of blue rolling mist with a flecked sky overhead ... The air is extraordinarily clear and still. Every little Cottage perched on the hill is sending up a column of blue wood smoke'. The overall impression reminded him of past trips to Bavaria. He spent the day firing from the side of a steep hill; one bird that he shot, 'fell nearly 200 ft down the side of the hill'.²⁶

Only his dreams spoiled this rural idyll. One night Head:

had a nightmare that some evil power was overwhelming me with things to be done. Telegrams to answer, people to be found, ladies to be dealt with and far above you sat enthroned. I could see your head and your hair dressed with a band and your red coral earrings. By these I knew it was you but I could not get at you and you did not see me.²⁷

Even in these Arcadian surroundings, Head could not entirely escape the worries that awaited him in the metropolis. He took comfort, however, from the knowledge that Ruth was also awaiting him in London.

Head enjoyed another shooting holiday at the Bradford's Welsh residence in December 1912. This time he engaged in some ethnographic observation that recalled the reflections on the state of the Bohemian peasantry he had recorded in 1886 (see Chapter One, pp. 46–47). The Bradfords were wont to play lord and lady by distributing gifts to some 130 local residents on Christmas Eve. Head cast a cold, clinical gaze upon these natives: 'On the whole they are small and weedy – a dying race'. But he was taken aback by the spirit of this physically unprepossessing people. When a fiddle was produced:

the whole audience sang the rough Coachhouse was changed to a place of vivid emotion. First they sang an old Welsh hymn and then the National Anthem 'Land of my Father'. No people brought up amongst outbursts of such intense emotion could ever live happily in our great Cosmopolitan world. At such moments they are at home, in their own Country, Lords of the World and we are outsiders, inferiors. It was the High art of the conjured and ineffective race but high art that could not be communicated and was their own secret.²⁸

On Christmas Day, after a morning spent shooting on the rainswept hills, Head returned 'wet but happy'. In the evening he attended a carol service at a nearby church that again gave him the chance to appreciate the musical talents of the local population. Members of the congregation would step out spontaneously to take up a tune – 'Then away they went with that most remarkable singing with odd intervals and harmonies which must go back to the time before our scale came into use'. Once more, Head felt that he had been privileged to glimpse an ancient way of life that had all but vanished from the earth. He had witnessed: 'a survival from the days before religion was divided into sects. I would not have missed it for anything'.[29]

Describing a subsequent shooting expedition, Head dwelt on how different these atavistic inhabitants of Merionethshire were from their English equivalents, and upon the fact that these natives seemed somehow an integral part of their environment. 'The Welch', he noted,

> are silent beaters; they don't bellow and tramp like our English labourers. They are keen to kill but [not] to follow up game and often you do not know that they are close except for the gentle tapping of their sticks which seems a gentle woodland noise and not the oncoming of an invading host.[30]

During these years Head and his wife also took holidays in Cumbria and at Malvern. On occasion they ventured further afield. Thus in July 1910, Head embarked on an extended stay in France. He set out alone; Ruth was to follow later. He had a difficult channel crossing, and his train into Paris was delayed thanks, in part, to his fellow passengers: 'The Americans with their enormous trunks made it difficult for us to get away and I was not in my hotel till nearly 6.30'.[31]

Once he had deposited his luggage, Head was quick to take advantage once more of Parisian high culture. After an al fresco meal on the Champs Elysée, he attended a performance at the Renaissance Theatre. The play was Fernand Wicheler and Frantz Fonson's *Le mariage de Mlle. Beulemans* – a study of the cultural gulf between the French and the Belgians. 'All Paris', Head noted, 'has rushed to hear the Belgian French'. He was fascinated by the issues of national and class difference around which the plot revolved. On the one hand, were the rich, but vulgar, Belgians: 'On the other, the young Frenchman of today, somewhat timid! Very highly educated, talking extremely fast but with an exquisite enunciation. It was exactly the contrast [?] between the Londoner, professionally educated and Lancashire or Yorkshire business people'.[32]

For most of his vacation, however, Head eschewed the metropolis in favour of a taste of provincial France. He returned to the Loire valley where he had previously enjoyed a cycling tour. Head stayed with a local family at the Chateau de Dieuzie at Rochefort sur Loire. His hosts took an indulgent attitude to the shortcomings of Head's conversational French. 'Our life here is', he rejoiced, 'a very simple one and I have already reached the stage when I cannot bear to be in a town for long'. Head rose at 7.30 and took: 'a douche in the basement. During the process I am absolutely exposed to anyone on the terrace but as no one ever come there at that hour the sun and water both combined only add to one's pleasure'. After coffee, he read for a while then took a 'short bright ride' before lunch.[33] He found his surroundings enchanting: 'The country', he marvelled, 'is a real poem. The Loire always reminds me of a great quiet highway which passes by towns villages and spreads out round village greens'. Indeed, he assured Ruth: 'this country is beautiful enough to be the setting even of our love'.[34]

Head found the nights especially alluring; they were: 'marvellous with a dark blue sky and stars but of course no moon. Bats and owls flit across the silence which is intense. For both our arms of the Loire are slow silent, streams. Occasionally, in the distance, one hears a bell like note of the edible frog'.[35]

Head had brought some work with him and he made himself known to the local medical dignitaries. But he was more intent on enjoying some of the major social events that the area had to offer. He had evidently developed an interest in horseracing and sought out a local meeting where he: 'spent a day of completely delightful boredom'. In the morning, 'I watched the racehorses gallop, talked to the trainer from Paris, his wife and his Belle mere'. In the afternoon he fell in with a group of other visitors and was refreshed by their lack of intellectual pretension. There was no talk of 'ideas'; instead: 'we talk about horses, dogs, jockey, the garden, the table, the household endlessly'.[36]

He returned to the races on 15 August when he attended a meeting held in the grounds of a local chateau. Head took the opportunity to indulge in some social observation, noting that both the upper class men and women present were dressed in clothes indistinguishable from those seen in England. He did, however, discern something characteristic in the physiognomy of these French aristocrats — namely the 'discovery' that:

> that curious colouring, the arched eyebrows the very unenglish sad look of Miss Courner [?] is exactly what you see over and over amongst the aristocracy in France. A kind of almost deprecating sadness associated with intense self possession. Two women in the Duchess's party today might have been her sisters.[37]

He was 'terribly upset' when a favourite horse cut its back while jumping. Assuming his clinical mantle, Head: 'went into Angers this afternoon and after Consultation with the Pharmacier brought back a fluid that may make his cut bearable, but I fear he must be scratched'.[38]

When he learned that there were to be some 'interesting fêtes' at Nantes, Head determined to witness the promised: 'Tournaments, a "Cour d'Amour" a Torchlight Procession and other historical representations'. After arriving in Nantes, Head passed the time making light conversation with the local ladies: 'we talked all the time about the infamy of other people of the books of the actresses'. One woman, on learning Head's profession, 'asked me what we did for

Rheumatism from which she was a sufferer and I recommended a diet, in the manner of the French physician!'[39]

Head maintained that he owed his conversational facility to his father, who had possessed an: 'extraordinary power of talking to everyone'. Head claimed that: 'Nobody bores me'. If a man with whom he was conversing became fixated on a single subject, 'I change the conversation and generally can get him going on something I want to know'. In the case of: 'middle-aged women, the only ones who address me, I find the simplest plan is to say something outrageous but true and immediately we become on a footing intimacy. At the same time, I learn quantities of words and have begun to make phrases'.[40]

Finally the fête began. Around 6p.m., 'a most amusing fool, "blageur" of the first water carried me off to the tournament …' Head discovered that some Welsh performers had made their way to the event, much to the amusement of the natives. These: 'poor welsh people wandered round like little lost sheep mocked by the real French or looked on like Red Indians. One poor creature had come all the way from Wales to wear a tall hat of Welsh costume … This made my French lady shout with laughter for at bottom the Parisians are really cruel'.[41]

Head did not entirely forego serious intellectual engagement during this stay in France. He spent a morning touring the hospital at Angers and lunched with the chief surgeon and his wife and nephew. 'We talked', he informed Ruth, 'of the philosophy of [Henri] Bergson who appears to have an enormous influence over the [young people]'.[42] Head was sufficiently intrigued by what he learned of Bergson's thought later to seek to learn more by reading one of his English expositors.[43]

Not all Head's excursions were so agreeable. After an outing with his hosts and some of their friends to an island on the river, he exclaimed: 'I had no conception that any party could be more unpleasantly arranged'. While the men sat and fished all afternoon, Head was left to sit under a nearby apple tree in the company of the ladies of the party. While he found the arrangement less than ideal, Head claimed that the experience had provided him with novel insights into the psychology of the French female. For five solid hours, his companions: 'mocked as only a French woman can mock her mankind'.[44]

While the men sat oblivious, immersed in their angling, 'We discussed every possible subject with that astounding frankness given only to the French woman'. Head confessed that: 'If I were younger and in my present stage I should be terrified of the Frenchwoman'. As it was, he felt sufficiently self-assured to enter into their discourse; he knew how to 'say the thing which "goes one better" and I have caught the trick of saying it so as to interest her'. Even though the 'middle aged French woman despises all men and mocks them all the time' (including, one presumes, their English visitor), they could be manipulated by a psychologist as adept as Head. The French woman:

rises at once to our unusual phrase, to some observation which strikes her as unexpected. In order to interest her, the answer must be different from that which she expects. Then she turns herself with your arm and there is nothing you may not hear. The three women and myself had a most amusing time and I succeeded in being paradoxical without being absurd.⁴⁵

While he enjoyed the horse races and other diversions of the neighbourhood, Head was excited by the prospect of one event in particular. On the 27 July, he told Ruth that: 'at the beginning of next week. I shall attend an aviation meeting at Nantes. At last I shall see real flying'.⁴⁶

For Head, as for many of his contemporaries, the advent of powered flight marked an epoch in the history of mankind.⁴⁷ Aviation represented a union of technological modernity and of the individual élan of the pioneer pilots that Head found as uplifting as any of the works of art in which he rejoiced. Unlike H. G. Wells and others who were worried by the destructive potential of the aeroplane, Head was among those who viewed this conquest of the skies as a triumph of the human spirit as well as the touchstone of the modern era.⁴⁸ His imagination had been excited by the accounts of pioneering flights that he had read; his visit to France provided him with the opportunity to witness these momentous developments at first hand.

In the account of the beautiful summer day when he saw his first flight he composed for Ruth's benefit, Head strove to convey the epic nature of the event:

> Imagine a space about half as big as the Regent's park – perfectly flat and covered with grass. In the centre an enormous mast like the fighting mast of a man of war covered with signals and with the white flag as the mast head. This meant 'flying is probable'. This vast space is surrounded by a palisade and at various points stands and restaurants had been erected. At the north end were the great green sheds of the aeroplanes each like a miniature Charing Cross. One good omen was that all but two were open in front so that one could see the head of a monoplane or a biplane. This meant that all but two intended if possible to fly.⁴⁹

Although the weather was fine, there was rather too much wind for comfortable flying. The audience was as result obliged to wait three hours before there was any movement. Then, around 6p.m.,

> a monoplane was slowly wheeled across the interminable plain. It was [Léon] Morane [1885–1918], the crack flyer and on looking at the mast, we saw he was off for the Cross County Prize. Three 'mecaniciens' pushed the tiny monoplane which looked rather like a large Cockchafer for at least a kilometre, then it was slowly turned and through my glasses I saw the three men wait for orders. A motor car dashes across the plain. A man and woman arm in arm walk slowly to the instrument. He draws her to him and kisses her firmly on the mouth. Then he slips into the body of the instrument, the mechanician in front gives the screw a whirl, a cloud of petrol vapour and

the thing has risen. It rises rapidly against the wind and dashes away to the eastward at a terrific height. At one moment he actually rises above the clouds that hang like [?] far in the Eastern sky. We are left gasping and applauding.[50]

Once Morane was airborne, several other pilots took off, allowing Head to appreciate the attributes of the various models of aeroplane on display. 'The biplanes', he noted, 'are clumsy slow and fly low'. The monoplanes, on the other hand, made more of an aesthetic appeal; they were: 'exactly like the Eagles or the Rook. They fly high, swiftly swoop and turn like a bird'. Head was also attentive to the techniques these aviators employed, noting, for instance, how they released a flight of pigeons – 'in order that the aviators might watch the birds dealing with the wind'.[51]

Around 7p.m., as Head was preparing to leave,

> a black speck appeared in the sky now reddened by the after glow of the setting sun. It looked like a speck, like a rook, but no rook could fly as high. Then the curious beetle like appearance of the Bleriot monoplane became evident and we yelled with delight 'Morane, Morane'. It was Morane back again from his Cross County Race. He came too much to the North at a terrific height. Cut off his engine and swooped like a rook into the lower levels, threw in his engine again and landed in just over an hour having cleared over 50 miles. The swoop from the clouds to the lower levels was the most splendid and birdlike thing I have ever seen.

His letter, Head admitted, was 'full of the flight of men'. Yet he saw a link between his ecstasy at the sublimity of the events he had recounted with the enduring passion he felt for his correspondent: 'somehow my mind has warmed [the letter] up with our joy. In fact all joy is now bound up with my love for you'.[52]

During the war, the Heads were to witness the other – terrifying – side that Wells had foreseen of the new flying machines. By 1915, German zeppelins were conducting regular raids on London exposing its inhabitants to a new and unprecedented menace that took a psychological as well as material toll. In June of that year, Ruth confided to Janet Ashbee that: 'It may be the Zeppelins or thunder in the air or cold winds bustling under leaden skies but I feel desperately slack and stupid'. The previous night, 'A whole boot factory was destroyed'. She found herself wondering: 'Will it be Montagu Square tonight [?]'[53]

There was, however, something of the sublime even about these monsters of modern warfare. Another raid later the same month was: 'the most thrilling experience'. Ruth and Henry had been busy with their correspondence before retiring to bed when:

> the most awful clap of what I took to be thunder rent the air. H. said: That's them and the bombardment began. We hang out of the study windows quite pleased and detached till Pratt en déhabille appeared saying 'Excuse me, sir, the view is so much better from the servants' bedrooms'. So up we went and there was the great golden

cloud into which the Zeppelins had just retired. The light came however not <u>from</u> it but from the searchlights of our antiaircraft men. Oh, the pity that they did not catch the brute, for the wreckage has been considerable at Bishopsgate Station, in Broad St: and Wood Street ... and in Queen Square where the Epileptic Hospital is but where no patient had so much as a fit. Bus No. 10 was blown up completely with all its contents.[54]

The Head household was obliged to adapt to this new normality. Ruth reported that: 'We have all clothes suitable for an air-raid laid out by our beds every night, buckets of water on every landing and an uncanny feeling of expecting the worst which yet does not prevent our all sleeping serenely'.[55]

The war impacted on the Heads in may other ways – spiritual as well as material. Both Ruth and Henry had for most of their adult lives been ardent Germanophiles. They now felt obliged to repudiate those former allegiances in the most forceful terms. Ruth wrote to Janet Ashbee that: 'I feel about the Enemy just as the Paris fishwives felt about the Aristocrats in the Revolution – I really do: bloodthirsty and bacchic'.[56] On learning later in the war of alleged atrocities committed by German soldiers, Ruth revealed how her outrage was coupled with a real sense of loss: 'It is impossible not to mourn even now over the fairy dreams of Germany as it was ... But any way we have vowed never to take one of them by the hand any more nor ever to forget these bitter wrongs'.[57]

The Heads were almost as incensed by those of their countrymen who questioned the justice of the war – including such long-time acquaintances as Bertrand Russell. '<u>No</u>', Ruth wrote in December 1916, 'we both say we can't do with Mr. B. Russell or any of his works. H.is <u>more</u> irrident even that I am. He really has done England an immensity of harm in America'.[58]

Moreover, as the toll of casualties rose, the true cost of the war became ever more apparent. As early as November 1914, Head reported that: 'I met poor Norman Moore in the street this afternoon looking distraught. He has lost his second boy at the front. What a toll the professional classes are paying! Herringham has lost the charming boy I met there at dinner in June'.[59] Many of Ruth and Henry's relatives had joined the armed forces adding to their anxieties. In February 1916, Ruth lamented that: 'A splendid young cousin of mine, a Captain in the Oxon and Bucks has been horribly shattered. He will live they say, but as soldiering is his dream and romance, his life as a crippled civilian will be unendurable to him, I fear'.[60]

The war struck, moreover, at the heart of Henry's family. In 1915, his younger brother Bernard: 'fell in that death-hole in the Dardanelles'. Bernard had been: 'the baby of Mrs Head's big family the youngest of her eight boys'. The report came as Head was recovering from an illness. Nonetheless, he took the news stoically, insisting that: 'he never for a moment thought Bernard would be spared'. Ruth allowed herself to mourn the loss of one who had brought: 'so much life and gaiety and youthfulness into the house'. She also reflected that of the eight boys that Head's mother had borne only three survived. It seemed: 'such a pity. They were such a fine stock and there is only one tiny frail male Head from all those children'.[61]

Head had no religious faith upon which to rely in the face of such adversities. But, when in March 1915 Ruth's mother became gravely ill, he urged her to seek succour in the resources of a secular faith: 'Oh! Let us pray for Courage and that Knowledge of Living which dominates even death'.[62] Head displayed a similar stoicism when in 1922 his long-time collaborator and friend, W. H. R. Rivers died suddenly. Siegfried Sassoon accompanied Head to Rivers's funeral service at Saint John's College in Cambridge. Sassoon broke down during the service and felt that: 'No doubt Head would have liked to do the same thing, but he is master of his lachrymal glands'.[63]

After the service, Head: 'talked about the necessity for our world to reconsider its traditional attitude towards the facts of death'. When confronted with the loss of one of his closest friends, Head was moved to reaffirm a secular ethic, insisting that:

> Nothing exists except life. Rivers has not died with his body. *'Death is swallowed up in victory'*, says *Corinthians*. But *'when this mortal shall have put on immortality'* must be applied to this and not to the mystical idea of any 'next world'. The burial service asks us to accept an escape from life. But life is the only real creative evolution. The Church's 'resurrection of the dead' is a divine dose of dope for people who have not the courage to face the facts of death.

Sassoon concluded that the time he spent with Head after Rivers's funeral was: 'an evening of perfect human sympathy. There was no false vibration. Death was swallowed up in victory'.[64]

Head recognized that the war had offered him opportunities to pursue his researches in ways that had been impossible in peacetime (see Chapter 3 above). Some of the suffering and loss was thus redeemed by service to science. Moreover, thanks to his work with traumatized soldiers, he made significant new acquaintances who were to loom large in his later life. The most prominent of these were Robert Nichols and Siegfried Sassoon. Through Sassoon, Head also met Robert Graves. Head's relationship with Graves was, however, never as close as that he enjoyed with those other two 'war poets' – although the two did share some intellectual interests (see Chapter Five, pp. 218–19).

Nichols had originally been a patient (see Chapter Two, p. 88), but was soon treated as a friend. When in March 1917, Head needed a quiet and secluded place to complete a paper, he went to stay with Nichols and his friend, the architect, Henry T. Lyon, in the village of Ilsington in South Devon. Head's journey was enlivened by sharing a carriage with a nurse and her charges. One of the children, Head noted, 'was a fascinating infant with a long head and the wildest curls but passionate to a degree that was frightening'. The nurse was concerned about the girl's mental state and plied Head with questions. He found the case interesting:

> For this child was a complete victim of words. Evidently visual impressions made little effect on her but you could alter her attitude by a word. She would refuse Cabbage absolutely and told me 'cabbage is dirt'; but call it greens and she would eat what she had previously refused. She absolute refused a sort of milk jelly that had been brought for her midday meal but liked it when it was called a soup and asked for more.

When they parted at Newton Abbot Station, Head claimed that the nurse: 'was truly grateful, for she confided to me that she had dreaded the long journey'.[65]

Nichols and Lyon met him on the platform and conveyed Head to the cottage where they were staying. Taking stock of his former patient, Head was pleased to note that: 'Nichols is marvellously normal ... His enthusiasm is as great but he has become orderly and reasonable to a degree I could never have expected'.[66] The tedious work of completing the paper was relieved by long walks with his companions. The cottage was, according to Nichols, 'a pleasant house: high up on the hills. Dartmoor is outside the porch, so to speak'.[67] Nichols also entertained Head in the evenings by reading aloud to him 'most beautifully' the poems of John Donne.[68]

Head was able to relate to Nichols and Sassoon as a fellow poet long after they ceased to be 'cases'.[69] He felt that these war poets had: 'written some astoundingly good things'.[70] They, in their turn, valued Head's talents as a literary critic as well as the insights he might offer as a trained psychologist. Thus Sassoon was sufficiently encouraged when he learned that: 'Head likes my "psycho-analytic" verses', to seek to publish them.[71] As part of these efforts to produce verse that was informed by psychological insight, Sassoon tried to get to grips with Freud's ideas. His initial conclusion was that: 'Freud can't see straight about sex, but he has discovered a lot about the mechanism of the human mind'. A walk through London with Head in June 1922 gave Sassoon the opportunity to discuss: 'Freud (or rather he elucidated my Freudian perplexities as far as possible in a distance of half a mile's perambulation)'.[72]

Head, for his part, seemed to take genuine joy in the role of mentor in which he was cast. He told Sassoon that he and Ruth were: 'were delighted and full of pride at the receipt of your beautiful little book of poems'. Head had seen most of the poems in proof; one of them:

> I kept ... in my desk and often quoted it – especially
> 'But thoughts are kingfishers that haunt the pools
> Of quiet; seldom seen; and all you need
> Is just that flash of joy above your dream'.[73]

Ruth confirmed the strong emotional impact that Sassoon's verse had on her husband. Henry read aloud to her 'Grandeur of Ghosts': 'in the strange muffled voice of one who reads through tears he fights with – "Lamps for my gloom, Lamps guiding where I stumble"'. Another poem in the collection: 'made him choke again'. Sassoon's fruition as a poet evidently gave both Ruth and Henry great satisfaction: 'We are quite swollen with pride in you, dear Siegfried'.[74]

In later years, even while striving to finish *Aphasia and Kindred Disorders*, and with his health failing, Head found the time to assure Sassoon that:

> I am delighted with your Poem, 'A Short Story'. Technically, I think you have managed to vary the rhythm of the lines in a wonderful way in consonance with their emotional value. With regard to your method of telling the story I admired your economy of words, and the selection of a moment outside the door of a room as its central point. Psychologically, you are certainly right, for after the talk of the night before, there is little doubt but that she would wake early in the morning with the aftermath of the emotional stress.[75]

Over and above such shared intellectual and literary interests, it is clear that both Nichols and Sassoon felt a genuine loyalty and affection for Head. Sassoon exclaimed after one 'very jolly conversation' with Head: '*What* an extraordinarily comforting man!'[76] Sassoon was always glad of Head's company. In June 1923, while wandering aimlessly by the banks of the Serpentine, he 'caught sight of Henry Head on the sky-line, tripping along in his baggy blue suit and looking more like an ex-sea-captain than a famous scientist. This was very pleasant, as I was feeling lonely ...' The two walked together and talked about Walter de la Mare's new book, *The Riddle*. 'Henry Head', Sassoon admitted, 'made me feel a callow youngster again'.[77]

In the years following the war, Sassoon, the Heads, and their mutual friends met frequently at Montagu Square. Sassoon gave the flavour of Head's table talk over dinner at one such gathering – at which his host was: 'very skilful and urbane' in managing the wayward utterances of a fellow guest. After holding forth on the psychological differences between the sexes, Head launched into a critique of the Freudian conception of the springs of human action. 'You react', he declared, 'with your totality (sex, intellect, everything). No good talking about an emotion being "due to sexual causes". If it *is* an emotion (and not crude automatic lust) you react with your totality to the phenomenon of causation'.[78] This notion of 'reacting with our totality' was one that Sassoon took to heart.[79]

The Heads were sometimes also received at Sassoon's residence in Tufton Street. Ruth and Henry mischievously hinted at the character of these gatherings when they sent Sassoon a postcard purchased at the British Museum depicting: 'Two Yogis (ascetics), with two other devotees' – with the inscription: 'A Symposium at 54 Tufton Street'.[80]

Sassoon in September 1924 joined the Heads at their holiday home at Malvern; he noted that his hosts were 'bubbling with delight at my arrival'. During the mornings, Henry stayed upstairs – 'working on his big book'. But he found time to read a German translation of Sassoon's poem, 'Concert Impression', which Head found to be: 'brilliantly literal'. One afternoon the two drove out to the Iron Age fort known as British Camp where Head: 'told me a lot about its postulated antecedents'. The three then had 'a homely sort of evening'. Head showed his guest a book on woodcarvings in churches; they then discussed the novelist, Anatole France. The next day Head took Sassoon around Malvern's Priory Church and showed his erudition by 'explaining the stained glass and wood-carvings etc'. Over dinner, the conversation turned to the topic of the supernatural. Head was: 'of course scientifically rational about it all. Very wholesome'.[81]

In the post-war period, Head continued to take an interest in the performing arts. In June 1921, together with Sassoon and Rivers, he attended a performance of Clemence Dane's play 'A Bill of Divorcement'.[82] The following July Sassoon wrote of a Schubert recital at the Queen's Hall. 'Most of the musical Teutons of this metropolis', he noted, 'were there', as well as: 'a lot of other people ... including Henry Head, who always looks more German than any German band that ever tootled'.[83]

Head's own account of attending a performance of Robert Nichols's *Guilty Souls* in December 1924 is preserved in a letter he sent to the author. After the performance, Head met the actor Ernest Milton, who played 'Paul Vyson', the lead role in the play, and 'and thanked him in your name for a most remarkable performance'. Head was particularly impressed with what he regarded as the psychological accuracy of Milton's acting: 'In the first two Acts of your Play, he made Vyson a typical neurotic of a highgrade type; later he brought out the man's implacable and obsessional ideas of spiritual revenge'.[84] In the following February, Ruth told Nichols that: 'The only really naturally acted play we have seen recently was VORTEX by Noel Coward, that wonderful boy of 24, and that he produced entirely himself, besides acting the hero'.[85]

The diaries that Sassoon kept in the 1920s show Head moving in the same literary circles as his poet friends. Thus when in January 1921 the novelist Arnold Bennett (1867–1931) held a farewell dinner for Nichols, who was about to take up a teaching post in Tokyo, the guests included Head and Aldous Huxley (1894–1963).[86] Huxley, along with John Masefield and the Sitwells, was also present at a wedding in July 1922, at which, Sassoon noted, 'the Heads were beaming'.[87] In March 1921, the Heads met: 'Lytton Strachey of the Eminent Victorians fame at dinner at his Publisher on Monday – such a nice shy, soft-voiced dear Thing'.[88] Ruth and Henry were later to presume upon this acquaintance to send Strachey a recondite query about the connection of Queen Elizabeth I to the Visconti.[89]

The Heads were sufficient of a presence in literary London to attract the hostility of at least one eminent man of letters. Sassoon averred that Head preferred to avoid such animosity, seeing it as self-defeating: 'As Head says, when you hate people you put yourself in their power'.[90] The poet and critic, Edmund Gosse (1849–1928) was, however, to oblige Head to make an exception to this rule.

Gosse was known for his abrasive ways. Sassoon noted that: 'E.G. cuts Turner at the Savile. He can't bear H.G. Wells. He has been rude to the Heads …'.[91] Two incidents in particular seem to have sparked the quarrel between Head and Gosse. Head had invited Gosse to dinner in order to meet the Belgian poet Emile Verhaeren (1855–1916), who was staying at Montagu Square. After the meal, as Head was showing him out, Gosse remarked: 'What a pity it is that Verhaeren doesn't know the right people!' This was, according to Sassoon, 'a remark which H.H. can't forget – naturally'. Sassoon was, however, also close to Gosse and had some sympathy with the sentiment, if not with the manner in which it was expressed. He could imagine how 'dear Ruth got on [Verhaeren's] nerves with her loquacities'. On another occasion Gosse and the Heads had been dining together at Gosse's house (presumably before the Verhaeren incident) when:

> Ruth, chatty and exuberant, upset her wine-glass and only saved it from spilling by dexterity. 'How lucky!' she exclaimed to E.G., who chillingly replied 'Yes, indeed. Most providential'. Poor Ruth looked crushed; and Henry, of course, would have been furious at such rudeness.[92]

So incensed was Head that Sassoon was astonished to hear his 'kind and sagacious mentor' declare: '"I should like to write *the book of the year*, so that I could cut Gosse". (The only time I've heard him speak bitterly of anyone.)'[93]

Because he wished to remain on good terms with both parties in this feud, Sassoon was sometimes placed in an awkward position. He recounted a farcical incident when he mistakenly delivered the copy of his book *Lingual Exercises* intended for the Heads to Gosse's house. Sassoon attempted to retrieve the book by stealth and substitute the copy dedicated to Gosse. The attempt at deception, however, failed miserably.[94] Sassoon finally dropped the correct volume through the Heads' letterbox and stole away unnoticed. Henry remained oblivious to the manoeuvres and machinations that had surrounded this seemingly simple transaction. He was entranced by the content of the gift, proclaiming: 'I am astounded at the way in which you have expressed complex emotional states in what appears at first sight to be the simplest language; not a word is out of place, not a word is redundant. I hope the inspiration is still flowing along these lines'.[95]

During the war, Head had become something of a man of affairs. He mixed with senior government officials and served on government bodies such as the Medical Research Committee. He was, in November 1916, appointed to an official committee to enquire into the position occupied by natural science in the British educational system; the committee reported in 1918.[96] During the 1920s he retained something of the aspect of a public figure, dining with the Japanese ambassador and mobilizing the Royal Society in an effort to replace books destroyed when fire ravaged the Imperial Library in Tokyo.[97] In December 1926, Head's status as a member of the establishment was formalized by the bestowal of a knighthood by Stanley Baldwin's government.[98] The official announcement stated that Head had been thus honoured: 'For public services. Has made distinguished contributions to knowledge of the nervous system'.[99]

Invitations to lecture on the Continent began to flow again. In May 1921, Head spoke to a group of around fifty people at the University of Leyden; he was pleased that they: 'seemed to understand and take the points'. After the lecture, he drove around the town, noting that: 'The tulips are over but the whole country is high with white and red may, white and red chestnuts and trees of every colour and greenness. We passed the village where Spinoza lived and ground his glasses'.[100] The following day Head gave a second lecture to an audience of around 150 in Amsterdam. He: 'threw over the manuscript and spoke nearly the whole, with I think advantage'. The experience again brought home to him the gap that seemed to separate British neurology from that of other nations. 'I now see', he wrote, 'how extraordinarily different is our view from those of continental workers. What seems to me the simplest and most obvious facts and modes of thought seem to them entirely new and original'.[101]

Head had more than forty years before passed through the Netherlands on his way to Halle. It seems that after 1914 he never returned to Germany – the country he had once regarded as his second home. The initial anger and sense of betrayal he had felt at the outbreak of war appears, however, to have abated. He even saw signs of hope in post-war Germany. In February 1924, Head told Robert Nichols of 'the movement in Germany called the "Jugendbund"'. Head found this movement:

> to be a very interesting one, especially in Germany, where they are founding schools apart altogether from the State and devoted entirely to new methods of teaching. Amongst the younger members there is a complete revolt against all the older ideas of Nationalism, Militarism, and all that we used to look upon as the German System. The young people travel about the country, clothed in the simplest manner with a rucksack, and their one passion seems to be, open air, music and companionship.[102]

Ruth Head, on the other hand, appears to have emerged from the war embittered. Her rage, as expressed in letters addressed to Janet Ashbee, was not directed solely at the Germans; the Irish and even the Poles attracted her indignation. Nor did the Dutch escape her contempt. When she learned of Henry's proposed visit to the Netherlands, Ruth expostulated: 'Why make up to that odious race? I asked Mr Walter of the Times. "They are so rich" he answered. Pigs'. She was no less incensed by the rising aspirations of the labour movement in Britain, declaring that: 'the greedy miners are odious people and the Proletariat generally rouse my bitterest ire. Protesters are also detestable'.[103] Ruth seems to have expressed such views – as well as her concerns about the state of the household finances – only to her trusted female friend.

For others the Head home appeared as an oasis of humour, erudition and welcome. While working in Tokyo, Nichols often found himself pining for the delights of Montagu Square. He painted an evocative word-picture of what he was missing:

> Along Edgware road a haze of ... [?] light ascends, Henrietta kindles her last pre-dinner cigarette & Henry, going to the window, remarks four more sheets of the Magnum Opus completed while both forbear to switch on the lamp ...
>
> Henry & Henrietta, the quiet, the laborious, the tranquilly hopeful, have just switched on the light. Aye, there is the soul, or rather all that is best in the metropolitan soul for me: the quiet study, the lighted lamp, cigarette smoke, sheets of the day's labour laid aside & after dinner talk yet to come! My darlings when shall I see Sig. and you again?[104]

In the years between 1904 and 1926, Head lived what in a letter to Ruth he had called 'a life of high passion'.[105] His commitment to his clinical and scientific work was undiminished. But this was combined with exultation in the beauties of the natural world and in the products of human art and intellect. Head felt that he lived at the dawn of a new age of unprecedented achievement, for which the aeroplane was an outstanding metonym. Even the trauma of war failed to dim this enthusiasm. A more personal calamity was, however, to bring to an end the life that Ruth and Henry had with so much toil and such ardour made for themselves.

This Foul Disease

In January 1920, Ruth wrote to the Ashbees in response to an invitation to visit them in Palestine. 'We <u>long</u> to come to Jerusalem', she affirmed, 'but our future lies all misty before us. Were you here I should enlarge on this point and confide largely, but I will be discreet on paper'.[106] This was an early hint that, beneath the apparent exuberant activity of the Head household, all was not well. An indication of the nature of the cause for concern can be found in a letter from Head dated 31 August 1921. He told Ruth that, after a busy day, 'I enjoyed an excellent dinner and [George] Riddoch has been sitting with me since'. The meeting with

Head's protégé had not been entirely social. Head confided that: 'I have discovered all about my little illness which I will tell you when we meet. Meanwhile it is rapidly passing away and I shall sleep well tonight'.[107] Head did not specify the nature of this illness; the fact that he chose to consult a fellow neurologist, however, hints at its character.

It is evident that Head was trying to make light of the condition. But such attempts to allay Ruth's concerns served only to heighten her fears. In January 1924, she admitted to Ashbee that: 'I am very anxious about dear H.H., but it is not a thing I can ever talk about, because he himself never mentions his condition'. When she pressed her husband as to what was ailing him, Henry would call 'it vaguely "the disabilities of old age". But he is only 63 so it seems hard'. Despite her frustration at this evasiveness, Ruth affirmed: 'I do firmly believe that strength comes to bear whatever burdens are put upon us and, wholeheartedly, I live only to make his lighter'. She advised Ashbee that:

> You can, and will I know help by ignoring his weaknesses and acting in his presence comme si de ne rien n'était [as if nothing was the matter], above all by not discussing it with outsiders to whom I very firmly and consistently lie! You see his salvation at present is by carrying on both his private practice and his own research work and if it gets about that he is ill (which in the actual sense of the words he is <u>not</u>) he would be put at once on the shelf, which would be very bad for him.[108]

Despite such pleas for discretion, in 1926 Ruth was forced to confront 'false reports' about her husband's health that were circulating through London medical society.[109]

By the mid-1920s it was in fact becoming impossible to hide the fact that something was wrong even from friendly eyes. When staying with the Heads in Malvern in September 1924, Sassoon had noticed that Henry 'is getting rather stiff and seems unable to lift his feet'.[110] By the following year, Sassoon had to admit that while his affection for Ruth and Henry 'increases with the years', he was 'saddened ... by Henry's increasing inactivity'.[111]

It was Ruth who was obliged to come to terms with the consequences of the progression of the disease. Henry had always been physically active, taking regular walks in Regent's Park for exercise. By 1926, however, it was a matter for remark when: 'H.H. trotted out, by himself to buy the new Bridges volume'.[112] As a rule, Ruth was now obliged to accompany her husband whenever he ventured out of doors.

Ruth confided to Head's niece that:

> Our little walks in pouring rain or beating wind are the most dismal things. He has to walk and hardly can. He gets very hot and has to walk without a great coat. He has to sit on some seat to rest and often cannot walk slowly and we rush along like dements. Isn't it horrible?

According to Ruth, Head took these impediments like a 'perfect angel', while she was obliged: 'to camouflage and play the giddy goat when I feel like tearing my hair'. She sometimes thought:

> another woman would bear it all more quietly, but as you know I have always idolized Uncle H. and I cannot get used to the feeling that he has been treated unjustly. If only I could be the invalid instead. I love being petted and having a Nurse and wearing pretty wraps and not having to order dinner and settle about the sweep and pay the beastly bills, all of which would presumably have been removed from my shoulders if only I could bear his burden.[113]

By the spring of 1926, Henry's condition had deteriorated further. All attempts at unassisted walking had been abandoned; he was effectively housebound. Head spent much of his time seated in an armchair, 'having to have his legs shifted frequently and taking little walks up and down the drawing-room occasionally to alleviate the horrible stiffness'. Moreover, 'poor H.H. cannot write letters now with his poor trembling hands, so I write instead'.[114]

There could no longer be any doubt about the identity of Head's disease. Ruth told Nichols:

> It is a kind of Paralysis Agitans, called 'Parkinson's disease'. Your doctor will tell you all about it – a terrible progressive malady, which never attacks the brain, does not kill, but cuts the patient off from nearly everything that makes life worth living. Dear Robert and Norah, it is such a tragedy. To see him with a nurse, confined mostly to an invalid chair, tired by even the most charming visitor after half an hour or so and walking only on my arm with bent back and dragging feet.

The only hope that Henry's doctors could offer was: 'that with country air and lots of sunshine he will find life more bearable'. Ruth was accordingly: 'moving heaven and earth to find a pleasant house where we can live the quiet invalid life which is all he is fit for'.[115]

Ruth had in fact been seeking for some time to persuade her husband to give up their London home so that they could retire to the country. By the summer of 1925, Head was so ill that he had decided to give up his practice as of 1 October. *Aphasia and Kindred Disorders* was at last finished, marking the end of his activity as a scientist. Head still tried, however, to indulge in some of his favourite recreational activities, though this became ever more difficult. Thus, when he went to the theatre: 'Harry can manage a seat in the front row of the stalls although getting cramped and miserable unless the play is very well worth while'.[116]

In the summer of 1925 the Heads took a holiday at Lyme Regis – staying in: 'the ancestral home of the Listers, known as High Cliff'. The house was within easy reach of such friends as the Hardys and the Granville-Barkers. They were joined by Sassoon whose company, Ruth noted, had a therapeutic effect on her husband:

> Siegfried is a wonderful tonic for Henry who has revived like a rescued fledgling under the tactful administrations of our guest. You can imagine the mingling of anecdote, reminiscence, psychology, physiology, philosophy, with the theme of prosody as a side-track, varied by readings from the poems of Thomas Hardy ...[117]

The three of them visited Hardy, 'and had the merriest time, though the old man is a little doubtful of the powers of Miss Frangcon Davies to represent his Tess, most beloved by him of all his heroines'. Hardy, moreover, 'twice entrusted' his wife to the Heads, letting Florence stay at High Cliff for a few days. Head noted that Robert Graves also stopped by, although he was called away by a family crisis. Head had been keeping abreast of Graves's latest writing – 'a new version of the story of Elisha and the Shunnamite woman'. He felt that it was: 'written with tremendous go and vividness, but I think his explanations are too rationalistic, and he has left too little of the fairy tale element in the story'.[118]

Ruth took the opportunity of the holiday to try to sound out 'Harry on his future plans, longing to sell 52 [Montagu Square] and its now useless multiplicity of sitting rooms, but it is still his beau idéal of a residence ...' Nor would Head consider taking an opportunity that had arisen to spend some time with friends in Italy: 'I think he dreads the fatigue of the journey and the crushing contrast between his eager sight seeing as a young man and the restrictive field of his capabilities to-day'.[119]

Head's resistance to leaving London was to prove short-lived. Indeed, he increasingly became the passive object of decisions made by others. Ruth assumed the de facto role of head of the household. She was, for example, obliged to take full control of the family finances. Previously, although she had paid the bills, Ruth had been reliant on a monthly 'allowance' from her husband. Now Henry was incapable of even writing a cheque. She found this responsibility somewhat daunting, but rose to the challenge:

> I find it so difficult to trust to the supply of money from the Bank, having all my married life seen it flow in uninterrupted streams through the channel of Harry's trouser pockets. It seems quite miraculous that I can fetch out what I want without carrying bags of cheques and notes to pay in. But I shall soon get used to the phenomenon. At present it makes me uncommonly careful; hence the three year old coat (and a deaf ear to bargains at the dressmaker's.)[120]

Despite Ruth's sense of a need for parsimony, the Heads were, thanks to a bequest from Henry's father, by the later 1920s financially well endowed. This affluence was of immeasurable help in managing the consequences of Head's illness.

By January 1926 the decision to sell the house at 52 Montagu Square and move out of London had been taken. Initially Ruth sought a suitable property in Lyme Regis, where the Heads had holidayed. She told Hester Marsden-Smedley that the process of 'chasing houses' was: 'nightmarish. As soon as we think we have secured one it melts in our fingers and Uncle H. gets desperate – just as though we were within a month of June instead of five months as we luckily are'.[121] In April Ruth was still house hunting. She identified one possible property near Lyme Regis, but was deterred when she realized the likely cost of its upkeep. The current owner: 'keeps 3 men and a boy for the grounds and their united wages would be a huge yearly sum'. She had enlisted the aid of the 'the best Lyme Regis agent', and also urged Hester to: 'keep an open ear if you hear of any good house to let'.[122]

Eventually, Ruth informed Harvey Cushing, they settled on an 'old Dorset Manor house built we are told in 1693'[123] in Forston near Dorchester. She asked Hester to take care of Henry while she travelled to Dorset to complete the arrangements.[124] The choice of Forston was determined in part by the proximity of both Max Gate, where Thomas Hardy and his wife Florence lived, and the home of the Granville-Barkers. The Heads were clearly anxious to maintain close contact with these friends during their retirement.

Living The Life Of Others

The Heads were to remain at Forston until 1929. The building they occupied was, according to Janet Ashbee, 'a beautiful W^m & Mary house – Dutch influence – formal garden – panelled rooms ...' Ashbee visited Forston in March 1927 and sent her husband an account of life at the new Head household:

> It is all <u>unspeakably</u> tragic, & it is wonderful we can all laugh in between as we do – Ruth is marvellous I think in adapting herself to the life of a nurse or warder in what is practically a gilded prison – every comfort & luxury ... a £900 car & perfect chauffeur, & there he is unable to sit down or rise without 3 or 4 minutes help & unstiffening of his limbs. <u>Almost</u> unable to take a spoon to his mouth, or dress himself – poor dear![125]

Ruth confided that Henry would tolerate few visitors – partly because he found company tiring, but also because: 'except in a few cases, he can't <u>bear</u> to be seen in this state, especially by <u>men</u> – after his sane & vigorous intercourse all his life'. The range of activities available to Head grew ever more narrow. He had long since been unable to write. Even dictating a letter to Ruth taxed his strength. Moreover, 'he <u>does not read</u> now – except what she reads aloud to him'. Head was now so feeble that he lacked the strength to hold a book or turn its pages.[126]

Ashbee noted that nonetheless: 'His brain is as keen alert & balanced as ever'. It was: 'a joy as ever to hear him talk & he is as yet brimful of ideas of new wonderful theories & generalizations'. She and Henry:

> had short but thrilling talks on all possible subjects from the Long Barrows all over the Dorset hills – Mycaenean Civilization – to the modern narrow girl whose pelvis is growing smaller & smaller like men's, so that 'Labour' has to be 'induced' at <u>8 months</u> if her children are to be born alive! 'Of course women with pelvises like that <u>ought</u> to die in childbirth' – he concluded!

Ashbee feared, however, that: 'his mind will slowly ossify like his muscles'.[127]

Among the few recreations still available to the Heads was to go for drives in the surrounding countryside. Ruth told Harvey Cushing that Henry: 'finds great relief in motoring so we have got a fine car and drive out daily, the country round is enchanting'.[128] The vibration of the motor seemed to relieve Henry's rigidity. Ashbee reported that during her stay, they had: 'taken long drives all through the lovely unspoilt Hardy country ... We went to Bridport ... Another day round beautiful Sherborne––, Winterbourne Abbas – Cerne Abbas – Charminster – all such lovely names'.[129] Often the Heads would drive out to visit such neighbours as Thomas and Florence Hardy. 'Max Gate', Ruth revealed, 'continues to be our great consolation'. In April 1927, Henry: 'called there on Easter day and found T.H. meditating upon Pontius Pilate'.[130]

Despite Henry's reluctance to entertain many visitors, a steady stream of select guests made their way to visit what Ruth referred to as 'the two solitaries of old Forston'.[131] Among the first was Sassoon who in January 1927 motored over with a: 'a charming young actor friend from town. He is devoting a few days to his elderly friends, T[homas].H[ardy]. and H.H'.[132] Ruth wrote to tell Robert Nichols: 'how sympathetic and entirely enchanting Sieg. was. He is quite absorbed in his work and only gave himself this holiday for the sake of his invalid friends to whom he brought a real breath of renewed life'. While Sassoon was with them, the Heads enjoyed 'a debauch of talk'. Once he and his friend had departed, however, Ruth confessed to feeling 'slightly depressed ... we are left: the candles are burnt low and the logs are all ashes'.[133]

Sassoon returned in March, this time with his friend Leslie Wylde, known familiarly as 'Anzie'. The two: 'motored down from Bray in the new Schuster[134] Rolls Royce (£2380, just bought) had tea at Max Gate and dined here, leaving us at 10 and sleeping at Torquay 85 miles away!' Although he was 'terribly shaky', Henry had not been too tired by these visitors. For Ruth, 'It was the fresh Springwater to hear so much talk and I quaffed deliriously'.[135]

Ruth wrote to assure Sassoon that her husband had: 'enjoyed every minute of your Angel's visit last week. How charming Mr 'Anzie' was, fitting in so well to our ways and bringing such a pleasant atmosphere of health and sanity with him'. Ruth only regretted that she had not had the opportunity for a private conversation with Siegfried about Henry's condition. 'You will have seen how much weaker he is', she wrote, 'but his London doctor who has been to see us says it is only the natural progress of this foul disease'. She and Henry would: 'need all the help your constant friendship gives us'.[136]

There was something almost fervent in Ruth's expressions of gratitude for the attention that Sassoon and his companions paid to them. She feared that, if such friends abandoned her and Henry, then their existence would become truly hermetic. Ruth wrote in a letter of May 1927, at a time when Sassoon's literary reputation was on the rise and he planned a trip to Germany: 'Soon you will disdain the company of the likes of us. Altogether we feel left behind'.[137]

Such anxieties were misplaced. The loyalty of the Heads' friends to them in the years following their enforced retirement to the country is striking. Nichols also remained in close touch and visited when he could. Janet Ashbee, the Oppés, the Granville-Barkers, Walter de la Mare, Peter Derwent, and John Jay Chapman (1862–1933) were also among their guests. Medical visitors from overseas, including Harvey Cushing and Adolf Meyer, also found their way to the Head residence.

On the other hand, during these years, the Heads lost the support of one of their 'dearest friends' – Alfred North Whitehead. Ruth told Janet Ashbee that through the mid–1920s Whitehead and his wife: 'stayed with us up to the last, they knew how bravely H. was fighting his inevitable disease, they were perfect in sympathy'. When, however, Whitehead took up a post at Harvard University, 'that was the end. Silence, and we hear that everybody in their circle of devoted English friends has been treated the same'.[138]

George Riddoch was a regular guest. Ruth described him as: 'the young neurologist who loves H.H. as a son and was his first pupil'. Riddoch came, not only as a friend, but also in the capacity of a physician to both Henry and Ruth. In February 1927, he examined his former mentor and: 'found to his sorrow that the horrible disease had made terrifying progress since his last visit to us at Whitstable in August'. Riddoch also interviewed a prospective nurse for Henry, and 'gave good advice and lively sympathy' before returning to London.[139]

Other visitors' attempts to help were less welcome. The American essayist John Jay Chapman seems to have suggested that Head's condition might be curable using some unorthodox course of treatment. It was left to Ruth to compose a tactful response; she confessed that she had been: 'puzzled how to reply'. Finally, she explained that:

> as himself a Physician Harry was most interested in all you told us, but personally unconvinced. He believes his disease to be incurable and progressive. This he has faced nobly and he has, as you saw, achieved patience and even to a certain degree

happiness. He does not wish to try anything by way of a cure or even of alleviation. He has planned out his own course and will continue in it.[140]

The Heads also enjoyed calls from their favourite niece Hester, although Ruth had sometimes to ask her to come without her children lest they disturbed the afternoon rests that Henry was obliged to take. Hester left a number of accounts of such visits. For instance in July 1938 she took the actor Cyril Maude (1862–1951) to see her aunt and uncle. She wrote that Maude had been: 'staggered by Aunt Ruth's gayness and looks – and especially her walk – the actor in him picked it out for the graceful, easy movement it is and the scene of your garden paths made it all the lovelier'.[141]

The appearance of Hester's mother: 'his disliked & tiresome sister Lady Pinney'[142] – had, however, quite another effect. Ruth confided to Sassoon that Henry: 'was fresh as a Sandboy till tiresome Lady Pinney came yesterday and bored and preached and squabbled and tired us both out'. Head had effectively disowned Hester when she married (see Chapter Five, pp. 195–196). Now he sought revenge for these intrusions by urging Sassoon not to spare Reginald Pinney's generalship in the war memoirs he was composing.[143]

Lady Pinney persisted in discharging her sisterly duty in later years. In November 1932, Ruth was moved to exclaim: 'Well, SHE has come and is gone and all is smiles and peace'. The elder Hester's company, she explained:

> has a queer effect on me. When she bounced in on Sat: she was the Town Lady, very grand in new stays about which I heard the moment she sat down! and indeed she looked very nice in green as I told her promptly, but she made me feel such a very small Country Mouse that I went on dwindling and dwindling till when I came to undress there was hardly any Me left to put to bed. Do you know that feeling after a few hours of Her. I made up the following epigram:––
> 'Hester P. is so busy claiming kinship with the Muses that she entirely neglects cultivating the society of the Graces'.[144]

In addition to visits – welcome or otherwise – Ruth and Henry relied heavily on letters from their friends to relieve the isolation and tedium of their existence. As Ruth admitted as they prepared to take up residence at Forston: 'We cultivate vicarious enjoyment as indeed we must and learn to look at happiness through other men's spectacles'.[145]

Robert Nichols responded to this need for news of the outside world by forwarding voluminous, if often illegible, screeds on a weekly basis. These letters were especially important in maintaining contact with the Heads during Nichols's extended stays in Japan and California. Ruth and Henry delighted in the trivia and gossip that Nichols purveyed, often begging him for more detail. Henry especially valued the 'glimpses of the beau monde littéraire'[146] that Nichols could provide. Thus in October 1926 Ruth declared:

> I can't tell you adequately with what joy H.H. read your letter of Oct:25 with the thrilling account of the historic dinner party. He hopes you will from time to time put down for him such events as seen by you and we will keep them carefully. They are a really valuable record of the big men of our Age as seen by big men of the next ...

In the same letter, she begged Nichols: 'Do tell us all about your visit to dear Miss Dickinson who writes me the most touching notes from her bed. Also notes on the Colefax dinner, please. You see we are insatiable'.[147] Head was: 'thrilled over his breakfast-sausage' by Nichols's account of a chance meeting with George Bernard Shaw in a London street. 'You tell the story so dramatically', Ruth exclaimed, 'we followed it step by step as only exiles do and when you emerged at Leicester Square, we both gasped'.[148] Even an account of a wedding that Nichols attended was found enthralling: 'H.H. listened to it with the fascinated attention of the Wedding Guest in the Ancient Mariner, although luckier than he was'.[149]

A member of the 'beau monde littéraire' whom Ruth and Henry found especially fascinating was Aldous Huxley. Nichols knew Huxley well and sent the Heads extracts from his letters; these intrigued Henry. He and Ruth both found Huxley: 'extraordinarily important and H. delights in tracing links with the Huxley grand pere, while I who knew his mother well years before A. was thought of love to see the Arnold heredity in him'.[150] When Nichols forwarded a sample of Huxley's 'table talk', Ruth expanded on why 'Aldous interests me intensely'. She saw: 'the Arnold in him extraordinarily clearly, the violence, the fundamental instability, the power of attracting all to whom they show favour in themselves as slaves, the discontent with the world, the great pride and arrogance, the surprising streaks of humanity, the critical faculty'.[151]

No intelligence of the outside world was, however, too mundane for the Heads. Henry was always eager for news of Nichols's 'terribly interesting family'.[152] When Nichols's sister decided to elope, Ruth and Henry became immersed in what they called the 'Anne Saga'. Successive letters served as instalments in an unfolding serial entertainment. They appreciated the literary quality with which Nichols endowed the narrative. 'You tell the whole thing so graphically', Ruth declared, 'with everyone's reaction it is thrilling for us ...' Later she assured him that: 'We drink in every development of the Anne-Henry story greedily and could have cried when you stop at their arrival chez vous'.[153]

The Heads differed somewhat, however, in the moral they derived from the story: 'H. thinks all marriages of people over 25 should be runaway, more or less and the way so many newly weds really grasp at presents and have big weddings to secure them sickens him. I am more romantical and think it can be a very pretty lovable ceremony'. They begged for more information about the reaction of Nichols's father to the affair –' That would interest us beyond anything'.[154]

The Heads' concern with the affairs of Nichols and his family was not entirely passive and voyeuristic. On occasion Henry would seek to offer guidance. Thus, when Nichols was considering buying a house in Winchelsea, Head was eager to tender advice. He had visited the area years before to visit Henry James in a professional capacity and had: 'used those sharp little eyes of his to some purpose and saw what a desirable neighbourhood it was in every respect'. Head was firmly of the opinion that: 'the place would be <u>ideal</u> for his former Patient'.[155]

Ruth and Henry also offered Nichols counsel on his personal relationships. They encouraged him to broach sensitive subjects, assuring him that: 'We never read your letters before [Henry's] men, but always wait for entire privacy'.[156] When Nichols and his wife, Norah, separated, the Heads expressed the hope that: 'you will stick courageously to your brave plans for the future. It would be disastrous we feel for you to return to Norah from sheer weariness and loneliness and to make a pretence of picking up threads some of which are irretrievably broken ...'. This advice was, somewhat incongruously, combined with an appreciation of Nichols's 'masterly' summation of his wife's character: 'She is more like one of Ibsen's strange heroines – The Lady from the Sea for instance. Something intangible I find in her and yet so extraordinarily practical – vague yet for ever planning'.[157]

Janet Ashbee was incredulous at the intensity of the Heads' interest in what seemed to her the trivia of daily life. She could not: 'help reflecting on the strange circumstance, really a miracle, that you 2 kind devoted friends should want to bother among all your friends & legions of letters, to want to read the reported facts of our family & its growth, & the pictures of its various infringements on each other!'[158] Nonetheless, she sought to satisfy this demand by forwarding to the Heads letters written to her by her daughter who was embarked on foreign travels. Ashbee even sent old letters that she had herself written while staying in Paris in the 1890s. Henry: 'dwelt long on them in his Rest period'.[159] Ruth was duly grateful for these efforts, assuring Janet that: 'We gloat over your letters'.[160]

Henry showed a special interest in Sassoon's doings. Ruth informed Siegfried that: 'H. constantly goes into every detail of your life'.[161] Sassoon took particular care to keep in touch with the Heads when he was touring abroad. He spent part of the spring of 1929 residing at Haus Hirth, a hotel in Garmisch, Bavaria where the Heads had also stayed. Ruth assured him that she and Henry were delighted with a 'magnificently long letter' Sassoon sent them while on this holiday. Indeed, she: 'had to limit the number of times I would read it aloud to H. like the children with a favourite fairy-tale, he was always saying "Again."'[162]

As well as letters, Sassoon sent Ruth and Henry gifts and photographs that were meant to revive memories of happier times. In response, Ruth wrote: 'Your Blue and White perfect Handkerchief with its bravery of Bavarian colours filled us with reminiscent joy'. A photograph of a deal table at which Head had been wont to write while staying in Garmisch also gave him special pleasure. There was, however, a bitter-sweet quality to these recollections: 'H. says of Haus Hirth it makes him understand that tragic passage "Then Christian with desire fell sick", because he can't get there--'[163]

Sassoon was accompanied on this vacation by Stephen Tennant (1906–87). Siegfried had in 1928 introduced Tennant to the Heads,[164] and Ruth now regarded the two as 'Twin Stars' of which she was the fortunate 'Earthly observer'.[165] Tennant would attach his own addenda to the letters Sassoon forwarded to the Heads. Moreover, even after his relationship with Sassoon ended, he remained a faithful correspondent.

Indeed among all the Heads' correspondents during their years of exile, it was Tennant who most clearly grasped the importance that such letters played in their lives. He strove to convey by means of words vivid impressions of the locations from which he was writing. Thus writing from the 'blissful haven' of Haus Hirth in February 1929, while Sassoon was in England, Tennant described how:

> sunshine has poured down everyday, hot, positive sunlight, lying in great pools on the carpet, glittering on my shells & books, and washing my January-in-London wizened heart, at the end of each burning day. The Twilight is lilac and momentary, the glittering sugar-mountains are dyed flamingo-pink & effulgent tangerine against the lilac fading to dove-grey, --& then the stars! I've never seen such a firmament! – such enormous, potent, flashing stars, ice-blue & green and pink, a blinding scintillation, --then I think the mountains are at their most beautiful (there must be some moon) they look marine & fabulous,-- and Italy beyond them seems doubly the Promised land, that must lie beyond all mountains that they may have their fullest significance?
>
> The snow is deep & crisp & dry, the air is cold & powdery, the peal of sleighbells is constantly on the still air,--there is no lovelier sound is there?[166]

In a later letter composed while the 'Twin Stars' were staying in Italy, Tennant stipulated that:

> I am writing this on a sun drenched terrace acacia shaded, --overlooking orchards--& orchards & orchards of blossoming almond & peach – among which are Greek temples & olive groves & then the sea, which today – is palest sapphire, breaking silver on the sands – the panorama is huge – all with – a bloom of distance and sunlight on it – while I write, lizards are near on crumbling patio ...[167]

Moreover, Tennant had some talent as an artist. He would sometimes illustrate his letters with pen drawings or even watercolour paintings in an attempt to provide Ruth and Henry with a virtual visual experience of places that it was now impossible for them ever to see at first hand. Some of these illustrations – such as Tennant's depiction of the parrot that shared his room at Haus Hirth – were humorous in nature. Others, such as the pictures of irises that adorned a letter Tennant sent from Sicily, were meant to convey something of the exoticism of the setting. A drawing of a 'white goat on sunny steps' thus served to: 'exhale the SOUTH'.[168]

The fact that so many of Head's friends were authors provided him with another source of relief and stimulation. He and Ruth were afforded the opportunity to read and critique works that were on their way to the press. Thus Sassoon sent the Heads the manuscript and proofs of the various volumes of the *Memoirs* that he composed in the interwar period. When an early instalment in the form of a 'beautiful Typescript' arrived in March 1929, Ruth assured Siegfried that: 'We shall digest it slowly. I have learnt to make our pleasures last'.[169]

In September 1936, Ruth confided that: 'H.H. was made so very happy this morning by your parcel and its contents and especially by the inscription which he found very moving'. The progression of Henry's illness, however, made the act of perusing these pages ever more arduous: 'it involves his man standing by his side and turning each page. Can you conceive of reading in such adverse circumstances?'[170] When the labour became too great for Henry, Ruth would read the drafts to him. On one occasion, Ruth's brother, Arthur:

> a man of great discretion was staying with us and we gave him the honour of reading it to us, a fact for which I had cause to be thankful as I could not have done it: it was too poignant and a choking voice tho' a tribute to your powers, would have annoyed H.H. who was listening with the open-eyed attention of a two-year old child joined to the critical mind of a Professor hearing a thesis from a candidate for the membership of the Royal Society.[171]

Henry's criticisms were in the event minor – 'the merest flea-bites'. His main comment concerned Sassoon's: 'use of "visualize" which is not always academically correct. He will enlarge on this should you wish'. The manuscript, and a subsequent letter from the author, had, however, done much to lift the invalid's mood:

> He chuckles every time he thinks of 'Whincop' – 'General Sir Archibald Whincop K.C.B'. he repeats, having spent his waking moments for several nights thinking out a suitable Christian name. We both like Archibald, for his wife who is a nagging woman says 'Archie' in a voice that makes the General tremble.[172]

'Whincop' was Sassoon's pseudonym for Henry's bête noire, Reginald Pinney.[173]

The Heads were also glad of the chance to read Charles Ashbee's memoirs in manuscript. Ruth remarked that Ashbee's: 'period comes in exactly between Harry's and my own and we would exclaim with delighted recognition at the cherished vistas of our youth opened again by your reminiscences'.[174] In later years Ashbee continued to provide Ruth and Henry with a 'Private View' of his journals. Ruth was especially enthusiastic about her friend Janet's contribution to the work: 'Everything she writes is so full of sparkle and vitality that we always read her with avidity'.[175] She felt, however, that the 'Egyptian' volume of Ashbee's memoirs: 'is not up to the other volumes in any way and in its conception is so totally different that it spoils the "set"'.[176]

More generally, books were among the few of their artistic and intellectual interests that Henry and Ruth were able to pursue in retirement. They kept abreast of new fiction; Ruth told Janet Ashbee that in his retirement, 'H. seems to like a good modern novel better than anything'.[177] He and Ruth developed a special interest in the works of Virginia Woolf (who had, briefly, been Head's patient). They thought: 'The Lighthouse more perfect than Mrs Dalloway and are so glad to have our opinion reinforced by Lytton Strachey. It is the most wonderful book and we talked of little else all the time I was reading aloud'.[178]

Robert Nichols shared his friends' enthusiasm for Woolf's work declaring her a genius, and noting that: 'she is particularly good at neurasthenics'. Nichols speculated that certain characters in *Mrs Dalloway* (1925) were based upon actual individuals, suggesting that: 'Peter Walsh is probably partly drawn from E.M. Forster'. He was, moreover, convinced that:

> Sir W. Bradshaw is partly drawn from Dr [Maurice] Craig. The touch about him & his being 'interested in' art was good. Real people aren't 'interested in' art. They either love it or leave it. I'm sure some people will think Sir W.B. a caricature. But he's not. By no means – alas! You, Henry, probably spend a third of your physician's time trying to undo what he & his perpetrate. My God how I hate the fellows & their clumsy yardsticks.[179]

Henry dismissed Nichols's identification of Walsh and Forster. However,

> you are right, H. says about Sir Maurice Craig (as he now is) and for some months we have steadily avoided mentioning the caricature, for fear of its getting round to Craig's ears. For caricature it really is, cleverly as she has reproduced the consulting room and the general atmosphere of his place of business.

The novel, which depicted a day in the life of a woman in post-war London, had a particular resonance for Ruth. She was moved to exclaim:

> But what a book it is and how inimitably she gives a woman's way of thinking and draming [sic] backwards as she pursues her little life filled to the brim with minute, inglorious actions. When she comes home to the small house in Westminster and stands for a moment in her hall, hearing the maid singing in the kitchen, proud of her preparations for her party and is filled with gratitude to her husband, the provider of all these good things, I positively started – just so have I often stood and brimmed over with thankfulness to Henry, giver of all things good for me also.[180]

Thanks to the advent of the gramophone and of the radio, the Heads were also able to listen to music and to spoken word broadcasts; live performances were out of the question for Henry and Ruth was rarely able to leave him alone.

The music of Josef Haydn was a particular favourite. Ruth revealed that: 'We have records of his London Symphony and likewise of the Surprise – too enchanting for sad hearts, and last night we had the Drum roll on the Wireless – a beautiful Queen's Hall Concert – Thibaud the Soloist Violin. These are our joys'.[181] The operas of Mozart, rather than Wagner, seemed to have had more appeal for Head during this period of his life. In 1936, he and Ruth heard the first Act of *Don Giovanni* 'on the Wireless last night, the first broadcast from Glyndebourne. Were we not in luck?'[182]

Henry derived some intellectual stimulation from listening to Goldworthy Lowes Dickinson's (1862–1932) broadcasts. Dickinson and Head had been contemporaries at Charterhouse. The two also shared an enthusiasm for the works of Goethe. Henry was so impressed with Dickinson's reading of one of Goethe's poems that he instructed Ruth to find his copy of the book so that she could emulate the performance.[183] There are hints that Ruth sometimes found her invalid husband a hard taskmaster. When he missed a broadcast of one of Nichols's poems, Henry was 'inclined to blame me for not knowing which I try to bear meekly'.[184]

In January 1929, Ruth was granted a brief respite from her custodial duties. Janet Ashbee temporarily took her place at Forston. She found the experience of joining Head for breakfast somewhat remarkable: 'It does seem very funny being suddenly thrust into this intimacy––& very flattering for the ex-patient!' She was left in no doubt of the continued vigour of her former physician's mind; it was: 'impossible to recapture all his dicta or suggestive remarks on all kinds of subjects. He was very thrilling at breakfast on the subject of vocational advice (psychology – balance & stability – etc.)'. Moreover, when the conversation turned to the pronouncements of the theologian, Canon Burnett Hillman Streeter (1874–1937), Head's revealed something of his old intellectual fire:

> H. said 'Whenever I hear a parson beginning about 'Evolution' etc. in the pulpit I begin to look out. This man Streeter got up & on the title 'Is Religion a Neurosis or a Psychosis?' thought he was going to delight us scientific men by being scientific – I stood it as long as I could & then got up & said – 'Neither the one nor the other. Religion, Science & Art are three of the ways that different types of people try to adjust themselves to the problems of the Universe & all 3 are necessary & essential & not in the least partaking of the nature of Pathology (which his title implied)'.[185]

Head retained something of his interest in developments in medical science. When commiserating with Janet Ashbee about a friend who had contracted cancer, Ruth revealed that: 'H. says that the Research people are on the wrong track, only going for the one definite thing, instead of working widely on paths that would lead them to their goal one day quite unexpectedly. This is one of his favourite doctrines. Too high and remote for me'.[186]

For Head, however, science had always been a communal enterprise, dependent upon contact with colleagues. Once he had withdrawn from London, maintaining such communication became ever more difficult. In a wistful letter he sent to Charles Sherrington via Ruth in January 1927, Henry admitted: 'I have been living greatly in the past since my retirement and the details you give me in your letter have added fixity to many happy memories'. Head urged Sherrington to visit Forston; it is unknown whether he took up this invitation.[187] When, however, he heard in a radio broadcast that Sherrington had been awarded a Nobel Prize, Head sent his congratulations. The honour, he declared: 'is long overdue. For you have in the course of your wonderful series of researches entirely transformed our views of the activity of the Nervous System'. Henry added the poignant postscript: 'May you long be spared in health and vigour to throw further light on the problems which you have made so entirely your own'.[188]

In November 1928, at Robert Nichols's suggestion, Ruth forwarded a copy of Head's Croonian Lecture 'On the Release of function in the Nervous System' to the biologist, Julian Huxley (1887–1975). Nichols had informed the Heads that Huxley had an interest in the subject of regression, which Head had essayed 'from a more psychical point of view' in his lecture. The tone of Ruth's letter was almost apologetic; but she excused the intrusion on Huxley's time by explaining that her husband was: 'pining of news of the world of thought from which his sad illness exiles him'. She ended with the plea: 'We remember so vividly your visit to us in London, and if as we hope to get nearer civilization next year, perhaps you would be good enough to see Harry there'.[189]

Ruth's reference to a perceived need 'to get nearer civilization' alluded to a decision to move house the Heads took towards the end of 1928. Despite its many advantages, Forston was deemed too remote from London; in the West Country, the Heads feared that they were indeed in danger of becoming Solitaries. They determined to find a new dwelling in the Home Counties. Ruth told Sassoon that: 'I yearn for Kent. Surrey is not real country to me and you would come sometimes en route for your part of the Weald, would you not: we should be 8 miles from Sevenoaks'.[190] The house in Kent that Ruth thought would be ideal however escaped her. She was obliged to search elsewhere – a process that she admitted was exhausting. In April 1929, she was: 'quite distraught househunting. Memoirs of a House-hunting Woman would be sorrowful reading'.[191]

By June, she had finally secured 'a suitable house tho' not, alas, an ideal one from many points of view'. Ruth described Hartley Court to Sassoon as:

3 miles from Reading, 20 from Eton where my favourite brother dwells, 40 from London, 33 from Winchester and 23 from Oxford... It is a pretty place of 26 acres with a small Park of its own and a roomy old house sitting in shady gardens with the traces of an old moat straggling through the wilder parts of the grounds. At present our man of law is struggling thorough the mazes of a complicated lease with the agent of the owner, a Captain Paul. This latter by the way asked for our references and among others H.H. gave Sir Archibald W[hincop]. K.C.B.[192]

Since the sale was completed without mishap, the vendor's solicitor apparently failed to notice that one of the Head's referees was a fictional character.

Among the first visitors at Hartley Court was, in July 1930, Harvey Cushing. Cushing confessed that he had 'not the slightest conception what Hartley Court would be like but merely knew it would be somewhere on the outskirts of Reading'. He had anticipated a modest residence and was taken aback at the grandeur of the Heads' new home. The house was, he reported, situated in:

A lovely park-like place ... An old house set a lovely grove of tress with a pasture newly mown extending before it – water wag-tails flitting about on the lawn, some magnificent oaks, old as the history of England – a clump of ancient Scotch firs, a huge linden or two. The house has some magnificent old Jacobean panelling and was made over, as the arch in the hallway and the windows of the bay make evident, either by [John] Soane himself or one of his pupils.[193]

It was in this setting that Ruth and Henry were to spend their last years.

My Little Narrowed Life

In his youth and prime Henry Head had been a supremely articulate individual. He projected representations of himself by means of his letters, journals, and poems. Through his scientific publications he hoped to make a permanent contribution to human knowledge. One of the most grievous aspects of the disease to which he fell victim was that it effectively denied Head the means of self-expression. By the late 1920s he was incapable of holding a pen; his career as an author was thus at an end. At first, he tried to dictate to his secretary or to Ruth. Soon, however, even this proved too tiring. Although Ruth would sometimes sign letters in Henry's as well as her own name – or use such devices as 'Henry+etta' to designate joint authorship – her voice increasingly dominated the communications that issued first from Forston and later from Hartley Court. Head thus became perforce the passive object of what others chose to say or write about him.

Much of Cushing's account of his visit to Hartley Court was taken up with his impressions of Head. His first sight of Henry was at breakfast:

> He was sitting at the end of the table – a typical advanced Parkinsonian, practically rigid except for the movements of which I will tell. He spoke in his customary cheery voice and when I sat down beside him said: 'Don't sit there, I can't turn my head, sit over there opposite to me'.

Cushing dwelt on the elaborate steps that had to be taken to get food into Head's mouth. Some of these seemed designed to create the illusion that: 'he [was] feeding himself'. After breakfast, 'his chair was pushed back, the attendant drew him out of it onto his feet and in a crouching attitude he was led out and the attendant holding both hands backed up the winding stairs, H.H. in festinating fashion following him'.[194]

Cushing later joined the Heads on their morning drive during which Henry: 'talked incessantly and brilliantly – about his pupils, George Riddoch in particular – about recent trends in neurology – about our accidental and happy friendship'. Head's memory seemed unimpaired; he recalled incidents that Cushing had forgotten, including what had transpired at the Budapest Congress. After lunch, Cushing noted that Ruth held the door open for Henry as he left the room – 'and as he tottered by kissed him on the cheek, and he said, "Thank you, my darling"'.[195]

Hester Marden-Smedley's young daughter, Henrietta, gave a more naïve view of life at Hartley Court. After a 1938 visit to her great-uncle:

> Henrietta chattered gaily of Uncle Harry 'with a flower in his coat' and 'sitting in a chair with great, normous [sic] sides' and Aunt Ruth giving an ice and sitting in a little house with a picture on the wall, 'and a big hole in the ground with a bucket going down into' and men with white coats 'and lots of rooms with lots of doors'.[196]

Ruth was inevitably, however, the most prolific source of such representations. She stressed the patience and fortitude with which Henry faced his fate, drawing upon familiar religious imagery – her own professed lack of belief notwithstanding. Thus Henry was: 'a Saint ripening for a better World', who would otherwise: 'have been very fractious and irritable cut off from all contact with the outer world'.[197] Christian notions of redemption through suffering served to give some meaning and dignity to her husband's affliction.

As Henry became ever more dependent upon her, Ruth began to depict their relationship as akin to one of mother and child. In November 1926, she wrote of how she strove to keep her charge comfortable: 'I cover H. with shawls and keep him as warm as I can, my cherished Baby'.[198] Later she declared to Head's niece: 'H's illness makes me so fierce in getting him pleasures: you know how you would feel if one of your little boys were ill and longing for a treat which did not arrive'.[199] Sometimes, like any mother, Ruth was, however, obliged to 'scold' Henry when he misbehaved.[200]

Ruth's descriptions of her husband were almost always compassionate. Such epithets as 'sweet' and 'heart-rending' often figure in her depictions of Henry. She recognized, moreover, 'what this inaction is to his active brain', and what: 'his increasing weakness brings to him of daily irritation and despair'.[201] Henry was: 'very bent and sad'.[202] Despite his fortitude, he could be: 'very feeble and low in his mind, seeing difficulties round every corner and averse, for the time from visitors'. Ruth was obliged to find Henry constant diversion because he was: 'too apt to droop or be very artificially entertained on the subject of his vile body and its pitiful needs'.[203] Indeed, 'H. grows daily more misanthropical'.[204]

There was sometimes an amused, condescending, although affectionate aspect to the representations that Ruth shared with her correspondents. In particular, as he assumed the role of invalid, she came to see ever more of the Quaker in Henry. When reporting his reluctance to divulge his opinions of the psychologist William MacDougall, she ascribed such caution to her husband's family background: 'One is not Quaker born and bred for nothing!'[205] Henry's financial parsimony was also the mark of: 'a prudent Quaker'.[206] Henry was similarly: 'the modest Quaker', who was shocked by Robert Nichols's use in one of his letters of the word 'lavatory'.[207]

The invalid Head was thus infantilized and rendered an object of pity and amusement. But at the same time Ruth's letters reveal him retaining some vestiges of his old dignity and authority. While forced to abandon medicine, some traces of his former clinical persona clung to Head. He was hungry for information on the state of health of acquaintances. When in October 1927 Janet Ashbee was informed of a '"non-virulent swab"', Ruth declared: 'The Oracle says now all is well, so well I trust it is'. She then turned to the health of Charles Ashbee – whom she referred to as 'Invalid II' – noting that: 'H. is deeply interested in all the coil of symptoms, so be medically explicit in your letters, won't you, nor think to spare us detail. He dotes on it'. Ruth was aware of the pathos inherent in the former consultant physician's hunger for medical gossip. It was: 'so sad that he should have to exist on hear say, having so long sat at the Hub'.[208]

On the basis of the snippets of information fed him, Henry was ready to offer his medical opinion. Thus he identified the complaint of one of Ashbee's children as: '"cystitis" – from your description he believes that is her complaint ....'[209] When the illness was one with which he was especially familiar, Head's pronouncements had particular weight. One relative's case was bound to be: 'horribly painful and slow to be healed. You know that Uncle H. is the authority on Shingles and has written the standard booklet on it?'[210] Head took a childish pleasure when one of his stabs at diagnosis was vindicated. He was: 'frightfully pleased to have made the right diagnosis of Robert's internal condition ...'. Moreover, he: 'highly commends the last Kiddie order for subsequent rest to let the treatment settle'.[211] Frank Kidd was the name of Nichols's actual doctor.

When, in 1930, Stephen Tennant required treatment for tuberculosis,[212] Head took an especially close interest in the case demanding news of the patient's condition and treatment from Sassoon, who would convey the information to the Heads by telephone. Ruth told Siegfried that Henry: 'wishes you had been his Clerk at the Hospital. He never saw anything more careful and exact than your résumé of Stephen's "care"'.[213] Head paid special attention to the records of Tennant's temperature that Sassoon transmitted.

According to Ruth, Head could not fathom the 'tactics' of one of Tennant's attendants: 'This annoys my H. who as a rule solves psychological problems with the ease of long practice'. What made the doctor's behaviour: 'more mysterious is that when at the London, H. always found Dr. [Frederick George] C[handler]. what he calls a "fussy man"'. In Tennant's case, Chandler's actions struck Head as negligent.

Head was, however, gratified when another of Tennant's doctors (Arthur de Winton Snowden) recognized his name: 'The dear creature is always naïvely surprised when such crumbs of comfort are brought to his table. He believes himself to be like a "dead man out of mind"'.[214] There were, in fact, a number of indications that such fears that Head had been forgotten by his professional brethren were misplaced. In January 1938, Hester Marsden-Smedley revealed that when she mentioned her relationship with Head to Egon Plesch – a Hungarian doctor who had settled in London – 'he simply leaped with excitement'.[215] Hester arranged for Plesch to make a visit to Hartley Court in order to make the acquaintance of someone who had previously been just a legendary name to him.

When the young son of his niece, Hester, became seriously ill, Head again provided advice and comfort through Ruth's letters. Henry was: 'much interested to hear he has been taking glucose which is the sugar antidote to acidosis, which is, as you know the condition of not making sugar enough for your bodily needs. Acidosis is a new discovery of the Bio-chemists since H's day, but he has heard all about it'. Henry was able to answer some of the parents' questions, explaining: 'the question whether a person does make too little, enough, or too much sugar is one that can only be solved experimentally and only declares itself by illness. This is to answer one of Basil's questions – why the acidosis had not declared itself earlier'. He also sought to provide Hester and Basil with a measure of comfort: 'Uncle H. says you are both model parents from a Specialist's point of view and he would have probably been lured into Basil's study and have discoursed to you on Reflexes for hours together!'[216]

Relics of Head's former persona as a psychic healer can also be glimpsed – for instance, in the counselling he continued to provide in retirement to his 'pet neurasthenic', Robert Nichols. In addition to his physical ailments, Stephen Tennant sometimes also appealed to this aspect of Head's medical character. During a stay in the United States, Tennant wrote of a 'dark tunnel of neurosis' from which he was only now emerging.[217] In the previous year he had confided:

> it may interest Sir Henry – I always expect to be disliked––& tremble before entering a room: ––psychologists always seem puzzled at this. I <u>am</u> a little troubled by the <u>amount</u> I mind, when people do dislike me. I go to my room & lie on my bed with ostensible sense of icy oppression I mean a more than natural distress that often lasts hours. As you can well understand my appearance often causes rage & I often feel I am to blame by my outré clothes etc.––& that for this I deserve no sympathy: ––& this is true – for I court & <u>love</u> attention & when I get the wrong sort should not mope.

Tennant gave an indication of the 'outré' look in question when he revealed how he planned on the following day to go abroad wearing: 'a lovely new blue & white check shirt (big checks like a dishcloth), blue peasant trousers & I shall have in my hands a huge bunch of mimosa & violets'.[218]

Head had in 1932 already sought to help with Tennant's psychological problems by referring him to the care of George Riddoch.[219] In response to this latest account of the mental distress to which Tennant was prone, Head evidently offered his own services as a therapist. Tennant's reaction was effusive: 'I cannot tell you how kind and beautiful I think it of Sir Henry to offer to help me, thank him please in all the nicest ways you can think of: ––I am so grateful: ––yes indeed I will visit you and it will be a great joy'.[220] It is unknown if this quasi-clinical visit ever took place.

Unlike Henry, during the years of exile Ruth remained capable of projecting representations of self. It sometimes seemed, however, that she was allowing her own identity to become subsumed by that of her husband. Thus she often signed letters 'Henrietta', 'Ruth-Henrietta', or 'Henry+etta'. As the nature of Henry's affliction became apparent, Ruth never questioned that it was her responsibility to remain with her husband to the end. From the outset, she had few doubts about how this role of carer would limit the scope of her own existence. As early as December 1925 Ruth felt that she was: 'sinking into a kind of Hospital Nurse existence'. Her 'chiefest pleasure' was when: 'H. has a good night and greatest pride when I make a really successful pillow of lamb's wool to put somewhere about the poor darling's stiff and useless limbs'.[221]

Ruth was, in fact, required to be much more than a 'Hospital Nurse'. She took over as de facto head of the household with all the trials and anxieties that entailed. She therefore had to manage the staff that the Heads employed at Forston and Hartley Court. As well as the usual compliment of household servants, this included a chauffeur and three male attendants who saw to Henry's personal needs.

In the past, Ruth had been assisted in the management of the household by an ancient Head family retainer, known only as 'Pratt'. She noted how during the war, food shortages notwithstanding: 'the admirable Pratt takes all the burden of shopping off my mind and Harry and I feed like Princes'.[222] When in the summer of 1925 the Heads holidayed at Lyme Regis, 'Pratt [was] a perfect Puss in Boots servant, always bringing home titbits from the town to lay before Harry at dinner time'.[223] Pratt seems, however, to have retired by the time the Heads moved to Forston. In November 1926, Ruth was:

> saddened this week by the sudden death of dear old Pratt, the family Butler. I expect you remember him from No. 4. Such a dear man he was and such a link with the happy carefree past. He had been for 40 years in the Head family and had seen the death of both Harry's parents and two of his brothers.[224]

The servants upon whom Ruth was obliged to rely in later years rarely came up to the standard of the saintly Pratt. Lamenting the unreliability of her maids, Ruth was moved to assert that: 'Robots would be bliss'.[225] Throughout all these trials, she felt obliged to appear: 'calm, serene, unalarmed, sweet-tempered and smiling to H'.[226]

A husband and wife duo, to whom Ruth referred as the 'owlets', served for a time as butler and housekeeper. The owlets, however, were to grow: 'too lazy and tyrannical for words – he unctuous this side the baize door, on the other a real heathen Pasha, she bitter, acid and suspicious to me and nagging to the maids'. When all the maids threatened to leave in the face of this tyranny, Ruth was obliged to dismiss the pair: 'Peace attend their ashes'.[227] Ruth was further vexed when it transpired that an accident had occurred because their chauffeur was drunk. He had been: 'nipping quietly for ages – whiskey and crème de menthe'. On this occasion, George Riddoch, who was visiting the Heads, took it upon himself to dismiss the miscreant. Nonetheless, the incident took its toll of Ruth: 'After 4 years of service during which we trusted him completely. It has been a horrible time. I feel, and look 10 years older'.[228]

Ruth was also obliged to try to maintain the fabric of the houses and grounds that she and Henry occupied. In a letter to Sassoon she listed the problems facing her on moving into Forston House: 'Lawsuits, Lawyers' letters, Accidents, "bad legs", swarming bees in the drawing-room, a general breakdown of windows, ball-cocks and taps all over the house, plague us not a little but offer some variety to our daily hum-drum'.[229] Hartley Court provided challenges of its own, including the infestation of the garden by rabbits. When Ruth ordered wire designed to exclude these pests, the tradesman provided her with a shipment with far too wide a mesh: 'I asked him if he thought I intended to keep Kangaroos?'[230]

The upkeep of Hartley Court was to prove a serious drain on Ruth's strength and patience as well as upon her purse. She told Hester:

> I was called to see our Landlord's Agent and the Clerk of the Works about repairs to our estate and I had a fearsome time while the long roll of things to be done this

autumn was unfolded before my horrified eyes, house, gate, garage, greenhouse to be repainted, front drive regravelled etc. etc. Hundreds of pounds to be laid out, contracts to be made, estimates to be considered and so on. I was really plunged into the depths and all this at a time when the new garden expenses will be doubled and all seems in confusion and upheaval. Well all will be put right in time but I would wish to feel more fit and less giddy and head-achy ...[231]

Ruth often wrote of her mental and physical states. As she first came to terms with the reality of Henry's illness, she manifested a mixture of anxiety and anger. Ruth's ire was, in particular, directed at those who she felt were insensitive to the true nature of her husband's condition – in particular those so-called: 'friends who write they hope "Harry is all right" make me sick and that after they have stayed with us at the sea and have seen the poor darling walk. Sickness and trouble do show one who are the real friends and who the pretence'.[232] Other 'best friends' were so obtuse as to assert: '"Why do you mind so much what happens to Harry's body, since the disease will never attack his mind?"'[233]

The consequences of Henry's condition extended to Ruth's own conception of self; she too was imprisoned by the limitations of his diseased body. In August 1925, she told Janet Ashbee: 'I plan to and plan and live just from moment to moment, which sounds paradoxical, but is many a woman's fate'.[234] By November the chill weather found a counterpart in Ruth's state of mind:

> here it is terribly cold and my ideas are frozen and my heart sick with poor darling Harry's disabilities. It is too sickening to write day after day and decline all sorts of invitations for him – visits to Cambridge, honourable proposals to give lectures etc. However I pray to keep a stiff lip and a cheerful outside ...[235]

After the move to Forston, Ruth sought to reconcile herself to what she called: 'my little narrowed life'.[236] She took some comfort in the beauty of the surroundings and in such joys of rural life as waking: 'every morning to a world of bubbling song. And we have willow-wrens too who warble under their breath too bewitchingly and with the bass of the rooks and the croonings of the wood pigeons and the occasional scream of the jay make a full orchestra for our delight'. She also delighted in the garden and its flowers – when they bloomed, 'a heavenly large bunch [of violets] [was] blushingly presented' to her each day by the 'Old Gardener'.[237]

Moreover, on the day of their wedding anniversary, she could maintain: 'that in spite of woe and death gay is life and sweet is breath. For really H.H. does grow lovelier to me every day and looks and tones that, where all else is hopeless, can still make a green garden of the future'.[238] Ruth managed to sound a note of optimism after the first Christmas at Forston: 'with broad sunshine over the brown ploughed fields and bare downs and remembrance from many many friends and a kind of inner content which will go on growing in us I do hope and believe'.[239]

In the coming years Ruth felt obliged for Henry's sake to maintain an outer facade of cheeriness. However, even from the early days of the exile, Ruth sometimes revealed to her close friends a growing sense of isolation as well as a recognition that her role as carer threatened to overwhelm all other aspects of her existence. As she told Hester: 'I have never a moment to myself now. Poor Uncle H. is terribly dependent on me all the time'.[240] To Janet Ashbee, Ruth expressed her fear that: 'I shall in time forget to talk to my own kind, for to Nurse I talk as one talks to one's Nurse – which is hardly to be called conversation, is it?'[241]

In April 1929, when the task of finding a new house was added to her usual responsibilities, Ruth confided in Sassoon:

> My own brains are disintegrating fast under the pressure of household cares at Forston combined with this evasive future Home. Indeed I feel half demented privately, although in public with H.H. I have to appear composed, especially as next Sunday is our little Silver Wedding Day and I must show him a shining bridal face on that beloved anniversary.[242]

She was in effect cast in a maternal role that threatened to become all-consuming to a child who would never grow up, but could with time only grow ever more needy. By 1933, Ruth confessed to Ashbee that she felt in: 'a state of cat-like weakness'. She was suffering from a bad cold at the time, but maintained that the true source of her debilitated condition: 'is not a cold'. The letter ends abruptly with the telling sentence: 'The Oracle speaks'.[243] In April 1934, although Henry was 'fairly well', Ruth confessed that: 'I feel, very feeble and so easily tired by even the nicest visitors'.[244] The comforts of Hartley Court notwithstanding, her home came increasingly to seem like a prison. Ruth admitted to Ashbee that she sat outside as much as possible because: 'I feel drowning for want of air in the house now'.[245]

At the end of 1935 Ruth's health was giving sufficient concern for her to seek professional help, though with little benefit. She told Janet:

> the visit of H's dear Dr Riddoch and his examination of my vile body has failed to show anything organically wrong and both doctors have left me to get intestinally comfortable by the light of my own reason. I am making experiments on diet and finding out by trial and error what will cure my ills – so far not very successfully ...[246]

Ruth came to realize that there was only one exit for her from the dreary and wearing existence to which her loyalty to Henry had committed her. She wrote in 1933 that her birthday had been 'such a merry day'. But this cheery sentiment had a sombre undertone: 'for I felt unexpectedly light-hearted and indeed every sign-post past lightens the load of responsibility and brings the desired Harbour nearer'.[247] Long before, while she and Henry were still in their prime, Ruth had already mused on the prospect of death:

> All my life I have been expecting Happiness to stand waiting for me at the end of every long lane, and lately it has been tragic to me to think that I have lived certainly half my life and it has never come. Now I know it is Death not Happiness who is waiting, but the thought brings no terror only peace.[248]

Ruth finally found this peace in May 1939 when she succumbed to pneumonia. *The Times* newspaper reported that: 'Lady Head, wife of Sir Henry Head, M.D., the physiologist famed for his researches into nervous diseases, died at Hartley Court, near Reading, on Monday'. Ruth's achievements as an author and educationalist were duly recorded. But the obituary dwelled mostly on the role she played in her later years, noting that after her husband became an invalid, 'she gave up all social life in order to look after him and be his amanuensis'.[249]

Gordon Holmes provided Ruth a fuller epitaph when he declared that Head had married a woman:

> who was in every respect a perfect wife to him and gave him a complete community of tastes and temperament. She shared his interests, stimulated his enthusiasms, criticized his writings and relieved him of many of the petty worries of life. In his later years she was his constant companion, and her philosophical outlook, her joy in life and her encouragement helped him to bear an illness which otherwise would have been an intolerable fate to one of his active mind and body.[250]

Little is know of how Henry spent the last lonely months of his life. Hester Marsden-Smedley tried to fill the gap left by Ruth's death. She was, however, beset with many other family and professional responsibilities. After the outbreak of war, moreover, Hester spent much of her time in Brussels, where she worked as a reporter. Another female relative moved with her sons into Hartley Court as a stopgap. But in September 1939, Hester wrote to her husband to tell him of: 'the appalling Hartley news; one of these boys had been struck down with infantile paralysis. Hester could:

> hardly bear to think of it for her for the boy for Uncle H.--and all the complications of that nightmare household with an extra invalid --I gather Uncle H. is being remarkable. They have a London Hospital nurse and the men are helping too --Riddoch in charge.[251]

From Belgium, Hester tried to keep Head informed of events on the continent. She wrote in April 1940 of the arrival in Brussels of staff evacuated from the British Legation in Copenhagen following the German takeover of Denmark. She described the various indignities to which their captors had subjected these diplomats; they were: 'rounded up, photographed from every angle and then driven to a barracks through the street in an open horsedrawn dray ... There was no doubt whatever that the English were treated more roughly that the other two nationalities. The military attaché ... was handcuffed for a while ... We are certainly THE enemy!'[252] It is sad to reflect that this account of how low the nation in which he had in the past invested so much admiration had sunk was among Head's last insights into the outside world.

In April 1940, Janet Ashbee passed on the news that Head: 'sounds sadly weaker'.[253] The neurologist, Russell Brain (1895–1966) claimed that he saw Head a few months before he died, noting that by this time he was immobilized and had great difficulty in speaking:

> But his mind was still alert, and he was eager to know what was happening in neurology. He talked about his own disabilities, and the physical obstacle they put in the way of speech, the part played by the auditor in speech, the body-image, and sensory perception.[254]

This is perhaps the last image of Henry Head in life that we possess.

George Riddoch observed that Ruth's death: 'was also the cause of his death'. Although the routine at Hartley Court continued, 'with the necessary changes required by his failing condition, [Head] longed to rejoin her, and his final illness was a happy and welcome release'.[255] Harry died on 8 October 1940.

Hartley Court passed to Head's favourite niece Hester Marsden-Smedley – by now a journalist and active in politics. She had always taken great delight from her visits to the house, describing one particular weekend as: 'just uninterrupted bliss'. Hester found it difficult to analyse whence this bliss derived:

> because its made up of so much, the talk, the sleep, the companionship, the beautiful things in your beautiful house, the knowledge that here are two to share even when they disagree – even should they disapprove – the exchange of impressions old and new, just everything. If I ever think of the future ... I have only one dread – to which air raids, revolutions, age and even the end of the world are as nothing, and that is the stopping of Hartley.[256]

In July 1941, Janet Ashbee paid a visit to a house that was now transformed. She marvelled at what Hester: 'is able to do with Henry's house & money'. Ashbee however sensed the shades that still haunted the place: 'how very strange it is here without Ruth or Henry! I am sleeping in Henry's room (never before seen!), & whether it is a still present spirit of Pain, or just potent ghosts, I slept very badly & woke at 3 – waiting for the dawn'.[257]

But in the hours of daylight the contrast with her previous stays at Hartley Court could hardly be greater. The garden that Ruth had so cherished had 'run quite wild'; Hester, whatever her other aptitudes, was no gardener. Nonetheless, the conditions of wartime economy demanded that vegetables should be grown:

'the kitchen garden [is] in superb order --& producing not only enough for an <u>average</u> of <u>18</u> people (including 5 small children) but a great deal is sold ...' The house was, moreover, teeming with people: 'with 2 young cousins (soldiers) billeted here – 2 maids with <u>babies</u> (husbands prisoners in Germany), 3 other miscellaneous servants, 2 of the M-Smedley children (Henrietta 6, Christopher 10) besides other children & people coming & going'. Ruth's drawing-room was now: 'a lovely comfy muddly <u>nursery</u>; the dining room unchanged but table full length & 7 or 8 at each meal – often 12 or 14 ...' In short:

> The atmosphere absolutely ALIVE, such a contrast to the strange & necessary Hush-hush of the previous dynasty, with the real culture (books everywhere, 2 pianos, gramophone etc) all still there – but the soignée-ness & terrific regard for <u>things</u> all vanished. They are now subservient to <u>people</u> – it is pathetic in a way but yet you feel <u>right</u>.

Janet could not help but wonder what the 'previous dynasty' would have made of this transformation. She concluded that: 'Ruth I feel w^d shed a tear over her Empire sofa with dolls' tea sets on it!' On the other hand, 'Henry w^d have rejoiced'.[258]

Postscript: A Life as Fruitful as it was Beautiful

In February 1925, Ruth told Robert Nichols:

> I do want you to promise that the very moment you read in the Times that Henry is dead, you will sit down there and then and put that just as you put it to us for other people to know what he has done for O so many suffering people. I can talk quite cheerfully about this as I have made up my mind to die a few weeks later.[259]

Ruth's prediction of who would die first was to prove mistaken; it was Henry who was obliged to endure life for several months after he had lost her.

Nichols was, however, quick to put down his tributes to his former physician in a letter written two days after Head's death:

> He was just; he was good; he was learned; he was true; he was kind; he was wise. Shakespeare, Leonardo, Mozart, Goethe would all have loved as well as respected him – our dear Professor, our spirit of knowledge, our deep well of compassionate gentleness and true serenity![260]

On 15 October, Nichols sent a more considered appreciation to *The Times*, describing himself as: 'one who was for 25 years the friend of Sir Henry Head F.R.S., and who was, like so many others, helped back from the "borderline" by him in the last War'. He wrote that Head: 'possessed the fullest as well as the finest mind I have ever known'. Nichols compared Head to Leonardo da Vinci in the scope of his intellectual interests:

Nor did he resemble Leonardo in mind only. He had Leonardo's lofty human compassion, humility, patience, and profound serenity of spirit. No stupidity or even downright wickedness could ruffle him, who, an imaginative scientific genius, looked on man but as part of nature, the supreme and unfathomable artist.

In remembering Henry, his friends could have only one regret – that because of his illness this 'Quaker Prospero' had been unable: 'to place upon paper that mass of varied reflections which rose in his being when he contemplated other works of nature and of man than those particular studies to which he consecrated a life as fruitful as it was beautiful'.[261]

The response of other friends was more measured though no less affecting. Nathan Redcliffe Salaman, on hearing the news, declared: 'And so, at last, the end has come to a tragedy which ... has been illuminated by the glow of the courage of and brief sacrifice of both Ruth and Harry'. Their: 'Two lives ... have meant so much for so many of us'. Salaman confessed that it would: 'be impossible for me to estimate how much I owe to Harry. It was he who first put me on to a line of Research and for the first years of my professional life was my guide and mentor'. But his debt to the Heads went beyond the purely professional. Henry's long struggle against his affliction, 'with the ever increasing help of Ruth – was an inspiration to all of us who knew what it meant. That he still kept [---] of what life was left to him after her death – is not far short of a miracle'.[262]

George Riddoch, another protégé, also composed a tribute. He wrote that the five years he had collaborated with Head: 'was indeed a happy time for me, the happiest of my life, a voyage of discovery and adventure, under the leadership of a great mind'. Riddoch maintained that Head's inspirational qualities made him: 'the great teacher he was to medical students, aspiring neurologists, and senior colleagues alike'. He noted that the same attributes: 'brought intelligent patients into his fold; they could not resist the glamour and excitement of his work on themselves, and became his ardent collaborators'.[263]

Riddoch dwelt on the onset of Head's 'creeping palsy'. The disease had: 'showed itself first in his left arm and hand and in small ways, so that, being strongly right-handed, he was for a considerable time unaware of what was happening to him, in fact only trained observers noticed the tragic signs'. Riddoch surmised that Head must, however, have known that something was wrong – 'for he toiled with almost feverish energy at the completion of his final work on "Aphasia"'. When the book was complete, Head had discussed his condition with Riddoch:

> With his practical philosophy, he accepted it at once, and his amazing power of adaptation enabled him to plan the rearrangement of his life. With his excellent general health, he well knew what he had to face – long years of steadily increasing physical disablement, with his mind unimpaired except for its capacity for continuous effort, in the grip of a relentless foe which medical science could not restrain. Without a

word of self-pity, with distress only for his devoted wife's heavy burden, he set himself in his methodical way to make his plans, and nothing was omitted.

Even as his illness progressed, Riddoch maintained that Head remained: 'the scientist and the teacher'. He observed: 'the phenomena of his progressive disease upon himself – in fact it was his second personal experiment, as carefully described as his first'. Thanks to this careful observation of his own bodily decline, Head was able to supply Riddoch and other neurologists the 'most valuable information which had not been described before about this not uncommon malady'.[264]

In a letter written to Robert Nichols in 1941, Arnold Mayhew recalled the happier days when Ruth and Henry were in their prime presiding over their London salon. He praised Nichols for:

> your picture of the Montagu Square tea table – Ruth presiding with her cigarette clipper (and how clever of you to have noticed the Henry Irving touch) – and Henry triumphant from his consulting room 'with a jocular gleam in the eye of the rogue elephant' – how well I know it – it came so often when I gave him the reply he expected, the reply whose source in a part of myself previously unexperienced by me he proceeded to lay bare.[265]

But Mayhew also stressed the calm courage with which Head had faced the affliction that had blighted his last years. He was uncertain how far this 'quietism' derived from Head's Quaker background or: 'how far it was Stoic in character'. In his final years, indeed, Henry had reminded Mayhew of: 'the Roman Senator with folded arms – awaiting the next stage of his disease'.[266]

Mayhew mused further upon the ethic that had guided Head through life and sustained him through his final illness. One of Head's favoured axioms had been: 'there is no such word as "ought"'. This was at first sight a surprising dictum from one who had throughout his life shown an exceptional sense of duty and responsibility. Mayhew correctly discerned that Head was in fact denying the right of any external authority to impose normative standards. It was for each individual to determine his or her own 'ought'. Mayhew remarked that Head failed to see that such externally impressed standards of behaviour were in fact necessary for most individuals. Head was oblivious to: 'the significance of the distinction in the ordinary man's mind between "right" and "wrong"'. This imperception was due to the fact that Henry was:

> no 'ordinary' man. From the very start he had a dominating purpose in his life – he knew exactly what he wanted and he deliberately, without effort and without slackening, shaped his whole life to get what he wanted – opportunities for accurate observation and scope for clear thought and expression. The results of life's work were far more beneficial to the world than what most people, with a very acute conscience, have been able to achieve. But it was always inevitable for him – he 'could no other'.

> I don't think that during his whole life he even reflected on what he ought to do – or had any difficulty in deciding on his next step. There was never any conflict of loyalties or delicate balance of values. There must be hundreds that owe him health or happiness – or the chance of a start on the right lines. But his reason for helping them was not 'thou shalt love thy neighbour as thyself'. He helped them because he hated waste and wanted to see every one 'functioning normally'.

Knowing Head had, moreover, taught Mayhew one: 'supreme ethical value that many persons with highly developed consciences could never have taught me – the need for absolute honesty in thought speech and action. He was the most honest man I have ever met'.[267]

The sense of independence from external ethical codes that Mayhew discerned in Henry Head is one of the defining signs of the modern condition. Head's personal circumstances made this predicament especially acute for him. He had been brought up in a religious household and was thus aware of the demands that Christian belief had imposed upon an earlier generation, as well as of the comfort and sense of direction that religious faith could bestow. As a young man, however, Head concluded that the truths revealed by natural science were incompatible with Christianity and indeed with all forms of theism. Physics revealed a cosmos in which the earth and the human beings that inhabited it were but a tiny insignificant speck. Biology, moreover, had uncovered a nature that was cruel and chaotic, rather than the work of a benevolent deity. It was left to men and women to decide how – indeed *if* – they would live in this demystified world.

Head found his own answer to this challenge through the pursuit of scientific knowledge. As Ruth surmised, science was his most exigent mistress; even in his final, crippling, illness, he sought to serve her through the meticulous observation of the deterioration of his body. As this study has shown, however, Henry combined this commitment to science with an untrammelled passion for beauty, companionship, and the life of the mind. Those who knew him felt they had been rarely privileged. When Salaman sought to encapsulate that sense of gratitude he concluded: 'I can but repeat the old Jewish Formula – "May his memory be for a blessing"'.[268]

Figure 4. Henry Head and Ruth Head

NOTES

The following abbreviations have been used for archives that are frequently referenced in the notes:

BL: British Library,
CMAC: Contemprorary Medical Archives Centre, Wellcome Library for the History of Medicine, London,
CUL: Cambridge University Library,
KCC: King's College Cambridge,
LHA: London Hospital Archives,
YUL: Yale University Library.

Introduction

1. R. Brain, *Henry Head: the Man and his Ideas* (London: Macmillan, 1961) – originally a paper given at the Head Centenary celebrations held at the London Hospital. See also: R. A. Henson, 'Henry Head: Man of Culture, Compassion and Science', *Journal of Medical Biography*, 6 (1998) 15–20; M. Critchley, 'Henry Head, 1861–1940', in *The Black Hole and Other Essays* (London: Pitman, 1964), pp. 172–9; G. Holmes, 'Henry Head, 1861–1940', *Obituary Notices of the Fellows of the Royal Society*, 3 (1939–41) 665–89.
2. J. M. Wilson, *Siegfried Sassoon: The Journey from the Trenches* (London: Duckworth, 2003); A. Charlton and W. Charlton, *Putting Poetry First: A Life of Robert Nichols* (Norwich: Michael Russell, 2003).
3. P. Barker, *Regeneration* (London: Penguin, 1992), *The Eye in the Door* (London: Penguin, 1993), *The Ghost Road* (London: Penguin, 1996).
4. L. S. Jacyna, *Lost Words: Narratives of Language and the Brain, 1825–1926* (Princeton, NJ: Princeton University Press, 2000).
5. See: S. Greenblatt, *Renaissance Self-Fashioning: from More to Shakespeare* (Chicago, IL: Chicago University Press, 1984); C. Taylor, *Sources of the Self: The Making of the Modern Identity* (Cambridge: Cambridge University Press, 1989); J. Siegel, *The Idea of the Self: Thought and Experience in Western Europe since the Seventeenth Century* (Cambridge: Cambridge University Press, 2005); C. Lawrence and S. Shapin, *Science Incarnate: Historical Embodiments of Natural Knowledge* (Chicago, IL: Chicago University Press, 1998); R. Porter (ed.), *Rewriting the Self: Histories from the Renaissance to the Present* (London: Routledge, 1997).
6. M. Foucault, *The History of Sexuality*, 3 vols, (London: Penguin Books, 1981–4).

7. D. Ross (ed.), *Modernist Impulses in the Human Sciences 1870–1930* (Baltimore, MD: Johns Hopkins Press, 1994), p. 6.
8. Ibid., p. 7.
9. On this reaction to modernity see: H. S. Hughes, *Consciousness and Society: the Re-Orientation of European Social Thought, 1890–1930* (St. Albans: Paladin, 1974); F. Stern, *The Politics of Cultural Despair* (Berkeley, CA: University of California Press, 1974); C. Schorske, *Fin de siècle Vienna: Politics and Culture* (Cambridge: Cambridge University Press, 1979); T. J. Jackson Lears, *No Place of Grace: Antimodernism and the Transformation of American Culture, 1880–1920* (Chicago, IL: University of Chicago Press, 1994).
10. G. R. Searle, *Eugenics and politics in Britain, 1900–1914* (Leyden: Noordhoff, 1976); D. Pick, *Faces of Degeneration: a European Disorder, c.1848–c.1918* (Cambridge University Press, 1989); W. Greenslade, *Degeneration, Culture, and the Novel, 1880–1940* (Cambridge University Press, 1994); F. Mort, *Dangerous Sexualities: Medico-Moral Politics in England since 1830* (London: Routledge, 1987); J. R. Walkowitz, *City of Dreadful Delight: Narratives of Sexual Danger in Late-Victorian London* (London: Virago Press, 1992).
11. See: M. J. Peterson, *The Medical Profession in Mid-Victorian London* (Berkeley, CA: University of California Press, 1978); C. Lawrence, 'Incommunicable knowledge: science, technology and the clinical art in Britain, 1850–1914,' *Journal of Contemporary History*, 20 (1985) pp. 503–20.
12. Lears, *No Place of Grace*, pp. xv–xvi.
13. L. Ginzburg, *On Psychological Prose*, translated by J. Rosengrant, (Princeton. NJ: Princeton University Press, 1991).
14. B. Linker, 'Resuscitating the 'Great Doctor': the career of biography in medical history,' in T. Söderquist (ed.), *The History and Poetics of Scientific Biography* (Aldershot: Ashgate, 2007), pp. 221–39. For a more general discussion of the status of biography in the history of science see: M. Shortland and R. Yeo, 'Introduction,' idem (eds), *Telling Lives in Science: Essays in Scientific Biography* (Cambridge: Cambridge University Press, 1996), pp. 1–44.

1 The Making of a Neurologist

1. Letter of Henry Head senior to Henry Head, 2 August 1902, CMAC, PP/HEA/D3.
2. Letter of Henry Head to Redcliffe Nathan Salaman, 26 August c. 1901, Salaman Papers, CUL, MS Add. 8171, Box 3.
3. Letter of [Hester Head] to Henry Head, 7 April 1900, CMAC, PP/HEA/D2.
4. Letter of Henry Head to Ruth Mayhew, 22 September 1899, CMAC, PP/HEA/D4/5.
5. H. Head, 'Autobiography', CMAC, PP/HEA, A.1, pp. 4–5.
6. G. Cantor, *Quakers, Jews, and Science: Religious Responses to Modernity and the Sciences in Britain, 1650–1900* (Oxford: Oxford University Press, 2005).
7. Head, 'Autobiography', p. 4.
8. Galton Papers, University College London, MS 133/2H.
9. Gordon Holmes, 'Henry Head, 1861–1940,' *Obituary Notices of the Fellows of the Royal Society*, 1939–41, 3: 665–89. Holmes, p. 683.
10. Cantor, *Quakers*, pp. 233–40.
11. Head, 'Autobiography', pp. 1–2.

12. Letter of Ruth Head to Henry Head, 23 March 1917, CMAC, PP/HEA/D4/19 Box 7.
13. Letter of Henry Head to Ruth Mayhew, 30 August 1899' CMAC, PP/HEA/ D4/5.
14. It may be noted that Morris's father was himself 'something in the City.'
15. Head, 'Autobiography', p. 6.
16. *Ibid.*, p. 3. 'Spillikins' is a game played by scattering a collection of sticks onto a surface. The players then seek to pick up each of the sticks without disturbing the rest.
17. *Ibid.*, p. 4.
18. *Ibid.*, p. 5.
19. *Ibid.*, p. 7.
20. *Ibid.*, p. 7.
21. 'Charlie' Head was to die in February 1878 while at the school.
22. Letter of Henry Head to Hester Head, 8 February *c.* 1876, CMAC, PP/HEA/D1.
23. Letter of Henry Head to Hester Head, undated, *ibid.*
24. See: J. A. Mangan, *Athleticism in the Victorian and Edwardian Public School: The Emergence and Consolidation of an Educational Ideology* (Cambridge: Cambridge University Press, 1981).
25. Head, 'Autobiography', p. 9.
26. *Ibid.*, p. 15.
27. *Ibid.*, p. 10.
28. Letter of Henry Head to Ruth Mayhew, 24 June 1900, CMAC, PP/HEA D4/6.
29. Head, 'Autobiography', pp. 11–12.
30. *Ibid.*, p. 13.
31. Letter of Henry Head to Hester Head, 9 June *c.* 1876, CMAC, PP/HEA/D1.
32. Head, 'Autobiography', p. 12.
33. *Ibid.*, p. 9.
34. Letter of Henry Head to Hester Head, undated, CMAC, PP/HEA/D1.
35. Letter of Henry Head to Hester Head, 8 February *c.* 1876, *ibid.*
36. Letter of Henry Head to Hester Head, undated, *ibid.*
37. The awards are recorded in the 'Blue Books' kept at Charterhouse. I am grateful to Mrs Ann Wheeler for supplying this information.
38. Letter of Henry Head to Hester Head, 15 August 1880, *ibid.*
39. Letter of Henry Head to Ruth Mayhew, 1897, CMAC, PP/HEA/D4/2.
40. Letter of Henry Head to Ruth Mayhew, 30 June 1899, CMAC, PP/HEA/D4/4.
41. Letter of Gerald S. Davies to Henry Head senior, 17 April. [1879], *ibid.*
42. Head, 'Autobiography', p. 14.
43. Letter of Henry Head to Hester Head, undated but written in summer 1879, CMAC, PP/HEA/D1.
44. Head, 'Autobiography', p. 17.
45. *Ibid.*, p. 18.
46. T. M. Porter, *Karl Pearson: the Scientific Life in a Statistical Age* (Princeton, NJ, Princeton University Press, 2004).
47. Letter of Henry Head to Hester Head, 9 June [1878], CMAC, PP/HEA/D1.
48. Head, 'Autobiography', p. 18.
49. Henry Head, 'Halle Journal', 13 April [1880], CMAC, PP/HEA, Box 3.
50. *Ibid.*, 14 April [1880].
51. Head, 'Autobiography', p. 19.
52. Head, 'Halle Journal', 15 April 1880.

53. *Ibid.*, 16 April 1880.
54. Letter of Henry Head to Hester Head, 17 April 1880, CMAC, PP/HEA/D1.
55. Head, 'Halle Journal', 21 April 1880.
56. *Ibid.*, 21 June or July 1880.
57. Head, 'Autobiography', p. 20.
58. Head, 'Halle Journal', 25 [April, 1880].
59. *Ibid.*, 27 July 1880
60. *Ibid.*
61. Letter of Henry Head to Hester Head. Sunday [25 April 1880], CMAC, PP/HEA/D1.
62. *Ibid.*
63. Head, Autobiography, p. 27.
64. Head, 'Halle Journal', 25 [April 1880].
65. Head, 'Autobiography', p. 21. On the bicycle as an emblem of modernity see: S. Kern, *The Culture of Time and Space 1880–1918* (London: Weidenfeld & Nicolson, 1983), pp. 111–13.
66. Head, 'Halle Journal', 24 [May 1880].
67. *Ibid.*, 28 *c.* July 1880.
68. *Ibid.*, 21 *c.* July 1880.
69. Head, 'Autobiography', p. 22.
70. Letter of Henry Head to Hester Head, [25 April 1880], CMAC, PP/HEA/D1.
71. Head, 'Autobiography', p. 27.
72. See: T. N. Bonner, *Becoming a Physician: Medical Education in Great Britain, France, Germany, and the United States 1750–1945* (New York, NY: Oxford University Press, 1995), ch. 9.
73. Head, 'Autobiography', p. 23.
74. Letter of Henry Head to Hester Head, [25 April 1880], CMAC, PP/HEA/D1.
75. See: P. F. Cranefield , 'The philosophical and cultural interests of the Biophysics Movement of 1847,' *Journal of the History of Medicine*, 21 (1966), pp. 1–7; E. A. Seyfarth, 'Julius Bernstein (1839–1917): pioneer neurobiologist and biophysicist,' *Biological Cybernetics*, 94 (2006), pp. 2–8.
76. Head, 'Halle Journal', 13 [May 1880]. On the development of graphic inscription devices in the nineteenth century see: S. de Chadarevian, 'Graphical method and discipline: Self-recording instruments in nineteenth-century physiology,' *Studies in History and Philosophy of Science*, 24 (1993), pp. 267–91.
77. Head, 'Halle Journal', 14 [May 1880].
78. Head, 'Autobiography', p. 24.
79. *Ibid.*, p. 25.
80. Head, 'Halle Journal', 26 [May 1880].
81. *Ibid.*
82. *Ibid.*, 27 [May 1880].
83. *Ibid.*, 28 [May 1880].
84. On these distinctions see: S. Shapin, *A Social History of Truth: Civility and Science in Seventeenth-Century England* (Chicago: Chicago University Press, 1994), ch. 8.
85. Head, 'Halle Journal', 28 *c.* July 1880.
86. Head, 'Autobiography', p. 25.

87. See: L. S. Jacyna, 'A host of experienced microscopists: the establishment of histology in nineteenth-century Edinburgh,' *Bulletin of the History of Medicine*, 75 (2001), pp. 225–53.
88. See: A. M. Tuchman, *Science, Medicine, and the State in Germany: the Case of Baden, 1815–1871* (New York, NY: Oxford University Press, 1993).
89. Head, 'Autobiography', p. 25.
90. Head, 'Halle Journal', Friday 23 *c.* June 1880.
91. Head, 'Autobiography', p. 25.
92. Lette of Henry Head to Hester Head, undated, CMAC, PP/HEA/D1.
93. Head, Autobiography, p. 28.
94. See: O. H. Wangensteen and S. D. Wangensteen, *The Rise of Surgery: from Empiric Craft to Scientific Discipline* (Folkestone: Dawson, 1978), pp. 427–8; Michael Sachs, *Historisches Chirurgenlexikon: ein biographisch-bibliographisches Handbuch bedeutender Chirurgen und Wundärtzte*, 5 vols, (Heidelberg: Kaden Verlag, 2002), vol. 3, pp. 371–7.
95. Head, 'Autobiography', p. 29.
96. Letter of Henry Head to Hester Head, 21 April 1880, CMAC, PP/HEA/D1.
97. Head, 'Halle Journal', 31 [July 1880].
98. Letter of Henry Head to Hester Head, [31 May 1880], CMAC, PP/HEA/D1.
99. Letter of Henry Head to Ruth Mayhew, 21 July 1899, CMAC, PP/HEA/ D4/5.
100. Head, 'Halle Journal', Tuesday, [29 August 1880].
101. Head, 'Autobiography', p. 29–30
102. Letter of Henry Head to Hester Head, 2 November 1880, CMAC, PP/HEA/D1.
103. On the Cambridge 'grinding' system see: A. Warwick, *Masters of Theory: Cambridge and the Rise of Mathematical Physics* (Chicago, IL: Chicago University Press, 2003), especially pp. 84–94.
104. Head, 'Autobiography', p. 30.
105. Girton College was established in 1869 and Newnham in 1872. Women were allowed to attend lectures and to sit examinations during Head's time in Cambridge; they were not, however, awarded degrees until 1947.
106. Letter of Henry Head to Hester Head, 2 November 1880, CMAC, PP/HEA/D1.
107. Letter of Henry Head to Hester Head, 5 March 1884, CMAC, PP/HEA/D1.
108. The programme is in: CMAC, PP/HEA/A.5.
109. Letter of Head to Hester Head, 22 May 1881, CMAC, PP/HEA/D1.
110. Head, 'Autobiography', p. 34.
111. Ibid., p. 39.
112. P. A. Schilpp, *The Philosophy of Alfred North Whitehead* (Evanston: Northwestern University, 1941), p. 7.
113. *Ibid.*, p. 8.
114. Letter of Ruth Head to Robert Nichols, 15 April 1930, Nichols Papers, BL, uncatalogued.
115. Head, 'Autobiography', p. 31. The derision for Paley shown by Head and his friends can be contrasted with Charles Darwin's much more respectful reading of the *Natural Theology* as a Cambridge undergraduate some fifty years earlier. See: Aileen Fyfe, 'The reception of William Paley's *Natural Theology* in the University of Cambridge,' *British Journal of the History of Science*, 30 (1997) pp. 321–35.
116. H. Head and R. Mayhew, 'Rag Book', vol. 2B, p. 2, CMAC, PP/HEA/E3/41.
117. *Ibid.*, p. 24.
118. *Ibid.*, p. 25.

119. Letter of Henry Head to Hester Head, [August 1880]. CMAC, PP/HEA/D1.
120. Head and Mayhew, 'Rag Book', vol. II.A, p. 80, CMAC, PP/HEA E3/3.
121. Head, 'Autobiography', p. 34.
122. On the prevalence of socialist ideas in Cambridge during the 1880s see: Alan Crawford, C.R. Ashbee: Architect, Designer and Romantic Socialist (New Haven, CT: Yale University Press, 2005), pp. 10–21.
123. Head, 'Autobiography', p. 33.
124. On Foster's teaching, see: G. L. Geison, *Michael Foster and the Cambridge School of Physiology: The Scientific Enterprise in Late Victorian Society* (Princeton, NJ: Princeton University Press, 1978), pp. 118–9.
125. M. Foster, *A Text Book of Physiology*, 3rd edn, (London: Macmillan, 1879).
126. Head, 'Autobiography', p. 37.
127. *Ibid.*, p. 36.
128. Geison, *Michael Foster*, p. 34.
129. Head, 'Autobiography', pp. 37–8.
130. *Ibid.*, p. 37.
131. Letter of Henry Head to Ruth Mayhew, 20 August 03, CMAC, PP/HEA/D4/15.
132. *Ibid.*
133. Geison, *Michael Foster*, p. 220.
134. Quoted in: E. G. T. Liddell, 'Charles Scott Sherrington 1857–1952,' *Obituary Notices of Fellows of the Royal Society*, 8 (1952) pp. 241–70, p. 244.
135. *Ibid.*, p. 245.
136. Head, 'Autobiography', pp. 38–9.
137. *Ibid.*, p. 38.
138. *Ibid.*, p. 41.
139. *Ibid.*
140. *Ibid.*, p. 42.
141. Letter of Henry Head to Ruth Mayhew, 7 December 1899, CMAC, PP/HEA/ D4/5.
142. C. Baumann, 'Henry Head in Ewald Hering's laboratory in Prague,' *Journal of the History of the Neurosciences.*, 14 (2005), pp. 322–33.
143. Head, 'Autobiography', p. 46.
144. *Ibid.*
145. On the industrialization of physiological work in the later nineteenth century, see: S. Dierig, 'Engines for experiment: Laboratory revolution and industrial labour in the nineteenth-century city,' in S. Dierig, J. Lachmund, and J. Andrew Mendelsohn (eds), *Science and the City, Osiris*, 18 (2003) pp. 116–34.
146. Head, 'Autobiography', p. 45.
147. *Ibid.*, p. 58.
148. See: M. G. Ash, *Gestalt Psychology in German Culture, 1890–1967: Holism and the Quest for Objectivity* (Cambridge: Cambridge University Press, 1995), pp. 54–5. R. Steven Turner, *In the Eye's Mind: Vision and the Helmholtz-Hering Controversy* (Princeton, NJ: Princeton University Press, 1994). C. Baumann, *Die Physiologe Ewald Hering (1834–1918)* (Frankfurt: Dr. Hänsel-Hohenhausen, 2002); F. Hillebrand, *Ewald Hering: ein Gedenkwort der Psychophysik* (Berlin: Julius Springer, 1918).
149. Head, 'Autobiography', p. 58.
150. Head, 'William Halse Rivers Rivers, 1864–1922,' *Proceedings of the Royal society of London Series B*, 45 (1924) pp. xliii–xlvii, p. xliii.
151. Head, 'Autobiography', pp. 58–9.

152. See the essays in: C. Lawrence and S. Shapin, *Science Incarnate: Historical Embodiments of Natural Knowledge* (Chicago: Chicago University Press, 1998).
153. Head, 'Autobiography', p. 60.
154. F. Kavka, *The Caroline University of Prague*, (Prague: Universita Karlova, 1962) pp. 59–60.
155. Head, 'Autobiography', pp. 50–1.
156. *Ibid.*, p. 51.
157. *Ibid.*, p. 52.
158. Letter of Henry Head to Hester Head, [September 1884], CMAC, PP/HEA/D1.
159. Head, 'Autobiography', pp. 49–50.
160. Letter of Henry Head to Hester Head, [September 1884]. CMAC, PP/HEA/D1.
161. Head, 'Autobiography', pp. 49–50.
162. Letter of Henry Head to Hester Head, [undated; probably written summer 1886], CMAC, PP/HEA/D1.
163. *Ibid.*
164. Letter of Henry Head to Hester Head, 1886a, CMAC, PP/HEA/D1.
165. Letter of Henry Head to Hester Head, summer 1886, *ibid.*
166. Letter of Henry Head to Hester Head, 1886a, *ibid.*
167. *Ibid.*
168. See: C. E. Schorske, *Fin-de-Siècle Vienna,* ch. 3.
169. D. Ross (ed.), *Modernist Impulses,* pp. 6–7.
170. Head, 'Autobiography', p. 49.
171. *Ibid.*, pp. 48–9.
172. Cf.: D. P. Todes, *Pavlov's Physiology Factory: Experiment, Interpretation, Laboratory Enterprise* (Baltimore: Johns Hopkins University Press, 2002).
173. On Hering's leadership style see: R. Steven Turner, 'Vision studies in Germany: Helmholtz versus Hering,' *Osiris*, 8 (1993), 80–103, pp. 87–8.
174. Head, 'Autobiography', p. 57.
175. Ibid., p. 58.
176. Letter of Henry Head to Hester Head, undated, CMAC, PP/HEA/D1.
177. *ibid.*
178. K. E. K. Hering, *Die Selbststeuerung der Athmung durch den Nervus vagus.* Sitzungsberichte der kaiserlichen Akademie der Wissenschaften. Mathematisch-naturwissenschaftliche Classe, Wien, 1868, 57: 2, pp. 672–7.
179. Head, 'Autobiography', pp. 52–3.
180. *Ibid.*, pp. 53–4.
181. *Ibid.*, pp. 54–5.
182. *Ibid.*, p. 55. On Knoll's political activities: Gary B. Cohen, 'Jews in German liberal politics: Prague, 1880–1914,' *Jewish History*, 1 (1986), pp. 55–74.
183. *Ibid.*, pp. 55–6.
184. Henry Head, 'On the regulation of respiration,' *Journal of Physiology*, 10 (1889) pp. 1–70; pp. 279–90.
185. Letter of Henry Head to Ruth Mayhew, 12 May 1902, CMAC, PP/HEA D4 11.
186. Henry Head, 'Über die und positiven Schwankungen des Nervenstromes,' *Pflügers Archiv*, 40: (1887), pp. 1–70.
187. Letter of Henry Head to Hester Head, [1886], CMAC, PP/HEA/D1.
188. Letter of Henry Head to Ruth Mayhew, 13 April 1900, PP/HEA/D4/6.
189. Letter of Henry Head to Hester Head, [1886a], CMAC, PP/HEA/D1.

190. Henry Head, [Journal of stay in Italy, March–April 1892], CMAC, PP/HEA/E1, 23 April.
191. Letter of Henry Head to Ruth Mayhew, 6 June 1897, CMAC, PP/HEA/DR/2.
192. *Ibid.* The allusion is to the garden of illusions belonging to the sorcerer in Richard Wagner's opera *Parsifal*.
193. Head, 'Autobiography', p. 60.

2 The Poles of Practice

1. *The Times*, 15 May 1886, p. 14; 16 June 1888, p. 16.
2. G. Holmes, 'Henry Head, 1861–1940,' *Obituary Notices of Fellows of the Royal Society*, 3 (1939–41), pp. 665–89, p. 667.
3. See: S. T. Casper, 'The idioms of practice: British neurology, 1880–1960,' (PhD Dissertation, University College London, 2006), pp. 131–46.
4. H. Head, 'Page-May Memorial Lectures on the Afferent Nervous System at Institute of Physiology UCL, 1912,' CMAC, PP/HEA/B13, p. [5].
5. *Ibid.*, pp. [5–7]
6. J. B. Lyons, 'Correspondence between Sir William Gowers and Sir Victor Horsley,' *Medical History*, 9 (1965) pp. 260–7, p. 266.
7. H. Head, 'The evolution of neurology and its bearing on medical education,' *British Medical Journal*, 2 (1924), p. 718.
8. Holmes, 'Henry Head', p. 667.
9. Head later confessed that: 'I became very despondent after my 4th month at Rainhill Asylum: but on putting together my notes I saw the lines along which I must plod gained confidence and left at the end of a year with a most voluminous & valuable series of records which have stood me in good stead ever since'. Letter of Henry Head to Redcliffe Nathan Salaman, 20 July [1901], Salaman Papers, CUL, MS Add. 8171, Box 3.
10. Letter of Everard Feilding to Henry Head, 13 October [no year given], CMAC, PP HEA/B2/2.
11. The letter makes the implausible claim that Head had: 'always shown himself to be well disposed to the Catholic faith,' *ibid.*
12. The notes he made of these examinations can be found in: CMAC, PP HEA/B2/1.
13. Letter of George Bull to Everard Feilding, 11 July 1895, *ibid.*
14. Letter of Hester Head to Henry Head, 16 August 1903, CMAC, PP HEA/D1.
15. Letter of Ruth Mayhew to Henry Head, 7 November 1899, CMAC, PP HEA/D4.
16. Letter of Ruth Head to Janet Ashbee, 27 June 1920, Ashbee Papers, KCC.
17. See: J. Oppenheim, *Shattered Nerves: Doctors, Patients, and Depression in Victorian England* (Oxford: Oxford University Press, 1991), pp. 31–4.
18. M. Critchley, 'Henry Head, 1861–1940,' in *The Black Hole and other Essays* (London: Pitman,1964), pp. 98–107, p. 106.
19. Head and Mayhew, 'Rag Book'. vol. 1.A. Oct 1901–July 1902, CMAC, PP HEA, Box 9, p. 32.
20. Letter of Ruth Head to Janet Ashbee, 3 March [1921], Ashbee Papers, King's College, Cambridge. This case may be an example of what Critchley claimed was Head's extensive practice in cases of alleged 'adolescent psychiatric problems'. Critchley, 'Henry Head', p. 106.
21. Head and Mayhew, 'Rag Book', vol. 2A, CMAC, PP/HEA, Box 9, p. 33.

22. Letter of Henry Head to Simon Flexner, 9 July 1919, Flexner Papers, American Philosophical Society, Philadelphia.
23. N. D. Jewson, 'Medical knowledge and the patronage system in 18th century England,' *Sociology*, 8 (1974) 369–85; *idem*, 'The disappearance of the sick-man from medical cosmology, 1770–1870,' *Sociology*, 10 (1976) 225–44.
24. H. H. Bashford, *The Corner of Harley Street: Being some Familiar Correspondence of Peter Harding. M.D.* (London: Constable & Co., 1911), p. 73.
25. Letter of Jacques Raverat to Margaret Keynes, 11 November 1921, Add.MS.9209.6:9, CUL. I am grateful to Katrina Gatley for this reference.
26. Letter of Ruth Head to Janet Ashbee, 6 April [1921], Ashbee Papers, KCC.
27. Letter of Janet Ashbee to Charles Ashbee, 10 March 1927, *ibid*.
28. Letter of Henry Head to Ruth Mayhew, 16 March 1900, CMAC, PP/HEA D4/6.
29. Letter of Henry Head to Ruth Mayhew, 8 September 1907, *ibid*.
30. Letter of Henry Head to Ruth Mayhew, 14 March 1915, *ibid*.
31. *The Times*, 2 June 1920, p. 5.
32. *The Times*, 15 June 1923, p. 5.
33. Letter of Henry Head to Ruth Mayhew, 6 March 1901, CMAC, PP/HEA/D4.
34. A. O. Bell, *The Diary of Virginia Woolf. Volume III: 1925–30* (Harmondsworth: Penguin, 1982), p. 193.
35. S. Trembley, *'All that Summer she was Mad': Virginia Woolf and her Doctors* (London: Junction Books, 1981), ch. 5; H. Lee, *Virginia Woolf* (London: Chatto & Windus, 1996), pp. 329–30.
36. V. Woolf, *Roger Fry: A Biography* (London: Hogarth Press, 1940), pp. 146–8.
37. A. Crawford, *C.R. Ashbee*, p. 149.
38. Letter of Janet Ashbee to Henry Head, 22 January [1920], CMAC, PP/HEA, F1.
39. Woolf, *Roger Fry*, p. 146.
40. R. Head (ed.), *Pictures and Other Passages from Henry James* (New York, NY: Frederick A. Stokes, 1916). A letter from Henry James to Ruth Head is reproduced in the Preface.
41. Letter of Ruth Head to Robert Nichols, 22 February 1927. Nichols Papers, BL, uncatalogued.
42. Letter of Henry James to Ruth Head, 1 November 1906, Henry James Papers Houghton Library, Harvard, bMS Am 1094.1, Box 2 [Head].
43. W. N. P. Barbellion [i.e., B. F. Cummings], *The Journal of a Disappointed Man* (London: Chatto & Windus, 1919), p. 81.
44. W. N. P. Barbellion with a Preface by Arthur J. Cummings, *A Last Diary* (London: Chatto & Windus, 1920), pp. xxx–xxxi.
45. *Ibid.*, p. xxxii.
46. Letter of C. R. Ashbee to Mary Elizabeth Downer, 17 September 1910, Ashbee Papers, KCC Archives, vol. 12.
47. Letter of Henry Head to Mary Elizabeth Ashbee, 29 March 1911, *ibid*.
48. Letter of Henry Head to Ruth Mayhew, 23 February 1900, CMAC, PP/HEA D4. The Avenue Matignon was the Paris street where Heinrich Heine lived during his last illness.
49. Letter of Henry Head to Ruth Mayhew, 7 September 1900, CMAC, PP/HEA D4.
50. Letter of Henry Head to Ruth Mayhew, 1 July 1912, *ibid*.
51. Letter of Arnold Mayhew to Robert Nichols, 30 March 41, Nichols Papers, BL, uncatalogued.

52. *Ibid.*
53. H. Head, 'The diagnosis of hysteria,' *British Medical Journal*, 1 (1922) 827–9, p. 827.
54. *Ibid.*, p. 829.
55. Letter of Henry Head to Ruth Mayhew, 18 May 1898, CMAC, PP/HEA/D4.
56. Head and Mayhew, 'Rag Book', vol. 2A, CMAC, PP HEA, Box 9, p. 33.
57. J. F. Fulton, *Harvey Cushing: A Biography* (Springfield, IL: Charles C. Thomas, 1946), p. 293.
58. Letter of Ruth Head to Janet Ashbee, 20 February 1916, Ashbee Paper, KCC.
59. Letter of Ruth Head to Henry Head, 2 December 1905, CMAC, PP/HEA/D4.
60. Letter of Ruth Head to Janet Ashbee, 24 October 1915, Ashbee Papers, KCC.
61. Letter of Ruth Head to Janet Ashbee, 6. December 1915, *ibid.*
62. Letter of Ruth Head to Harvey Cushing, 20 December 1916. Cushing Papers, YUL, series 1 correspondence, folder 670.
63. Letter of Ruth Head to Janet Ashbee, 6 August 1918, Ashbee Papers, KCC.
64. Letter of Ruth Head to Janet Ashbee, 5 January 1917, *ibid.*
65. Letter of Ruth Head to Janet Ashbee, 13 January [1918], *ibid.*
66. Letter of Ruth Head to Janet Ashbee, 23 November 1918, *ibid.*
67. Letter of Ruth Head to Harvey Cushing, 18 October 1923, YUL, series 1 correspondence, folder 670.
68. Letter of Ruth Head to Janet Ashbee, 27 June 1920, Ashbee Papers, KCC.
69. Letter of Ruth Head to Janet Ashbee, 6 April [1921], *ibid.*
70. Letter of Ruth Head to Janet Ashbee, 3 March [1921], *ibid.*
71. On the London see: E. W. Morris, *A History of the London Hospital* (London: Edward Arnold & Co., 1926); A. E. Clark-Kennedy, *The London: A Study in the Voluntary Hospital System* (London: Pitman Medical Publishing Co. Ltd., 1963).
72. Letter of H. H. Bashford to R. Brain, 10 August *c.* 1940, CMAC, PP/HEA/A.
73. *Ibid.*
74. D. Hunter, 'Sir Henry Head, 1900,' London Hospital Archives, LH/X/145/2.
75. 3 May 1919, *Ibid.*
76. Head and Mayhew, 'Rag Book', vol. 1B, CMAC, PP/HEA E3, p. 96.
77. Hunter, 'Sir Henry Head,' 20 June 1947. For another version see: R. Graves, *Goodbye to All That* (London: Jonathan Cape, 1929), pp. 381–2. According to Harvey Cushing, Head was aggrieved by the inaccuracy of Graves's account of the incident in 'that awful book'. He gave Cushing his own version of the story, which agreed with that of Hunter in its main features. A few details were added: for example, the patient in question had suffered from herpes zoster and had been killed in a traffic accident. Head described the attendant who suffered the fit and subsequently became homicidal as: 'a big burly negroid individual'. 'A visit to Henry and Ruth Head, July 16, 1930,' Harvey Cushing Papers, YUL, series I Correspondence, Folder 670.
78. Letter of Geoffrey Slade to Redcliffe Nathan Salaman, 27 December 1904, Salaman Papers, CUL, MS Add. 8171, Box 3.
79. Letter of Henry Head to Ruth Mayhew, 20 January 1903, CMAC, PP/HEA//D4.
80. Head and Mayhew, 'Rag Book', vol. 1A, CMAC, PP/HEA E3, pp. 56–57.
81. Letter of Henry Head to Ruth Mayhew, 17 December 1898,CMAC, PP HEA/D4/14.
82. Letter of Henry Head to Ruth Mayhew, 15 March 1902, CMAC, PP/HEA D4/11.
83. On the social tensions of the period see: G. S. Jones, *Outcast London: a Study in the Relationship between the Classes in Victorian Society* (Harmondsworth: Penguin, 1984).

84. R. N. Salaman, 'The Helmsman Takes Charge,' LHA, LH/X/101. On the long-standing links between the London and the Whitechapel Jewish community see: Morris, *A History*, pp. 113–21.
85. Letter of H. Adler to Redcliffe Nathan Salaman, 25 January 1904, Salaman Papers, CUL, MS Add. 8171, Box 3.
86. Letter of Henry Head to Ruth Mayhew, 21 July 1899, CMAC, PP HEA/ D4/5. Salaman is not named in this letter, but appears to be the most likely candidate.
87. Salaman, 'The Helmsman,' p. 26.
88. Letter of Henry Head to Redcliffe Nathan Salaman, *c.* 20 July 1901, Salaman Papers, CUL MS Add. 8171, Box 3.
89. Head and Mayhew, 'Rag Book', vol. 1A, CMAC, PP/HEA E3, pp.57–8.
90. *Ibid.*, pp. 58–9.
91. Head and Mayhew, 'Rag Book', vol. 2A, p. 1, *ibid.*
92. Letter of Ruth Head to Janet Ashbee, 1 June 1915, Ashbee Papers, KCC.
93. Letter of Henry Head to Ruth Mayhew, 9 November 1900, CMAC, PP/HEA D4.
94. Letter of Henry Head to Ruth Mayhew, 5 September 1902, CMAC, PP/HEA/D4. A further example of the complexity of Head's attitudes to Jews is found in the fact that in 1912 he signed an open letter protesting against the revival in Russia of the so-called 'Blood Accusation' against Jews – i.e., the claim that they engaged in ritual murder. *The Times*, 6 May 1912, p. 7.
95. Letter of Henry Head to Ruth Mayhew, *c.* 1899, CMAC, PP/HEA D4/1.
96. Letter of Henry Head to Ruth Mayhew, 1 October 1900, CMAC, PP/HEA D4/7.
97. Letter of Henry Head to Ruth Mayhew, 21 September 1900, *ibid.*
98. Head and Mayhew, 'Rag Book', vol. 2B, CMAC, PP/HEA/E3/41, p. 80. On the same page Head copied W. B. Yeats's poem 'A Faery Song,' one stanza of which was perhaps especially apposite: 'Give to these children, new to the world / Rest far from men / Is anything better, anything better? / Tell us it then'.
99. Letter of Henry Head to Ruth Mayhew, 8 April 1900, CMAC, PP/HEA D4/6.
100. Letter of Henry Head to Ruth Mayhew,13 April 1900, *ibid.*
101. Letter of Henry Head to Ruth Mayhew, 5 January 1900, *ibid.*
102. Letter of Henry Head to Ruth Mayhew, 7 June 1900, *ibid.*
103. Letter of Henry Head to Ruth Mayhew, 15 January 1901 CMAC, PP/HEA/D4.
104. Letter of Henry Head to Redcliffe Nathan Salaman, 5 March 1905, Salaman Papers, CUL, MS Add. 8171, Box 3.
105. Letter of Henry Head to Ruth Mayhew, 28 September 1900, CMAC, PP/HEA D4/7.
106. Letter of Henry Head to Ruth Mayhew, 14 December 1900, CMAC, *ibid.*
107. Letter of Henry Head to Ruth Mayhew, 16 October 1903, CMAC, PP/HEA D4/16.
108. Letter of Ruth Head to Janet Ashbee, 21 February 1920, Ashbee Papers, KCC.
109. Letter of Henry Head to Ruth Mayhew, 8 February 1901, CMAC, PP/HEA/D4. The quite exhaustive report on diet that Head composed with Robert Hutchison can be found at LHA: 'Report on the Diets at the London Hospital,' LH/A/17/19.
110. Letter of Henry Head to Ruth Mayhew, 30 May 1901, CMAC, PP/HEA D4.
111. Head and Mayhew, 'Rag Book'. vol. 1A, Oct: 1901–July 1902, CMAC, PP/HEA E3/1, p. 15.
112. Letter of Henry Head to Redcliffe Nathan Salaman, 26 August 1901, Salaman Papers, CUL, MS Add. 8171, Box 3.
113. London Hospital Medical Council Minute Book, 1900–1901, Friday July 20 1900, p. 146.

114. 'Path. Institute. Rep^t by Dr Head. May 01,' London Hospital Archives, LM/5/9.
115. Letter of Henry Head to Ruth Mayhew, 30 May 1901, CMAC, PP/HEA D4.
116. London Hospital Medical Council Minute Book, 1900–1901, Friday May 31 1901, p. 192.
117. Letter of Ruth Mayhew to Henry Head, 10 June 1901, CMAC, PP/HEA D4.
118. Letter of Henry Head to Ruth Mayhew, 11 July 1901, *ibid*.
119. 'London Hospital Pathological Institute,' *The Times*, 11 July 1901, p. 10.
120. Letter of Henry Head to Redcliffe Nathan Salaman, 20 July [1901], Salaman Papers, CUL, MS Add. 8171, Box 3.
121. Letter of Henry Head to Ruth Mayhew, 31 January 1902, CMAC, PP/HEA/D4.
122. London Hospital Council Minutes, 1902, 3 February 1902, p. 16.
123. London Hospital House Committee Minutes 1900 to 1903, 10 February 1902, pp. 340–41.
124. Letter of Henry Head to Redcliffe Nathan Salaman, 11 February 1902, Salaman Papers, CUL, MS Add. 8171, Box 3.
125. Letter of Henry Head to Ruth Mayhew, 8 February 1902, CMAC, PP/HEA/D4.
126. Letter of Henry Head to Redcliffe Nathan Salaman, 23 February 1902, Salaman Papers, CUL, MS Add. 8171, Box 3.
127. Letter of Henry Head of Redcliffe Nathan Salaman, 11 February 1902, *ibid*.
128. Letter of George W. Ross to Redcliffe Nathan Salaman, 23 October 1905, *ibid*.
129. Letter of Henry Head to Redcliffe Nathan Salaman, 5March 1905, *ibid*.
130. H. M. Turnbull, 'Autobiography,' LHA, PP/TUR/1/1.
131. Letter of Henry Head to Redcliffe Nathan Salaman, 14 May 1905, Salaman Papers, CUL, MS Add. 8171, Box 3.
132. Clark-Kennedy, *The London*, pp. 207–8.
133. See: C. Lawrence, *Rockefeller Money, the Laboratory and Medicine in Edinburgh 1919–1930* (Rochester, NY: University of Rochester Press, 2005), chapter 3.
134. Letter of Walter Fletcher to Sydney Holland (Viscount Knutsford), 27 May 1919, LHA, LH/A/23/53.
135. Letter of Sydney Holland (Viscount Knutsford) to Henry Head, 24 May 1919, *ibid*.
136. Letter of Walter Fletcher to Sydney Holland (Viscount Knutsford), 27 May 1919, *ibid*.
137. *Ibid*.
138. Letter of Henry Head to Sydney Holland (Viscount Knutsford), 22 May 1919, *ibid*.
139. Letter of Sydney Holland (Viscount Knutsford) to Henry Head, 24 May 1919, *ibid*.
140. Letter of [Sydney Holland (Viscount Knutsford)] to Walter Fletcher, 29 May 1919, *ibid*.
141. Letter of Walter Fletcher to Sydney Holland (Viscount Knutsford), 27 May 1919, *ibid*.
142. Letter of Henry Head to Viscount Knutsford, 8 January 1919, LHA, LH/A/5/56
143. H. Head, 'The elements of the psycho-neuroses,' *British Medical Journal*, 1 (1920) pp. 389–92, p. 391.
144. *Ibid*., p. 390.
145. Letter of Ruth Head to Henry Head, 20 March 1917, CMAC, PP/HEA/D4.
146. Letter of Henry Head to Ruth Head, 22 March 1917, *ibid*.
147. Letter of Ruth Mayhew to Henry Head, 19 July 1899, *ibid*.
148. Letter of Henry Head to Ruth Mayhew, 24 June 1903, *ibid*.
149. Letter of Henry Head to C. R. Ashbee, 8 December 1913, Ashbee Papers, KCC.
150. Letter of Janet Ashbee to C. R. Ashbee, 4 December 1913, *ibid*.
151. Letter of Henry Head to C. R. Ashbee, 8 December 1913, *ibid*.

152. Letter of Janet Ashbee to C. R. Ashbee, 4 December 1913, *ibid.*
153. Letter of Henry Head to C. R. Ashbee, 8 December 1913, *ibid.*
154. Head and Mayhew, 'Rag Book', vol. 2A, p. 78, CMAC, PP/HEA/E3/3.
155. On shell shock see: E. Showalter, *The Female Malady: Women, Madness and English Culture* (London: Virago Press, 1987), ch. 7; A. Young, *The Harmony of Illusions: Inventing Post-Traumatic Stress Disorder* (Princeton, NJ: Princeton University Press, 1995); B. Shephard, *A War of Nerves* (London: Jonathan Cape, 2000); M. S. Micale and P. Lerner (eds), *Traumatic Pasts: History, Psychiatry, and Trauma in the Modern Age, 1870–1930* (Cambridge: Cambridge University Press, 2001).
156. Head, 'The elements,'; Shephard, *War of Nerves*, pp. 57–8.
157. Head, 'The elements,' p. 392.
158. Head, 'The diagnosis,' p. 827.
159. Letter of Robert Nichols to Henry Head, 7 July 1927, Nichols Papers, BL, uncatalogued.
160. On the establishment of the 'Special Hospital' at Palace Green see: S. Holland, *In Black and White* (London: Edward Arnold, 1928), p. 269.
161. A. Charlton and W. Charlton, *Putting Poetry First*.
162. Letter of Robert Nichols to Henry and Ruth Head, 26 August 1930, Nichols Papers, BL, uncatalogued..
163. Letter of Ruth Head to Robert Nichols, 11 Februay 1925, *ibid.*
164. Letter of Robert Nichols to Henry Head, 26 June 1925, *ibid.*
165. Letter of Robert Nichols to Henry Head, 3 May 1917, *ibid.*
166. Letter of Robert Nichols to Henry and Ruth Head, 21 July 1934, *ibid.*
167. Letter of Robert Nichols to Henry Head, 16 August 1927, *ibid.*
168. Letter of Robert Nichols to Henry and Ruth Head.,30 April 1934, *ibid.*
169. *Ibid.*
170. Letter of Robert Nichols to Henry Head, 14 March 1925, *ibid.*
171. *Ibid.*
172. Letter of Henry Head to Robert Nichols, 9 April 1925, *ibid.*
173. Letter of Ruth Head to Robert Nichols, 4 April 1925, *ibid.*
174. Letter of Robert Nichols to Ruth Head, September 1924, *ibid.*
175. Letter of Ruth Head to Robert Nichols, 16 December 1926, *ibid.*
176. R. Nichols, 'Notes on Henry Head,' *ibid.*
177. *Ibid.*
178. Letter of Robert Nichols to Henry Head, 14 May 1917, *ibid.*
179. Letter of Robert Nichols to Henry Head, 20 February 1918, *ibid.*
180. Letter of Robert Nichols to Henry and Ruth Head, 23 January 1933, *ibid.*
181. Letter of Robert Nichols to Henry and Ruth Head, 12 February 1932, *ibid.*
182. Letter of Robert Nichols to Henry and Ruth Head, 13 April 1930, *ibid.*
183. Letter of Robert Nichols to Henry and Ruth Head, 14 January 1933, *ibid.*
184. Letter of Robert Nichols to Henry and Ruth Head, 23 July 1933, *ibid.*
185. Letter of Robert Nichols to Henry and Ruth Head, 14 January 1933, *ibid.*
186. Letter of Robert Nichols to Ruth Head, September 1924, *ibid.*
187. Letter of Robert Nichols to Henry Head, 4 December 1927, *ibid.*
188. Letter of Robert Nichols to Henry and Ruth Head, 6 April 1930, *ibid.*
189. Letter of Robert Nichols to Henry and Ruth Head, 29–30 January 1932, *ibid.*
190. Letter of Robert Nichols to Henry and Ruth Head, 14 March 1932, *ibid.*

191. Letter of Henry Head to Harvey Cushing, 23 December 1914, Cushing Papers, Yale University, series I Correspondence, Folder 670.
192. In January 1915 Head sent a poem to Charles Sherrington that expressed these feelings of frustration. The opening lines read: 'How can I serve who am too old to fight? / I cannot stand and wait / With folded hands and lay me down at night / In restless expectation that the day / Will bring some stroke of Fate / I cannot help to stay'. Sherrington Papers, University Library of Physiology, Oxford.
193. Thus, in his diary Cushing noted that cases of injury to the spinal cord incurred at the front: 'are sent to Henry Head at the London Hospital, by whom they are subsequently followed'. J. W. Fulton, *Harvey Cushing: A Biography* (Springfield, IL: Charles C. Thomas, 1946), p. 400.
194. Letter of Henry Head to Ruth Head, 17 February 1919, CMAC, PP/HEA/D4/19.
195. Letter of Henry Head to Ruth Mayhew, 18 February 1919, *ibid.*
196. Letter of Henry Head to William Bulloch, 19 October 1914, Pearson Papers, University College London, MS 647/5.
197. Letter of Henry Head to Harvey Cushing, 8 September 1916, Harvey Cushing Papers, YUL, series I Correspondence, Folder 670.
198. H. Head, 'Some principles of neurology', *Lancet*, 1918: 657–60, p. 657.
199. H. Head, 'Disease and diagnosis', *British Medical Journal*, 1 (1919) 365–7, p. 365. Head's rhetoric reflected a more general mood. Reporting the annual meeting of the British Medical Association *The Times* declared: 'Every speaker seemed to be imbued with the idea that the old order was giving place to new...' *The Times*, 3 July 1920, p. 17. Head chaired the Section of Neurology and Psychiatry at this meeting: *British Medical Journal*, 2 (1920) 49–50. Head gave the opening paper at the session on 'Early symptoms and signs of nervous disease and their interpretation,' *ibid.*, pp. 691–3.
200. *Ibid.*
201. *Ibid.* p. 366.
202. 'National Council for Mental Hygiene', *British Medical Journal*, 1922, 2:766–7, p. 766. On the mental hygiene movement, see: M. Thomson, *Mental Hygiene as an International Movement* (Cambridge: Cambridge University Press, 1995).
203. *Ibid.*
204. *Ibid.*
205. *Ibid.*, p. 767.
206. A. Flexner, *I Remember: The Autobiography of Abraham Flexner* (New York, NY: Simon and Schuster, 1940), p. 137.
207. See, for example, B. V. White, *Stanley Cobb: A Builder of Modern Neurosciences* (Boston, MA: Francis A. Countway Library, 1984), pp. 101–3.
208. Letter of Ruth Head to Janet Ashbee, 10 November 1925, Ashbee Papers, KCC.
209. Letter of Henry Head to Ruth Mayhew, 18 December 2001, CMAC, PP/HEA D4/8.

3 'The Great Hard Road of Natural Science'

1. Head, 'Autobiography', p. 4.
2. 'Commission on vivisection. Evidence of Dr. Henry Head, F.R.S.,' *British Medical Journal*, 1 (1909), pp. 227–9, p. 228.
3. This was the idealized narrative of the relationship between theory and surgical practice in the case of germ theory. The reality was considerably more complex. See: C. Lawrence and R. Dixey, 'Practising on principle: Joseph Lister and the germ theory of disease,' in:

Notes to pages 100–7

4. Head, 'Autobiography,' pp. 4–5.
5. Letter of Henry Head to Ruth Mayhew, 21 September 1900, CMAC, PP/HEA/D4/7.
6. C. Lawrence, 'Incommunicable knowledge' pp. 503–20.
7. See: J. H. Warner, 'The history of science and the sciences of medicine,' *Osiris*, 10 (1995), pp. 164–95.
8. Head, 'Autobiography,' pp. 31–2, 37.
9. Head and Mayhew 'Rag Book', vol. 2B, CMAC, PP/HEA E3/41pp. 64–6.
10. *Ibid.*, p. 62.
11. *Ibid.*, p. 68.
12. Letter of Henry Head to Ruth Mayhew, 25 October 1902, CMAC, PP/HEA/D4/12. For Mackenzie's views of Head see p. 102.
13. Letter of Henry Head to Ruth Mayhew, undated, CMAC, PP/HEA D4/1.
14. Letter of Henry Head to Ruth Mayhew, 25 November 1900, CMAC, PP/HEA D4/7.
15. *Ibid*.
16. The full list of proposers is: W. H. Gaskell, E. A. Schafer, J. H. Jackson, H. C. Bastian, D. Ferrier, V. Horsley, J. N. Langley, J. R. Bradford, C. S. Sherrington, F. W. Mott. Royal Society Archives, EC/1899/08.
17. Letter of Henry Head to Ruth Mayhew, 7 May 1899, CMAC, PP/HEA D4.
18. Head and Mayhew, 'Rag Book,' vol. 1A, CMAC, PP/HEA E3/1, pp. 50–1.
19. *Ibid.*, p. 76.
20. The 'creature' in the first extract appears to have been Sydney A. Monckton Copeman who in 1902 received the Copley Medal for his researches in bacteriology and the comparative pathology of vaccination. Letter of Henry Head to Ruth Mayhew, 6 December 1902, CMAC, PP/HEA/D4/12.
21. See: L. S. Jacyna, 'Questions of identity: science, aesthetics, and Henry's Head,' in C. Lawrence and G. Weisz (eds), *Greater than the Parts: Holism in Biomedicine, 1920–1950* (New York, NY: Oxford University Press, 1998), pp. 211–33.
22. Letter of Henry Head to Ruth Mayhew, 11 July 1902, CMAC, PP/HEA D4/12. See also: H. Head, 'The conception of nervous and mental energy,' *British Journal of Psychology*, 1923–4, *14*: 126–47, p. 129. Head's participation in experimental work is noted in: H. C. Bazett and W. G. Penfield, 'A study of the Sherrington decerebrate animal in the chronic as well as the acute condition,' *Brain*, 1922, *45:* 185–265, p. 234.
23. Letter of Henry Head to Ruth Mayhew, 17 January 1902, CMAC, PP/HEA/D4/11.
24. Letter of Henry Head to Ruth Mayhew, 22 September 1897, CMAC, PP/HEA/D4/2.
25. See: S. Schaffer, 'Genius in romantic natural philosophy,' in A. Cunningham and N. Jardine (eds) *Romanticism and the Sciences* (Cambridge: Cambridge University Press, 1990), pp. 82–98.
26. Head did, however, acknowledge the tedious and exhausting, yet strangely addictive, aspects of research. Writing in November 1903 of his experiments on nerve regeneration he described: 'the sense of being chained to an oar for a period to be measured in years that leads to occasional revolt – And yet, like the drunkard, I could not keep away from it now, however much I wanted to do so'. Letter of Henry Head to Ruth Mayhew, 6 November 1903, CMAC, PP/HEA D4/16.
27. Head and Mayhew, 'Rag Book', vol. 1A, CMAC, PP/HEA/E3/1, p. 27.
28. Letter of Henry Head to Ruth Mayhew, 28 September 1900, CMAC, PP/HEA/D4/7.
29. Letter of Henry Head to Ruth Mayhew, 5 August 1902, CMAC, PP/HEA/D4/12.

30. Letter of Brenda Seligman to Russell Brain, 6 July 1961, CMAC, PP HEA Box 1.
31. Head and Mayhew, 'Rag Book,' vol. 1A, CMAC, PP/HEA/E3/1, pp. 13–15.
32. See: A. Harrington, 'Other "ways of knowing": The politics of knowledge in interwar German brain science,' in A. Harrington (ed.), *So Human a Brain: Knowledge and Values in the Neurosciences* (Boston, MA: Birkhäuser, 1992), pp. 229–44, especially p. 236; idem, *Reenchanted Science: Holism in German Culture from Wilhelm II to Hitler* (Princeton, NJ: Princeton University Press, 1996), p. 40.
33. Head and Mayhew, 'Rag Book,' vol. 1A, CMAC, PP/HEA/E3/1, p. 31.
34. Letter of Robert Nichols to Henry and Ruth Head, 8 October 1934, Nichols Papers, BL, uncatalogued.
35. Letter of Henry Head to Ruth Mayhew, 21 January 1899, CMAC, PP/HEA D4.
36. Letter of Henry Head to Ruth Mayhew, 27 November 1898 CMAC, PP/HEA/D4/3.
37. Letter of Ruth Mayhew to Henry Head, [5 November 1900], CMAC, PP/HEA/D4/7.
38. Letter of Henry Head to Ruth Mayhew, 9 November 1900, *ibid*.
39. Head does not discuss Nietzsche's work in his extant writings. There are, however, hints in his correspondence with Robert Nichols that he was familiar with Nietzsche's work. For references to Nietzsche, see, for example: letters of Robert Nichols to Henry Head, 10 December 1932, 14 January 1934, 30 April 1934, Nichols Papers, BL uncatalogued. On the currency of Nietzsche's ideas see: D. Stone, *Breeding Superman: Nietzsche, Race and Eugenics in Edwardian and Interwar Britain* (Liverpool: Liverpool University Press, 2002).
40. Letter of Henry Head to Charles Sherrington, January 1915, Sherrington Papers, Oxford University Laboratory of Physiology.
41. Head and Mayhew, 'Rag Book,' vol. 2B, CMAC, PP/HEA/E3/41, pp. 17–18.
42. *Ibid.*, p. 19.
43. *Ibid.*
44. *Ibid.*, pp. 20–1.
45. *Ibid.*, p. 22. The full quotation is: 'And what does the spirit need in the face of modern life? The sense of freedom. That naive, rough sense of freedom, which supposes man's will to be limited, if at all, only by a will stronger than his, he can never have again'. W. Pater, *The Renaissance: Studies in Art and Poetry* (1873; Oxford: Oxford University Press, 1998), p. 231.
46. *Ibid.*, p. 18.
47. Letter of Henry Head to Charles Sherrington, 8 October 1925, Sherrington Papers, Oxford University Laboratory of Physiology.
48. *Ibid.*
49. Letter of Henry Head to Ruth Mayhew, February 1903, CMAC, PP/HEA/D8. This note is written on the back of a letter from Gaskell to Head dated 6 February 1903 congratulating him on the award of the Marshall Hall prize. Gaskell had added: 'It is a very great pleasure to me that your work is being appreciated at its true worth at last'.
50. W. H. Gaskell, 'On the relation between the structure, function, distribution and origin of the cranial nerves; together with a theory of the origin of the nervous system of vertebrata,' *Journal of Physiology*, 10 (1889) pp. 153–211, pp. 190–1.
51. Henry Head, 'On disturbances of sensation with especial reference to the pain of visceral disease,' *Brain*, 16 (1889) pp. 1–133, p. 1.
52. Letter of Henry Head to James Mackenzie, 19 November 1893, Wellcome Library for the History of Medicine, Autograph Letters.
53. Head, 'On disturbances of sensation' p. 4.

54. *Ibid.*, p. 5.
55. I take the term from: S. Shapin and S. Schaffer, *Leviathan and the Air-Pump: Hobbes, Boyle, and the Experimental Life* (Princeton, NJ: Princeton University Press, 1985).
56. Letter of Henry Head to Redcliffe Nathan Salaman, 26 August [1901], Salaman Papers, CUL, MS Add. 8171, Box 3.
57. Head, 'Disturbances of Sensation', p. 4.
58. Letter of Henry Head to Charles Sherrington, 7 January 1927, Sherrington Papers, Oxford University Laboratory of Physiology. Head had maintained close communication with Sherrington while formulating his early theories on the nature of sensation as well as citing some of the latter's experimental work in these papers.
59. Head, 'On disturbances of sensation', pp. 5–6.
60. *Ibid.*, p. 7.
61. H. Head, 'Herpes Zoster Case Records, chiefly from the London Hospital 1891-9', Royal College of Physicians of London, MS 329–30.
62. H. Head and A. W. Campbell, 'The pathology of Herpes Zoster and its bearing on sensory localization,' *Brain* 32 (1900), pp. 353–523, p. 396.
63. Head, 'On disturbances of sensation', p. 38.
64. Jackson was among the clinicians who supplied Head with clinical materials for this study. See: *ibid.*, p. 49.
65. H. Head, 'On disturbances of sensation with especial reference to the pain of visceral disease: Part II – Head and neck,' *Brain* 17 (1894) 339 – 480, p. 390.
66. Head, 'On disturbances of sensation', p. 101.
67. Letter of James Mackenzie to Arthur Keith, 27 March 1906, quoted in: Alex Mair, *Sir James Mackenzie MD 1853–1925: General Practitioner* (London: Royal College of General Practitioners, 1973), p. 113. In 1904, Mackenzie tried to detach Redcliffe Nathan Salaman from the teachings on referred pain of 'my good friend Dr. Head'. Mackenzie claimed that Head: 'has not grasped the full significance of these hyperaesthetic areas to which I call attention'. Letter of James Mackenzie to Redcliffe Nathan Salaman, 4 August 1904, Salaman Papers, CUL, MS Add. 8171, Box 3.
68. Head, 'On disturbances of sensation,' p. 2.
69. Letter of Henry Head to Ruth Mayhew, 25 July 1900, CMAC, PP/HEA/D4/7.
70. Head, 'On disturbances of sensation' part 2, p. 344.
71. *Ibid.*
72. *Ibid.* p. 466. On the association between migraine and intellectual exertion see: E. G. Mussellman, *Nervous Conditions: Science and the Body Politic in Early Industrial Britain* (Albany, NY: State University of New York, 2006), pp. 101–3.
73. Head, 'On disturbances of sensation', p. 468.
74. *Ibid.*,, p. 124.
75. Letter of Henry Head to Ruth Mayhew, 5 January 1901, CMAC, PP/HEA/D4/8. Head did not always describe the processes of scientific composition in such lyrical terms. While writing another of his papers, he grumbled: 'It is a most toilsome process – I write 'Sensation' 'boundary' 'area' a hundred times and try to replace them by other words. In attempting to be precise I am monotonous. To change the turn of a phrase frequently make it say something I do not intend and yet some changes must be made for Euphony. I scratch out and re-write and swear at the toil involved to make the whole appear as if it was all as simple as an elementary lecture'. Letter of Henry Head to Ruth Mayhew, 14 August 1902, CMAC, PP/HEA D4/12.

76. Henry Head, 'Certain mental changes that accompany visceral disease,' *Brain*, 24 (1901), pp. 345–403, p. 346.
77. *Ibid.*, p. 347.
78. *Ibid.*
79. *Ibid.*, p. 348.
80. *Ibid.*, p. 349.
81. *Ibid.*, p. 352.
82. *Ibid.*, p. 354.
83. *Ibid.*, p. 353.
84. *Ibid.*, p. 356.
85. *Ibid.*, p. 359.
86. *Ibid.*, pp. 400–1.
87. *Ibid.*, p. 401.
88. On this discourse see: R. Smith, *Inhibition: History and Meaning in the Sciences of Mind and Brain* (London: Free Association Books, 1992), especially ch. 2.
89. Head, 'Certain mental changes', p. 401.
90. Head was reiterating some of the commonplace assumptions about femininity of contemporary medical science. See: O. Moscucci, *The Science of Woman: Gynaecology and Gender in England 1800–1929* (Cambridge: Cambridge University Press, 1990), ch. 1; C. Sengoopta, *The Most Secret Quintessence of Life: Sex, Glands, and Hormones, 1850–1950* (Chicago, IL: University of Chicago Press, 2006), especially pp. 12–15.
91. Head, 'Certain mental changes', p. 402.
92. *Ibid.*
93. *Ibid.*
94. Henry Head, *Studies in Neurology*, 2 vols, (London: Henry Frowde, 1920), vol. 1, p. 8.
95. Letter of Henry Head to Robert Nichols, 29 February 1924, Nichols Papers British Library, uncatalogued.
96. Head, *Studies*, vol. 1, p. 8.
97. On Wright's involvement with the Royal Victoria Hospital at Netley see: M. Worboys, 'Almroth Wright at Netley: modern medicine and the military in Britain, 1892–1902,' in M. Cooter, M. Harrsion and S. Sturdy, *Medicine and Modern Warfare*, pp. 77–97.
98. Letter of Henry Head to Ruth Mayhew, 1 March c. 1902, CMAC, PP/HEA D4/11.
99. *Ibid.*
100. Letter of Henry Head to Sydney Holland (Viscount Knutsford) 30 December 1918, LHA, LH/A/23/53.
101. H. Head, 'The consequences of injury to the peripheral nerves in man,' *Brain*, 28 (1905) pp. 116–299, p. 118.
102. *Ibid.*, p. 119.
103. Head, 'A human experiment in nerve division' in *Studies*, vol. 1, p. 225.
104. *Ibid.*, vol. 2, p. 741.
105. Head and Mayhew, 'Rag Book', vol. 2B, CMAC, PP/HEA/E3/41, p. 29.
106. *Ibid.*, p. 30.
107. Letter of A. P. Oppé to Henry Head, CMAC, PP/HEA/D9, 4 December, no year.
108. Head, 'A human experiment,' p. 226.
109. *Ibid.*
110. *Ibid.*, p. 242.
111. *Ibid.*
112. Letter of Henry Head to Ruth Mayhew, 6 May 1903, CMAC, PP/HEA/D4/14.

113. Letter of Henry Head to Ruth Mayhew, 15 May 1903, *ibid*.
114. Letter of Henry Head to Ruth Mayhew, 27 November 1903, CMAC, PP/HEA/D4/16.
115. Letter of Henry Head to Ruth Mayhew, 7 February 1904, CMAC, PP/HEA/D4/17.
116. Letter of Henry Head to Ruth Mayhew, 2 November 1903, CMAC/PP/HEA/D4/16.
117. Letter of Henry Head to Ruth Mayhew, 16 October 1903, *ibid*.
118. Letter of Henry Head to Ruth Mayhew, 20 July 1903, CMAC, PP/HEA/D4/15.
119. Letter of Henry Head to Ruth Mayhew, 21 July 1903, CMAC, PP/HEA/D4/15.
120. Letter of Redcliffe Nathan Salaman to Henry Head, 1 January 1906, CMAC, PP/HEA/D9.
121. Letter of Henry Head to Ruth Mayhew, 24 May 1903, CMAC, PP/HEA/D4/14.
122. Letter of Ruth Mayhew to Henry Head, 18 May 1903, *ibid.*.
123. Letter of Ruth Mayhew to Henry Head, 22 July 1903, CMAC, PP/HEA/D4/15.
124. Letter of Ruth Mayhew to Henry Head, 4 May 1903 CMAC, PP/HEA/D4/14.
125. Head, 'A human experiment', p. 243.
126. *Ibid.*, p. 244.
127. *Ibid.*
128. *Ibid.*
129. *Ibid.*, p. 245.
130. Letter of Henry Head to Ruth Mayhew, 6 April 1903, CMAC, PP/HEA/D4/13.
131. Letter of Henry Head to Ruth Head, 3 April 1908, CMAC, PP/HEA/D4/19.
132. *Ibid.*, p. 324.
133. *Ibid.*, p. 329.
134. *Ibid.*
135. On the interplay between social and biological thought see: R. M. Young, 'Malthus and the evolutionists: the common context of biological and social theory,' *Past and Present* 43 (1969), pp. 109–45.
136. H. Head and J. Sherren, 'The consequences of injury to the peripheral nerves in man,' *Brain*, 28 (1969), pp. 116–299, p. 284.
137. Salaman, in facetious mode, captured something of these connotations when in his imaginary Head coat of arms he included: 'Bracers. Dext. Ye Epicritic Fop. Sin. Ye Protopathic Bully'. Letter of Redcliffe Nathan Salaman to Henry Head, 1 January 1906, CMAC, PP/HEA, D9.
138. Head, *Studies*, vol. 2, p. 743.
139. On the relations between psychology and anthropology see: H. Kuklick, *The Savage Within: the Social History of British Anthropology, 1885–1945* (Cambridge: Cambridge University Press, 1991).
140. L. S. Jacyna, *Lost Words: Narratives of Language and the Brain, 1825–1926* (Princeton, NJ: Princeton University Press, 2000), pp. 151–6.
141. Head, *Studies*, vol. 1, p. 781.
142. Letter of Henry Head to Ruth Head, 25 March 1917, CMAC, PP/HEA/D4/19.
143. H. Head and G. Holmes, 'Sensory disturbances from cerebral lesions,' *Brain*, 34 (1911), pp. 102–254, p. 177.
144. *Ibid.*, pp. 190–1.
145. *Ibid.*, p. 136.
146. Letter of Henry Head to Ruth Mayhew, 24 May 1903, CMAC, PP/HEA/D4/14.
147. W. H. R. Rivers noted the close analogy when he wrote: '...Head working in conjunction with Holmes has discovered a relation between the cerebral cortex and the optic thala-

mus very similar to that existing between protopathic and epicritic sensibility'. *Instinct and the Unconscious: A Contribution to a Biological Theory of the Psycho-Neuroses* (Cambridge: Cambridge University Press, 1922), p. 27.
148. H. Head, 'Sensation and the cerebral cortex,' *Brain*, 41 (1918) pp. 57–253, p. 191.
149. Head and Holmes, 'Sensory disturbances', p. 191.
150. *Ibid.*, p. 71.
151. Head, 'Sensation', p. 191.
152. *Ibid.*, p. 162.
153. Head, 'Sensory disturbances', p. 106.
154. Head, 'Sensation', p. 58.
155. Henry Head, 'The grouping of afferent impulses within the spinal cord', *Brain*, 29 (1907), pp. 537–741, p. 643.
156. *Ibid.*, p. 639.
157. *Ibid.*, p. 644.
158. *Ibid.*, p. 645.
159. Head, 'Sensory disturbances', pp. 125, 172; Head, 'Sensation', pp. 82, 88–9, 91.
160. Head, 'Sensory disturbances', pp. 186, 189.
161. *Ibid.*, p. 183.
162. *Ibid.*, p. 105.
163. *Ibid.*, p. 105.
164. Head, 'Sensation,' p. 158. It will be recalled that Head found the newly introduced taxicabs of great help in travelling around London to see patients (Chapter 2, p. 61).
165. *Ibid.*, p. 188.
166. *Ibid.*, p. 183.
167. *Ibid.*, p. 192.
168. *Ibid.*, p. 193.
169. See: Kern, *The Culture of Time and Space*, especially pp. 45–6.
170. Letter of Henry Head to Ruth Mayhew, 22 January 1901, CMAC, PP/HEA/D4/8. This recognition extended to the international scientific community. During a visit to Germany, Head reported: 'I did not know I had made so many friends through my papers. One after another came up and introduced himself saying he was a 'Verehrer' of many years'. Letter of Henry Head to Ruth Head, [1904], CMAC, PP/HEA/D4/19.
171. Letter of Ruth Mayhew to Henry Head, 23 January 1901. CMAC, PP/HEA/D4/8.
172. This section is based in part on Jacyna, *Lost Words*, ch. 5 and *idem*, 'Starting anew: Henry Head's contribution to aphasia studies,' *Journal of Neurolinguistics*, 18 (2005), pp. 327–36.
173. Letter of Henry Head to Ruth Mayhew, *c.* 13 January 1901, CMAC, PP/HEA D4/8.
174. H. Head, 'Aphasia: an historical review,' *Brain*, 43 (1921), pp. 390–411, p. 403.
175. Letter of Henry Head to Constantin von Monakow, 11 September 1919, Monakow Papers, University of Zurich.
176. J. H. Jacklings 'Reprint of some of Dr. Hughlings Jackson's papers on affections of speech', *Brain*, 38 (1915), pp. 28–42.
177. Head, 'Sensory disturbances', p. 179.
178. Letter of Henry Head to Constantin von Monakow, 11 September 1919, Monakow Papers, University of Zurich.
179. See: Holland, *In Black and White* pp. 269–70; 'Special hospitals for officers,' *Lancet*, 1915, 2: 1155–7.

180. On the status of these officer patients in Head's research see: Jacyna, *Lost Words*, pp. 154–6.
181. H. Head, *Aphasia and Kindred Disorders* 2 vols, (Cambridge: Cambridge University Press, 1926), vol. 1, p. 221.
182. *Ibid.*, p. 146.
183. *London Times*, 18 October 1940.
184. Head, *Aphasia*, vol. 1, p.180.
185. Letter of Henry Head to Robert Nichols, 9 April 1924, Nichols Papers, BL, uncatalogued. For some of the reactions to Head's attempt to revolutionize understanding of aphasia see: H. Head, 'Discussion of aphasia,' *Brain*, 43 (1921), pp. 412–50.
186. Head, *Aphasia*, vol. 1, p. 431.
187. *Ibid.*, p. 269.
188. *Ibid.*, p. 397.
189. *Ibid.*, pp. 56–7.
190. Head, 'Aphasia,' p. 396.
191. *Ibid.*, p. 400.
192. *Ibid.*, p. 410.
193. Head, *Aphasia*, vol. 1, p. 166.
194. *Ibid.*, p. 165.
195. *Ibid.*, p. 257.
196. *Ibid.*, p. 142.
197. See: F. C. Bartlett, 'Critical review: *Aphasia and Kindred Disorders of Speech*. By Henry Head,' *Brain*, 49 (1926), pp. 581–7, p. 582.
198. Letter of Henry Head to Robert Nichols, 18 December 1923, Nichols Papers, BL, uncatalogued.
199. Letter of Ruth Head to Janet Ashbee, 26 April [1920], Ashbee Papers, KCC.
200. Letter of Ruth Head to Robert Nichols, 5 August 1922, Nichols Papers, BL, uncatalogued.
201. Letter of Ruth Head to Janet Ashbee, 7 December 1922, Ashbee Papers, KCC.
202. Letter of Ruth Head to Janet Ashbee, 18 August 1925, *ibid.*
203. Letter of Henry Head to Charles Sherrington, 3 January 1924, Sherrington Papers, Department of Physiology, Oxford.
204. Letter of Henry Head to Robert Nichols, 28 July 1925, Nichols Papers, BL, uncatalogued.
205. Letter of Janet Ashbee to Charles Ashbee, 10 March 1927, Ashbee Papers, KCC.
206. For representative examples of such assessments see: R. A. Henson, 'Henry Head's work on sensation,' in *Henry Head Centenary: Essays and Bibliography* (London: Macmillan & Co.), pp. 7–22; M. Critchley, 'Head's contribution to aphasia', *ibid.*, pp. 23–32.
207. Letter of Adolf Meyer to Henry Head, 25 March 1922, Meyer Papers, Alan Mason Chesney Medical Archives, Johns Hopkins University, 1/1633/1.
208. Adolf Meyer, 'Baltimore Working Notes. 1909–41,' *ibid.*, XII/1/20.
209. Adolf Meyer, 'New York Working Notes,' *ibid.*, XI/1/5.
210. See: E. L. Wolfe, A. C. Barger and S. Benison, *Walter B. Cannon, Science and Society* (Cambridge, MA.: Harvard University Press, 2000), pp. 167–8. Head and Cannon corresponded on their areas of shared interest. The only surviving trace of this correspondence appears to be an extract from one of Head's letters in: Letter of Walter Cannon to Philip Bard, 28 September 1928, Walter B. Cannon Papers, Francis A. Countway Medical Library, Harvard University.

211. W. H. R. Rivers, *Instinct and the Unconscious: A Contribution to a Biological Theory of the Psycho-Neuroses* (Cambridge University Press, 1922), especially ch. 6.
212. Letter of Henry Head to Robert Nichols, 28 July 1925, Nichols Papers, BL, uncatalogued.
213. Letter of Henry Head to Stanley Cobb, 1 January 1928, Cobb Papers, H MS c53, Box 4, f. 90, Francis A. Countway Library, Harvard University.
214. H. Head, 'Disorders of symbolic thinking and expression,' *British Journal of Psychology*, 11 (1920–1), pp. 179–93, p. 180.
215. Jacyna, *Lost Words*, pp. 194–203.
216. 'Disorders of symbolic thinking,' *Nature*, 1920, *106:* 197–8, p. 198.
217. 'British Institute of Philosophical Studies,' *British Medical Journal*, 1 (1927), p. 699.
218. Letter of Henry Head to Bertrand Russell, 16 May 1919, Bertrand Russell Papers, McMaster University.
219. A. N. Whitehead, *Adventures of Ideas* (Cambridge: Cambridge University Press, 1933), p. 274.
220. H. Head, 'The physiological basis of the spatial and temporal aspects of sensation,' in 'Symposium: Time, space and material: are they, and if so in what sense, the ultimate data of science?,' *Aristotelian Society. Supplementary Volume II* (London: Williams and Norgate, 1919), pp. 44–108, pp. 77–86, on pp. 77–8.
221. *Nature*, 1919, 104: 267. See also: 'New conceptions of psychology,' *Nature*, 1920, 105: 363–4.
222. See: Jacyna, *Lost Words*, pp. 163–5.
223. G. G. Campion and G. Elliot Smith, *Neural Basis of Thought* (London: Kegan Paul, 1934), especially pp. 105–6.
224. Letter of George Campion to Grafton Elliot Smith, 26 September 1921, Grafton Elliot Smith Papers, John Rylands Library, Manchester, Folder 1. Elliot Smith replied that Head was unlikely to cooperate: Letter of Grafton Elliot Smith to George Campion, 30 September 1921, *ibid*.
225. C. K. Ogden and I. A. Richards, *The Meaning of Meaning: A Study of the Influence of Language upon Thought and of the Science of Symbolism* (London: Kegan Paul, 1923), p. 350. Ogden and Richards were, however, critical of the way in which Head used the term 'meaning'.
226. Letter of Henry Head to Charles Kay Ogden, 4 November 1921, Ogden Papers, McMaster University.
227. Letter of Henry Head to Charles Kay Ogden, 3 December 1921, *ibid*.
228. 'The progress of science. Sensation controlled by the brain. Dr. Henry Head's Researches,' *The Times*, 25 October 1921, p. 8.
229. The title in question is: Head, 'Disorders of symbolic thinking and expression'.
230. 'Philosophers at Oxford. Henri Bergson. Semantic aphasia,' *The Times*, 6 October 1920, p. 8.
231. Letter of Henry Head to Ruth Mayhew, 7 March 1902, CMAC, PP/HEA/D4/11. Wright worked on this book, which was meant to embody a revised, modern system of logic, for much of his life but never completed the work. See: L. Colebrook, 'Almroth Edward Wright 1861–1947,' *Obituary Notices of the Royal Society*, 6 (1948), pp. 297–314, p. 309.
232. V. Lee, *Music and its Lovers: an Empirical Study of Emotion and Imaginative Responses to Music* (London: George Allen & Unwin, 1932), p. 130.
233. Letter of Violet Paget to Ruth Head, 8 October 1925, CMAC, PP/HEA/B16.

234. Letter of Henry Head to Robert Nichols, 29 February 1924, Nichols Papers, BL, uncatalogued.
235. *Ibid.*
236. Letter of Robert Nichols to Ruth Head, 2 June 1933, *ibid.*
237. Letter of Robert Nichols to Henry and Ruth Head, 23 July 1933, *ibid.*
238. Letter of Robert Nichols to Henry Head, 9 September 1927, *ibid.*
239. Hughes, *Consciousness and Society;* Ross (ed.), *Modernist Impulses* and J. Ryan, *The Vanishing Subject: Early Psychology and Literary Modernism* (Chicago, IL: University of Chicago Press, 1991).
240. See: J. Crary, *Techniques of the Observer: On Vision and Modernity in the Nineteenth Century* (Cambridge, MA: MIT Press, 1992), especially pp. 40–3.
241. G. C. Campion, *Elements in Thought & Emotion: An Essay on Education, Epistemology, & the Psycho-Neural Problem* (London: University of London Press, 1923), pp. 4–5.

4 Ruth and Henry

1. CMAC, PP/HEA/D4/2.
2. This conceit appears to have been in vogue among Head's acquaintances. A friend advised him of a letter about an appointment at the London Hospital from a 'Mrs Ladislaw.' This provided the opportunity for some humorous banter about Head's professional persona and possibly the hostility he had aroused in some quarters. Mrs Ladislaw had apparently favoured a candidate who: 'was universally recognized as the coming physician by all the best judges in London. He was "so broad minded that he would treat you homoeopathically or allopathically just as you wished and never missed an opportunity of speaking against the cruel & devilish practice of vivisection." Yet in spite of these sound views the corrupt clique that governs such elections chose a young man who had no practice and was only known for a disgusting new method of torturing animals and some foolish papers on the stomach ache that contained nothing everybody did not know before he was born. She is therefore going to write to you to tell you you must immediately elect her protégé to avoid a great public scandal. (By the by I don't know the fellow's name but I wonder if you are the person she is so hot against....)' Letter of John Waldegrave to Henry Head, undated, CMAC, PP/HEA/D9.
3. Letter of Ruth Head to Janet Ashbee, 24 October 1915, Ashbee Papers, KCC.
4. *Ibid.*
5. Letter of Ruth Head to Henry Head, 23 March 1917, CMAC, PP/HEA/D4/19.
6. R. Head, *Pictures,* p. v.
7. *Ibid.*, p. vi.
8. M. N. Cohen (ed.), *The Selected Letters of Lewis Carroll* (London: Macmillan, 1982), p. 86.
9. *Ibid.*, p. 88.
10. M. N. Cohen, *Lewis Carroll: A Biography* (London: Macmillan, 1995), p. 170.
11. Letter of Ruth Mayhew to Henry Head, 10 November 1902, CMAC, PP/HEA/D4.
12. Head, *Pictures*, pp. vi–vii.
13. Letter of Ruth Head to Janet Ashbee, [1933], Ashbee Papers, KCC.
14. This information is drawn from a draft letter in Head's hand to the Governors of St. Paul's School in London written on 17 February 1903 on the occasion of Mayhew seeking an appointment there. CMAC, D4/13.
15. Letter of Ruth Mayhew to Henry Head, 11 December 1897, CMAC, PP/HEA/D4/2.

16. Letter of Ruth Mayhew to Henry Head, 21 January 1901, CMAC, PP/HEA/D4/8.
17. Letter of Ruth Mayhew to Henry Head, 19 August 1899, CMAC, PP/HEA/D4/5.
18. Letter of Ruth Mayhew to Henry Head, 25 August 1902, CMAC, PP/HEA/D4/12.
19. Letter of Ruth Head to Thomas Hardy, 23 July 1921, Thomas Hardy Papers, Dorchester County Museum.
20. Letter of Ruth Mayhew to Henry Head, 19 August 1899, CMAC, PP/HEA/D4/5.
21. Letter of Ruth Mayhew to Henry Head, 17 May 1897, CMAC, PP/HEA/D4/2.
22. *Ibid.*
23. Letter of Ruth Mayhew to Henry Head, 18 February 1901, CMAC, PP/HEA/D4/8. After she married Head, Ruth recounted attending Church with his father. She confided that: 'much of the Service struck me as real nonsense: only we had O God our Help in Ages past which is a fine rolling impersonal kind of appeal to a great invisible Jovian kind of Deity, that I liked.' Letter of Ruth Head to Henry Head, 20 November 1904, CMAC, PP/HEA/D4/18.
24. Letter of Henry Head to Ruth Mayhew, 1 March 1901, CMAC, PP/HEA/D4/8.
25. Letter of Ruth Mayhew to Henry Head, 10 November 1899, CMAC, PP/HEA/D4/5.
26. Head and Mayhew 'Rag Book', vol. 1A., CMAC, PP/HEA E3/1, p. 10.
27. *Ibid.*, p. 42.
28. Letter of Henry Head to Ruth Mayhew, 1 March 1901, CMAC, PP/HEA/D4/8.
29. [Journal of Stay in Italy, March–April, 1892], CMAC, PP/HEA, BOX 9 E1.
30. Letter of Henry Head to Ruth Mayhew, 4 March 1899, CMAC, PP/HEA/D4/4.
31. Letter of Henry Head to Ruth Mayhew, [1897], CMAC, PP/HEA/D4/2.
32. Letter of Henry Head to Ruth Mayhew, 21 February 1899, CMAC, PP/HEA/D4/4.
33. Letter of Henry Head to Ruth Mayhew, [February 1899, fragment], *ibid.*
34. Ruth Mayhew, 'Journal intime', 18 September 1900, CMAC, PP/HEA/DR/2.
35. Letter of Henry Head to Ruth Mayhew, 4 March 1899, CMAC, PP/HEA/D4/4.
36. Letter of Henry Head to Ruth Mayhew, 19 March [1899], *ibid.*
37. Letter of Ruth Mayhew to Henry Head, 3 April 1899, *ibid.*
38. Letter of Henry Head to Ruth Mayhew, 4 March 1899, *ibid.*
39. Letter of Ruth Mayhew to Henry Head, 17 March [1899], *ibid.*
40. Letter of Henry Head to Ruth Mayhew, 22 March [1899], CMAC, PP/HEA/DR/2.
41. Letter of Ruth Mayhew to Henry Head, 12 April 1899, *ibid.*
42. Letter of Ruth Mayhew to Henry Head, 29 March 1899, *ibid.*
43. Letter of Henry Head to Ruth Mayhew, 31 March 1899, *ibid.*
44. Letter of Ruth Mayhew to Henry Head, 12 April 1899, *ibid.*
45. Letter of Ruth Mayhew to Henry Head, 15 April 1899, *ibid.*
46. Letter of Ruth Mayhew to Henry Head, 24 April 1899, *ibid.*
47. Letter of Henry Head to Ruth Mayhew, [Spring 1899] fragment, *ibid.*
48. Letter of Ruth Mayhew to Henry Head, 24 April 1899, *ibid.*
49. Letter of Ruth Head to Janet Ashbee, *c.* 1928, Ashbee Papers, KCC.
50. Letter of Henry Head to Ruth Mayhew, 19 November 1900, CMAC, PP/HEA/D4/7.
51. Letter of Ruth Mayhew to Henry Head, 17 November 1898, CMAC, PP/HEA/D4/3.
52. Letter of Ruth Mayhew to Henry Head, 11 August 1899, CMAC, PP/HEA/DR/2.
53. Letter of Ruth Mayhew to Henry Head, 24 November 1898, CMAC, PP/HEA/D4/3.
54. Letter of Ruth Head to Janet Ashbee, 7 November 1933, Ashbee Papers, KCC.
55. Letter of Ruth Mayhew to Henry Head, 11 March 1899, CMAC, PP/HEA/D4/4.
56. Letter of Ruth Mayhew to Henry Head, 27 January 1902, CMAC, PP/HEA/D4/11.
57. Letter of Ruth Mayhew to Henry Head, 19 March 1900, CMAC, PP/HEA/D4/6.

58. Letter of Ruth Mayhew to Henry Head, 27 July 1899, CMAC, PP/HEA/D4/4.
59. Letter of Ruth Mayhew to Henry Head, *c.* 1897, CMAC, PP/HEA/D4/2.
60. Letter of Ruth Mayhew to Henry Head, 14 September 1898, CMAC, PP/HEA/D4/3.
61. Letter of Ruth Mayhew to Henry Head, 17 July 1899, CMAC, PP/HEA/D4/4.
62. Letter of Ruth Mayhew to Henry Head, 19 July 1899, *ibid.*
63. Letter of Henry Head to Ruth Mayhew, undated, *ibid.*
64. Letter of Ruth Mayhew to Henry Head, 19 July 1899, *ibid.*
65. Letter of Henry Head to Ruth Mayhew, 8 September 1899, *ibid.*
66. Letter of Ruth Mayhew to Henry Head, 13 December 1899, *ibid.*
67. Letter of Ruth Mayhew to Henry Head, 1 January 1901, CMAC/PP/HEA/ D4/8.
68. Letter of Ruth Mayhew to Henry Head, 5 January 1901, *ibid.*
69. Letter of Ruth Mayhew to Henry Head, 28 January 1901, *ibid.*
70. Letter of Henry Head to Ruth Mayhew, *c.* 13 January 1901, *ibid.*
71. Letter of Ruth Mayhew to Henry Head, 10 January 1901, *ibid.*
72. Letter of Arnold Mayhew to Ursula [Whitehead], 26 July 99, CMAC, PP/HEA/D10.
73. Letter of Ruth Mayhew to Henry Head, 6 March 1901, CMAC, PP/HEA/D4/8.
74. Letter of Ruth Mayhew to Henry Head, 17 June 1901, CMAC, PP/HEA/D4/9.
75. Letter of Henry Head to Ruth Mayhew, 21 June 1901, *ibid.*
76. Letter of Ruth Mayhew to Henry Head, 14 July 1902, CMAC, PP/HEA/ D4/12.
77. Letter of Henry Head to Ruth Mayhew, 18 July 1902, *ibid.*
78. Letter of Henry Head to Ruth Mayhew, 20 August 1903, CMAC, PP/HEA/D4/15.
79. Letter of Henry Head to Ruth Mayhew, 18 September 1903, *ibid.*
80. Letter of Ruth Mayhew to Henry Head, 21 April 1903, CMAC, PP/HEA/ D4/13.
81. Letter of Ruth Mayhew to Henry Head, 9 July 1903, CMAC, PP/HEA, D4/15.
82. Letter of Ruth Mayhew to Henry Head, 10 July 1903, *ibid.*
83. Letter of Ruth Mayhew to Henry Head, 20 September 1903, *ibid.*
84. *Ibid.*
85. Letter of Henry Head to Ruth Mayhew, 22 September 1903, *ibid.*
86. Letter of Hester Head to Henry Head, 19 September 1903, CMAC, PP/HEA/D3.
87. Letter of Henry Head to Ruth Mayhew, 22 September 1903, CMAC, PP/HEA/ D4/15.
88. Letter of Hester Head to Henry Head, 23 September 1903, CMAC, PP/HEA/D3.
89. Letter of Henry Head to Anthony Mayhew, 27 December 1903, CMAC, PP/HEA/ D10. The operation in question was to sever the cutaneous nerves in Head's arm in order to further his researched into sensation – see Chapter 3, p. 125.
90. Letter of Ruth Mayhew to Janet Ashbee, 2 October 1936, Ashbee Papers, King's College, Cambridge.
91. *Ibid.*
92. Letter of A. L. Mayhew to Henry Head, 28 December 1903, CMAC, PP/HEA/D9.
93. Letter of Alfred North Whitehead to Ruth Mayhew, 3 January 1904, *ibid.*
94. Letter of Henry Head to Ruth Mayhew, 28 June 1904, CMAC, PP/HEA/D3.
95. Letter of Henry Head to Anthony Mayhew, 12 January 1904, CMAC, PP/HEA/D10.
96. Letter of Ruth Head to Hester Pinney, 21 November 1926, CMAC, PP/HEA/D6/1.
97. Letter of Henry Head to Ruth Mayhew, 22 September 1899, CMAC, PP/HEA/D4/4.
98. Letter of Ruth Mayhew to Henry Head, 28 January 1901, CMAC, PP/HEA/D4/8.
99. Letter of Henry Head to Ruth Mayhew, 20 November 1898, CMAC, PP/HEA/D4/3.
100. Letter of Henry Head to Ruth Mayhew, 25 May 1899, CMAC, PP/HEA/D4/4.
101. Letter of Henry Head to Ruth Mayhew, 25 May 1899, *ibid.*

102. Letter of Henry Head to Ruth Mayhew, 17 December 1898, CMAC, PP/HEA,D4/3.
103. Letter of Ruth Mayhew to Henry Head, undated, CMAC, PP/HEA, unclassified.
104. Letter of Ruth Mayhew to Henry Head, undated, CMAC, PP/HEA/D4/1.
105. Letter of Ruth Mayhew to Henry Head, 3 August 1900, CMAC, PP/HEA/D4/7.
106. Letter of Ruth Mayhew to Henry Head, 21 February 1901, CMAC, PP/HEA/ D4/8.
107. Letter of Ruth Mayhew to Henry Head, 11 February 1900, CMAC, PP/HEA/D4/6.
108. *Ibid*
109. Letter of Henry Head to Ruth Mayhew, 31 August 1900, CMAC, PP/HEA/D4/7.
110. Letter of Henry Head to Ruth Mayhew, 9 November 1900, *ibid*.
111. Letter of Ruth Mayhew to Henry Head, 12 November 1900, *ibid*.
112. Letter of Henry Head to Ruth Mayhew, 25 June 1899, CMAC, PP/HEA, D4/4.
113. Letter of Ruth Mayhew to Henry Head, 26 June 1899, *ibid*.
114. Letter of Ruth Mayhew to Henry Head, 7 October 1902, CMAC, PP/HEA/D4/12.
115. Letter of Ruth Mayhew to Henry Head, [20 November 1900], CMAC, PP/HEA, D4/7.
116. Ginzburg, *On Psychological Prose*, especially pp. 9–11.
117. Letter of Henry Head to Ruth Mayhew, 28 January 1900, CMAC, PP/HEA/D4/6.
118. Letter of Ruth Mayhew to Henry Head, 19 March 1901, CMAC, PP/HEA/D4/8.
119. Letter of Ruth Mayhew to Henry Head, 28 January 1901, CMAC, PP/HEA/8.
120. Letter of Ruth Mayhew to Henry Head, 3 July 1899, CMAC, PP/HEA/ D4/5.
121. Letter of Ruth Mayhew to Henry Head, 1 November 1899, *ibid*.
122. Letter of Ruth Mayhew to Henry Head, 7 November 1899, *ibid*.
123. Letter of Ruth Mayhew to Henry Head, 27 October 1898, CMAC, PP/HEA/D4/3.
124. Letter of Ruth Mayhew to Henry Head, 4 November 1899, *ibid*.
125. Letter of Ruth Mayhew to Henry Head, 24 March 1901, CMAC, PP/HEA/D4/8.
126. Letter of Ruth Mayhew to Henry Head, 2 October 1898, CMAC, PP/HEA/D4/3.
127. Letter of Ruth Mayhew to Henry Head, 11 February 1900, CMAC, PP/HEA/D4/6.
128. Letter of Henry Head to Ruth Mayhew, 4 July 1899, CMAC, PP/HEA/D4/5.
129. Letter of Henry Head to Ruth Mayhew, 18 January 1901, CMAC, PP/HEA/D4/8.
130. Letter of Henry Head to Ruth Mayhew, undated, CMAC, PP/HEA/D4.
131. Letter of Ruth Mayhew to Henry Head, 25 March 1901, CMAC, PP/HEA/D4/8.
132. Letter of Ruth Mayhew to Henry Head, 2 July 1900, CMAC, PP/HEA/D4/7.
133. Letter of Ruth Mayhew to Henry Head, 17 May 1897, CMAC, PP/HEA/D4/2.
134. Letter of Ruth Mayhew to Henry Head, 30 March 1901, CMAC, PP/HEA/D4/8.
135. Letter of Ruth Mayhew to Henry Head, 1 January 1901, *ibid*.
136. Letter of Ruth Mayhew to Henry Head, 19 June 1899, CMAC, PP/HEA/D4/4.
137. Letter of Ruth Mayhew to Henry Head, 22 April 1900, CMAC, PP/HEA/D4/6.
138. Letter of Ruth Mayhew to Henry Head, 20 July 1899, CMAC, PP/HEA/D4/4.
139. Letter of Ruth Mayhew to Henry Head, 21 January 1901, CMAC, PP/HEA/ D4/8.
140. Letter of Ruth Mayhew to Henry Head, 13 December 1899, CMAC, PP/HEA/D4/4.
141. Letter of Ruth Mayhew to Henry Head, 3 September 1898, CMAC, PP/HEA/D4/3.
142. Letter of Ruth Mayhew to Henry Head, 10 January 1901, CMAC, PP/HEA/ D4/8.
143. Letter of Ruth Mayhew to Henry Head, 9 May [1899], CMAC, PP/HEA/D4/4.
144. Letter of Henry Head to Ruth Mayhew, 6 September 1899, *ibid*.
145. *Ibid*.
146. Letter of Henry Head to Ruth Mayhew, 7 June 1900, CMAC, PP/HEA/D4/6.
147. Letter of Henry Head to Ruth Mayhew, 24 June 1900, *ibid*.
148. Letter of Ruth Mayhew to Henry Head, 24 February 1902, CMAC, PP/HEA/ D4/11.

149. Letter of Henry Head to Ruth Mayhew, 6 May 1903, CMAC, PP/HEA/D4/14.
150. Letter of Henry Head to Ruth Mayhew, 6 March 1898, CMAC, PP/HEA/unclassified.
151. Letter of Ruth Mayhew to Henry Head, 28 April [1898] CMAC, PP/HEA/D4/3.
152. Letter of Henry Head to Ruth Mayhew, 1 May 1898, CMAC, PP/HEA/unclassified.
153. Letter of Henry Head to Ruth Mayhew, 25 May 1899 CMAC, PP/HEA/D4/4.
154. Letter of Henry Head to Ruth Mayhew, 21 January 1899 CMAC, PP/HEA/D4/4.
155. Letter of Henry Head to Ruth Mayhew, 17 December 1898, CMAC, PP/HEA/D4/3.
156. Letter of Ruth Mayhew to Henry Head, 24 January 1898, CMAC, PP/HEA/D4/4.
157. Letter of Ruth Mayhew to Henry Head, 10 November 1898, CMAC, PP/HEA/D4/3.
158. *Ibid.*
159. *Ibid.*
160. Letter of Henry Head to Ruth Mayhew, 12 November 1898, CMAC, PP/HEA/D4/3.
161. Letter of Ruth Mayhew to Henry Head, 3 January 1900, CMAC, PP/HEA/D4/6. Emphasis added.
162. Letter of Henry Head to Ruth Mayhew, 22 September 1899, CMAC, PP/HEA/D4/4.
163. Letter of Henry Head to Ruth Mayhew, 30 August 1899, *ibid.*
164. Letter of Henry Head to Ruth Mayhew, 5 November 1899, *ibid.*
165. Letter of Henry Head to Ruth Mayhew, 7 May 1899, *ibid.*
166. Letter of Henry Head to Ruth Mayhew, 12 May 1899, *ibid.*
167. *Ibid.*
168. Letter of Henry Head to Ruth Mayhew, 1 October 1898, CMAC, PP/HEA/unclassified.
169. Letter of Ruth Mayhew to Henry Head, 4 October 1898, CMAC, PP/HEA/ D4/3.
170. Letter of Henry Head to Ruth Mayhew, 12 March 1900, CMAC, PP/HEA/ D4/6.
171. Letter of Henry Head to Ruth Mayhew, *c.* 1897, CMAC, PP/HEA/D4/2.
172. Letter of Henry Head to Ruth Mayhew, 3 December 1898, CMAC, PP/HEA/D4/3.
173. Letter of Henry Head to Ruth Mayhew, 27 April 1904, CMAC, PP/HEA/D4/18.
174. Letter of Henry Head to Ruth Mayhew, [1904], CMAC, PP/HEA/D4/19.
175. Letter of Henry Head to Ruth Mayhew, [1909], *ibid.*
176. Letter of Henry Head to Ruth Mayhew, 19 May 1907, *ibid.*
177. Letter of Henry Head to Ruth Mayhew, 20 May 1907, *ibid.*
178. Letter of Ruth Mayhew to Henry Head, 30 May 1905, *ibid.*
179. Letter of Ruth Head to Henry Head, 13 January 1905, *ibid.*
180. Letter of Ruth Head to Henry Head, 19 March 1904, CMAC, PP/HEA/D4/18.
181. Letter of Henry Head to Ruth Mayhew, 28 July 1910, CMAC, PP/HEA/D4/19.
182. Head and Mayhew, 'Rag Book', vol. 1B, CMAC, PP/HEA/E3/2.
183. Head and Mayhew 'Rag Book', vol. 2A, CMAC, PP/HEA E3/3, title page.
184. Head and Mayhew 'Rag Book', vol. 2B, CMAC, PP/HEA/E3/41, title page.
185. Head and Mayhew 'Rag Book', vol. 2A, CMAC, PP/HEA/E3/3, p. 52.
186. Head and Mayhew 'Rag Book', vol. 1A, CMAC, PP/HEA/E3/1, p. 66.
187. *Ibid.*, p. 79.
188. *Ibid.*, p. 2.
189. *Ibid.*, p. 28.
190. Head and Mayhew, 'Rag Book', vol. 1A, CMAC, PP/HEA/E3/1, p. 19.
191. Head and Mayhew, 'Rag Book', vol. 2B, CMAC, PP/HEA/E3/41, facing p. 6.
192. Head and Mayhew, 'Rag Book', vol. 1A, CMAC, PP/HEA/E3/1, p. 12.

193. Head and Mayhew 'Rag Book', vol. 2A, CMAC, PP/HEA/E3/3, pp. 41–2.
194. *Ibid.*, p. 43.
195. *Ibid.*, p. 10.
196. *Ibid.*, p. 78.
197. Head and Mayhew 'Rag Book', vol. 2B, CMAC, PP/HEA/E3/41, p. 15.
198. Head and Mayhew, 'Rag Book', vol. 2A, CMAC, PP/HEA/E3/3, p. 54.
199. Head and Mayhew, 'Rag Book', vol. 2B, CMAC, PP/HEA/E3/41, facing p. 28. The poem is evidently inspired by Rudyard Kipling's story, 'The Cat that Walked by Himself.'
200. Letter of Robert Nichols to Henry and Ruth Head, 14 June 1931, Nichols Papers, BL, uncatalogued.
201. Letter of Robert Nichols to Henry and Ruth Head, 14 August 1933, *ibid.*
202. Letter of Ruth Head to Janet Ashbee, 30, December *c.* 1916, Ashbee Papers, KCC.
203. Letter of Ruth Head to Janet Ashbee, 24 October 1915, *ibid.*
204. Letter of Ruth Head to Janet Ashbee, 3 January [1920], *ibid.*
205. Letter of Ruth Head to Janet Ashbee, 19/21 June 1915, *ibid.*
206. Letter of Ruth Head to Janet Ashbee, 19 November 1915, *ibid.*
207. Letter of Ruth Head to Janet Ashbee, 23 November 1918, *ibid.*
208. R. Head, *Pages from the Works of Thomas Hardy* (London: Chatto and Windus, 1922); *idem*, *Pictures*.
209. Letter of Ruth Head to Thomas Hardy, 20 July 1921, Thomas Hardy Papers, Dorchester County Museum.
210. Letter of Ruth Head to Thomas Hardy, 23 July 1921, *ibid.*
211. Letter of Ruth Head to Thomas Hardy, 16 August 1921, *ibid.*
212. Letter of Thomas Hardy to Ruth Head, 21 May 1922, *ibid.*
213. Letter of Ruth Mayhew to Henry Head, 1 May 1898, CMAC, PP/HEA/D4/3.
214. Letter of Ruth Mayhew to Henry Head, 9 May 1898, *ibid.*
215. Letter of Ruth Mayhew to Janet Ashbee, 24 July 1918, Ashbee Papers, KCC.
216. Letter of Ruth Mayhew to Janet Ashbee, 30 December 1916, *ibid.*
217. Letter of Ruth Head to Robert Nichols, 26 September 1924, Nichols Papers, BL, uncatalogued.
218. Letter of Robert Nichols to Ruth Head, 1 September 1924, *ibid.*
219. Ruth Head, *A History of Departed Things* (London: Kegan Paul, 1918), p. 2.
220. *Ibid.*, pp. 56–7.
221. *Ibid.*, p. 17.
222. *Ibid.*, p. 22.
223. *Ibid.*, pp.23–4.
224. *Ibid.*, p. 29.
225. These poems are to be found in: CMAC, PP/HEA, F1/1.
226. R. Head, *A History.*, p. 86.
227. *Ibid.*
228. *Ibid.*, p. 36.
229. *Ibid.*, p. 52.
230. Presumably a skit on the name of the French surgeon, Ambroise Paré (1510–90).
231. R. Head, *A History*, p. 117.
232. *Ibid.*, p. 149.
233. Letter of Henry Head to Ruth Mayhew, 22 May 1903, CMAC, PP/HEA/D4/14.
234. Letter of Ruth Mayhew to Henry Head, 25 May 1903, *ibid.*

235. R. Head, *A History*, pp. 184–5.
236. *Ibid.*, pp. 236–7.
237. The quotation comes from Stevenson's essay, 'Pulvis et Umbra'.
238. Letter of Ruth Head to Janet Ashbee, 19 November 1915, Ashbee Papers, KCC.
239. Letter of Henry Head to Ruth Head, 14 March 1915, CMAC, PP/HEA/D4/19.
240. Letter of Henry Head to Ruth Head, 28 September 1922, *ibid.*

5 The Cultivation of Feeling

1. Letter of Henry Head to Ruth Mayhew, 31 August 1900, CMAC, PP/HEA/D4/7.
2. Letter of Henry Head to Ruth Mayhew, 21 September 1900, *ibid.*
3. Letter of Henry Head to Ruth Mayhew, 25 November 1900, *ibid.*
4. Letter of Ruth Mayhew to Henry Head, 28 November 1900, *ibid.*
5. Letter of Henry Head to Ruth Mayhew, 25 November 1900, *ibid.*
6. Letter of Ruth Mayhew to Henry Head, 3 July 1899, CMAC, PP/HEA/D4/5.
7. Letter of Henry Head to Ruth Mayhew, 25 November 1900, CMAC, PP/HEA/D4/7.
8. Letter of Henry Head to Ruth Mayhew, 21 September 1900, *ibid.*
9. Letter of Henry Head to Ruth Mayhew, 2 June 1899, CMAC, PP/HEA/D4/4.
10. Letter of Henry Head to Ruth Mayhew, 31 March 1900, CMAC, PP/HEA/D4/6.
11. Letter of Henry Head to Ruth Mayhew, 22 May 1903, CMAC, PP/HEA/D4/14.
12. Letter of Henry Head to Ruth Mayhew, undated, CMAC, PP/HEA/D4/1.
13. Letter of Henry Head to Ruth Mayhew, 25 November 1900, CMAC, PP/HEA/D4/7.
14. Letter of Henry Head to Ruth Mayhew, 17 May 1901, CMAC, PP/HEA/D4/9.
15. *ibid.*
16. Head and Mayhew 'Rag Book', vol.1A. October 1901–July 1902, CMAC, PP/HEA E3/1, p. 29.
17. Letter of Henry Head to Ruth Mayhew, 21 July 1899, CMAC, PP/HEA/D4/4.
18. Letter of Henry Head to Ruth Mayhew, 11 December 1900, CMAC, PP/HEA/D4/7.
19. Letter of Henry Head to Ruth Mayhew, 14 July *c.* 1901, CMAC, PP/HEA/D4/9.
20. Letter of Henry Head to Ruth Mayhew, 28 June 1909, CMAC, PP/HEA/D4/19.
21. Letter of Henry Head to Ruth Mayhew, 3 August 1910, *ibid.*
22. In the 'Rag Book', Head entered the following quotation from D'Anunzzio's *Les vierges aux Rochers*: 'Le monde est la représentation de la sensibilité et de la pensée d'un petit nombre d'hommes supérieures qui l'ont crée, puis amplifié et orné dans le cours de temps, et qui continueront de l'amplifier et de l'orner toujours plus dans l'avenir. Le monde, tel qu'il apparaît aujourd'hui est un don magnifique dispensé par une élite à la multitude, par les hommes libres aux escalves, par ceaux qui pensent et sentient à ceaux qui doivent travailler'.
23. Letter of Henry Head to Ruth Mayhew, 31 August 1900, CMAC/PP/HEA/D4/7.
24. Letter of Henry Head to Ruth Mayhew, 14 July *c.* 1901, CMAC, PP/HEA/D4/9.
25. *Ibid.*
26. Letter of Henry Head to Ruth Mayhew, 7 February 1902, CMAC, PP/HEA/D4/11.
27. Letter of Henry Head to Ruth Mayhew, 14 November 1902, CMAC, PP/HEA, D4/12.
28. Letter of Henry Head to Ruth Mayhew, 28 September 1898, CMAC, PP/HEA/D4/3.
29. Letter of Henry Head to Ruth Mayhew, 28 November 1902, CMAC, PP/HEA, D4/12.
30. Letter of Henry Head to Ruth Mayhew, 4 July 1902, *ibid.*

31. Letter of Henry Head to Ruth Mayhew, 20 June 1903, CMAC, PP/HEA/D4/14.
32. R. Nichols, 'Notes on Henry Head', Nichols Papers, BL, uncatalogued.
33. Head and Mayhew, 'Rag Book', vol. 2A, CMAC, PP/HEA E3/3, p. 56.
34. Letter of Henry Head to Ruth Mayhew, 13 April 1900, CMAC/PP/HEA/D4/6.
35. *Ibid.*
36. Letter of Henry Head to Ruth Mayhew, 1 October 1898, CMAC, PP/HEA/D4/3.
37. Letter of Henry Head to Ruth Mayhew, 21 March 1901, CMAC, PP/HEA/D4/8.
38. Ann Banfield, *The Phantom Table: Woolf, Fry, Russell and the Epistemology of Modernism* (Cambridge: Cambridge University Press, 2000), pp. 11–12.
39. [Journal of Head's stay in Italy, March–April 1892], March 12, CMAC, PP/HEA/ BOX 9 E1.
40. [Ruth Head, 'Record of Honeymoon in Paris', April 28–May 10, 1904], May 6, *ibid*.
41. Letter of Henry Head to Ruth Mayhew, 1 October 1898, CMAC, PP/HEA/D4/3.
42. Letter of Henry Head to Ruth Mayhew, 1 October 1898, CMAC, PP/HEA/unclassified.
43. Head and Mayhew, 'Rag Book', vol. 1A. Oct: 1901–July 1902. CMAC, PP/HEA/E3/1, p. 10.
44. Head and Mayhew 'Rag Book', vol. 2A., CMAC, PP/HEA/E3/3, p. 29.
45. *Ibid.*, pp. 29–30.
46. *Ibid.*, p. 30.
47. Head and Mayhew 'Rag Book', vol. 1B, CMAC, PP/HEA/E3/2 Box 9, p. 114. See: Jonathan Crary, *Suspensions of Perception Attention, Spectacle, and Modern Culture* (Cambridge, MA: MIT Press, 2000).
48. *Ibid.*, pp. 112–4.
49. *Ibid.*, p. 114.
50. Letter of Henry Head to Ruth Mayhew, 1 October 1898, CMAC, PP/HEA/D4/3.
51. Letter of Henry Head to Ruth Mayhew, 12 October 1898, *ibid*.
52. See: Lionel Lanbourne, *The Aesthetic Movement* (London: Phaidon Press, 1996).
53. Head and Mayhew 'Rag Book', vol. 1B, CMAC, PP/HEA/E3/2 Box 9, p. 20.
54. *Ibid.*, p. 22.
55. Letter of Henry Head to Ruth Mayhew 12 October 1898, CMAC, PP/HEA/D4/3.
56. *Ibid.*
57. Letter of Henry Head to Robert Nichols, 18 December 1923, Nichols Papers, BL, uncatalogued.
58. Letter of Ruth Mayhew to Henry Head, 21 November 1897, CMAC, PP/HEA/D4.
59. Head and Mayhew, 'Rag Book', vol. 1B, CMAC, PP/HEA E3/2, p. 52.
60. Letter of Henry Head to Ruth Mayhew, 28 September 1898, CMAC, PP/HEA/D4/3.
61. An allusion to the play by Friedrich von Schiller.
62. *Ibid.*
63. Letter of Henry Head to Ruth Mayhew, 3 November 1898, CMAC, PP/HEA/D4/3.
64. Head and Mayhew preserved the prospectus of the Pharos Club and the programmes of several of its productions in their 'Rag Book'. 'Rag Book', vol. 2B, CMAC, PP/HEA/E3/41, p. 57
65. *Ibid.*, p. 56.
66. *Ibid.*, pp. 51–2.
67. *Ibid.*, p. 78.
68. *Ibid.*, p. 80.
69. *Ibid.*, p. 88.

Notes to pages 212–20   323

70. Letter of Henry Head to Ruth Mayhew, 11 December 1898, CMAC, PP/HEA/ D4/3.
71. Head and Mayhew, 'Rag Book', vol. 3A, CMAC, PP/HEA/E3/5, pp. 28–30.
72. *Ibid.*, p. 32.
73. Ruth Mayhew did, however, on one occasion dream of: 'reading a novel [Head] had written as a quite young man. It was the finest novel I had ever read'. Letter of Ruth Mayhew to Henry Head, 1 October 1902, CMAC, PP/HEA/D4/12.
74. Letter of Ruth Mayhew to Henry Head, , 20 July 1899, CMAC, PP/HEA/D4/5.
75. Head and Mayhew, 'Rag Book', vol. 3A, CMAC, PP/HEA/E3/5, p. 28.
76. Letter of Henry Head to Ruth Mayhew, 18 August 1899, CMAC, PP/HEA/D4/5.
77. [Journal of Head's stay in Italy, March–April 1892], CMAC, PP/HEA/E1.
78. Letter of Henry Head to Ruth Mayhew, 3 March 1901, CMAC, PP/HEA/D4/8.
79. Letter of Henry Head to Ruth Mayhew, 6 March 1901, CMAC, PP/HEA/D4/8.
80. Letter of Henry Head to Ruth Mayhew, 5 August 1899, CMAC, PP/HEA/D4/5.
81. Letter of Henry Head to Ruth Mayhew, 27 October 1899, *ibid.*
82. *Ibid.*
83. Letter of Henry Head to Ruth Mayhew, 5 January 1901, CMAC, PP/HEA/D4/8.
84. Mayhew and Head, 'Rag Book', vol. 3A, CMAC, PP/HEA/E3/5, p. 28.
85. *Ibid.*, p. 1.
86. Letter of Henry Head to Ruth Mayhew, 21July 1899, CMAC, PP/HEA/D4/5.
87. Letter of Henry Head to Ruth Mayhew, 9 January 1902, CMAC, PP/HEA/D4/13.
88. Letter of Robert Nichols to Henry Head, 26 June 1925, Nichols Papers, BL, uncatalogued. See Chapter 2, p. 88 above.
89. Robert Nichols, 'Notes on Henry Head'. It is tempting to speculate that Head's choice of a novel dealing with the conflict between duty and honour and the sense of self-preservation when seeking to help a shell-shocked officer was not accidental.
90. See L. Rothfield , *Vital Signs: Medical Realism in Nineteenth-Century Fiction* (Princeton, NJ: Princeton University Press, 1992), ch. 2.
91. Head and Mayhew, 'Rag Book', vol. 2A., CMAC, PP/HEA/E3/3, p. 13.
92. *Ibid.*, p. 4.
93. *Ibid.*, pp. 5–7.
94. *Ibid.*, p. 9.
95. *Ibid.*, p. 28.
96. Letter of Henry Head to Ruth Mayhew, 30 October 1903, CMAC, PP/HEA/D4/16. Head met James socially at least once: Letter of Henry Head to Ruth Mayhew, 23 February 1900, CMAC, PP/HEA/D4/6. After James suffered a stroke, Head also saw him professionally. See Chapter 2 above.
97. Head and Mayhew, 'Rag Book', vol. 2A, CMAC, PP/HEA/E3/3, p. 23.
98. See: R. B. Pippin, *Henry James and Modern Moral Life* (Cambridge: Cambridge University Press, 2000).
99. Letter of Henry Head to Ruth Mayhew, 22 February 1898, CMAC, PP/HEA, unclassified.
100. Letter of Henry Head to Ruth Mayhew, 21 July 1899, CMAC, PP/HEA/D4/5.
101. Letter of Henry Head to Robert Graves, 7 April 1925, Graves Papers, Morris Library, Southern Illinois University, 890.
102. *Ibid.*
103. *Ibid.*
104. Letter of Henry Head to Robert Nichols, 9 April 1925, Nichols Papers, BL, uncatalogued.

105. Nichols, 'Notes on Henry Head'.
106. I cannot see what flowers are at my feet,
    Nor what soft incense hangs upon the boughs,
    But, in embalmèd darkness, guess each sweet
    Wherewith the seasonable month endows
    The grass, the thicket, and the fruit-tree wild;
    White hawthorn, and the pastoral eglantine;
    Fast fading violets cover'd up in leaves;
    And mid-May's eldest child,
    The coming musk-rose, full of dewy wine,
    The murmurous haunt of flies on summer eves.
107. Nichols, 'Notes on Henry Head'.
108. *Ibid.*
109. Head and Mayhew, 'Rag Book', vol. 2A, CMAC, PP/HEA/E3, p. 75.
110. Head and Mayhew, 'Rag Book', vol. 1B, CMAC, PP/HEA/E3. p. 62.
111. *Ibid.*, p. 58.
112. Head told Harvey Cushing that he and Bridges had taken the examination at the Royal College of Physicians at the same time: 'A visit to Henry and Ruth Head, July 16 1930', Harvey Cushing Papers, Yale University Library, series I Correspondence, Folder 670.
113. Letter of Henry Head to Ruth Mayhew, *c.* 1897, CMAC, PP/HEA/D4/2.
114. Letter of Henry Head to Ruth Mayhew, 27 October 1899, *ibid.*
115. Letter of Henry Head to Ruth Mayhew, 10 November 1899, CMAC, PP/HEA/D4/5.
116. *Ibid.*
117. Head and Mayhew, 'Rag Book', vol. 2A, CMAC, PP/HEA/E3/3, p. 75.
118. Letter of Ruth Head to Siegfried Sassoon, 13 December 1929, Siegfried Lorraine Sassoon Collection, Harry Ransom Center, University of Texas, Austin.
119. Head and Mayhew, 'Rag Book', vol. 1B, CMAC, PP/HEA/E3/2, pp. 6–9.
120. *Ibid.*, p. 40.
121. See: George Saintsbury, 'W.E. Henley', in *The Cambridge History of English and American Literature*, 1907–21, http://www.bartleby.com/223/0656.html.
122. Letter of Henry Head to Ruth Mayhew, 13 April 1900, CMAC, PP/HEA D4/6.
123. C. Klonk, *Science and the Perception of Nature: British Landscape Art in the Late Eighteenth and Early Nineteenth Century* (New Haven, CT: Yale University Press, 1996).
124. Letter of Henry Head to Ruth Mayhew, *c.* 1899, CMAC, PP/HEA/D4/1.
125. Letter of Head Head to Ruth Mayhew, 18 March 1901, CMAC, PP/HEA/D4/8.
126. Letter of Henry Head to Ruth Mayhew, 18 August 1899, CMAC, PP/HEA/D4/5.
127. Letter of Henry Head to Ruth Mayhew, 27 October 1899, *ibid.*
128. Letter of Henry Head to Ruth Mayhew, 31 August 1900, CMAC, PP/HEA/D4/7.
129. Letter of Henry Head to Ruth Mayhew, 5 October 1903, CMAC, PP/HEA/D4/16.
130. Letter of Henry Head to Ruth Mayhew, 21 March 1901, CMAC, PP/HEA/D4/8.
131. Letter of Henry Head to Ruth Mayhew, 11 May 1902, CMAC, PP/HEA/D4/11.
132. *Ibid.*
133. Heand and Mayhew, 'Rag Book', vol 1A, CMAC, PP/HEA/E3/1, p. 4.
134. *Ibid.*, p. 3.
135. Head copied the quotation, taken from Henry James's *Hubert Crackenthorpe*, into the 'Rag Book'. Head and Mayhew 'Rag Book', vol. 2A, CMAC, PP/HEA/E3/3, p. 71

136. Head may have derived this understanding of aesthetics, in large part, from his reading of Pater. See: J. Matz, *Literary Impressionism and Modernist Aesthetics* (Cambridge: Cambridge University Press, 2001), especially pp. 56–7.
137. Letter of Henry Head to Ruth Mayhew, 12 March 1900, CMAC, PP/HEA/D4/6.
138. Letter of Henry Head to Ruth Mayhew, 20 November 1898, CMAC, PP/HEA/D4/3.
139. Letter of Henry Head to Ruth Mayhew, 25 August 1899, CMAC, PP/HEA/D4/5.
140. Letter of Henry Head to Ruth Mayhew, 9 October 1902, CMAC, PP/HEA, D4/12.
141. *Ibid.*
142. *Ibid.*
143. *Ibid.*
144. *Ibid.*
145. *Ibid.*
146. *Ibid.*
147. Letter of Henry Head to Ruth Mayhew, 11 October 02, CMAC, PP/HEA/D4/12.
148. *Ibid.*
149. *Ibid.*
150. *Ibid.*
151. Letter of Henry Head to Ruth Mayhew, 12 October 02, *ibid.*
152. *Ibid.*
153. *Ibid.*
154. *Ibid.*
155. Letter of Henry Head to Ruth Mayhew, 5, 6 and 7 April 1901, CMAC, PP/HEA/D4/9.
156. Letter of Henry Head to Hester Head senior, 7 December, year not given, CMAC,PP/HEA/F1/1. The words in square brackets formed part of an earlier version of the sonnet.
157. Letter of Ruth Mayhew to Henry Head, 20 July 1899, CMAC, PP/HEA/D4/5.
158. Letter of Henry Head to Ruth Mayhew, 25 June 1899, *ibid.*
159. Letter of Henry Head to Ruth Mayhew, 25 September [1897], CMAC/PP/HEA/F1/1.
160. Letter of Ruth Mayhew to Henry Head, 1 October 1897, *ibid.*
161. Letter of Ruth Mayhew to Henry Head, June 1897, CMAC, PP/HEA/D4/2.
162. Letter of Ruth Mayhew to Henry Head,15 September 1897, *ibid.*
163. Letter of Henry Head to Ruth Mayhew, 1897, *ibid.*
164. Letter of Ruth Mayhew to Henry Head, 1 October 1897, *ibid.*
165. Letter of Henry Head to Ruth Mayhew, 18 August 1899, CMAC, PP/HEA/D4/5.
166. Letter of Ruth Head to Arthur Mayhew, 5 July [1919], CMAC, PP/HEA/D10.
167. H. Head, *Destroyers and Other Verses* (London: Humphrey Milford, Oxford University Press, 1919), Dedication.
168. Letter of Henry Head to Ruth Mayhew, 31 May 1898, CMAC, PP/HEA/D4/3.
169. Letter of Henry Head to Ruth Mayhew, 24 November 1899, CMAC, PP/HEA/D4/5.
170. Mayhew and Head, 'Rag Book', vol. 1A, CMAC,PP/HEA/E3/1, p. 6.
171. Letter of Henry Head to Ruth Mayhew, 24 November 1899, CMAC, PP/HEA/D4/5.
172. In: CMAC, PP/HEA/F1/1.
173. *Ibid.* This was published in the 'Sun and Showers' series in: Head, *Destroyers*, p. 79 as a 'He' poem. Both of the titles that appeared in the manuscript were suppressed.
174. *Ibid.*
175. *Ibid.*

176. *Ibid.*
177. *Ibid.*
178. *Ibid.*
179. Some of Head's poems were previously published in periodicals such as: *The Yale Journal, The English Review,* and *The Dublin Review.*
180. Head, *Destroyers*, p. 11.
181. *Ibid.*, p. 12.
182. *Ibid.*, p. 10.
183. *Ibid.*, p. 16.
184. *Ibid.*
185. In: CMAC, PP/HEA/F1/1.

6 The Two Solitaries

1. G. Holmes, 'Henry Head 1861–1940', *Obituary Notices of Fellows of the Royal Society*, 1939–41, 3: 665–89, p. 670.
2. *Ibid.*, p. 667.
3. Letter of Henry Head to Macmillan & Co., 28 November 1910, BL, Add MS 55250 (26).
4. Letter of Henry Head to Macmillan & Co., 18 January 1913, *ibid.* (48).
5. A copy of the invitation is preserved in the Macmillan archive: *ibid.*, (36).
6. 'Famous doctors get Fordham degrees: Dr. Henry Head of London discovers a unique case in the Montefiore Home', *New York Times*, 12 September 1912, p. 2.
7. 'Present age decadent but happy, says Dr Henry Head', *New York Times*, 15 September 1912, p. 5.
8. *Ibid.*
9. *Ibid.*
10. Letter of Henry Head to Ruth Head, 22 April 1909, CMAC, PP/HEA/D4/19.
11. Letter of Henry Head to Ruth Head, 5 April 1908, *ibid.*
12. In his account of the Congress, Cushing indicated that Head was 'rather bored' by some of the banter at the social gatherings that followed sessions. J. F. Fulton, *Harvey Cushing: A Biography* (Springfield, IL: Charles C. Thomas, 1946), p. 294.
13. Letter of Henry Head to Ruth Head, 29 August 1909, CMAC, PP/HEA/D4/19.
14. Letter of Henry Head to Ruth Head, 31 August 1909, *ibid.*
15. Letter of Henry Head to Ruth Head, 30 August 1909, *ibid.*
16. Letter of Henry Head to Ruth Head, 7 April 1916, *ibid.*
17. *Ibid.*
18. *Ibid.*
19. Letter of Henry Head to Ruth Head, 25 September 1904, *ibid.*
20. Letter of Henry Head to Ruth Head, 19 May 1907, *ibid.*
21. Letter of Ruth Head to John Jay Chapman, 1 June 1922, Chapman Papers, Houghton Library, Harvard, Bms Am 1854.1 (330).
22. Letter of Henry Head to Ruth Head, 30 June 1912, CMAC, PP/HEA/D4/19.
23. Letter of Henry Head to Ruth Mayhew, *c.* 1899, CMAC, PP/HEA/D4/4.
24. Letter of Henry Head to Ruth Mayhew, *c.* October 1900, *ibid.*
25. Letter of Henry Head to Ruth Head, 30 December 1911, CMAC, PP/HEA/D4/19.
26. Letter of Henry Head to Ruth Head, 4 January 1910, *ibid.*
27. *Ibid.*

28. Letter of Henry Head to Ruth Head, 25 December 1912, *ibid*.
29. Letter of Henry Head to Ruth Head, 26 December 1912, *ibid*.
30. Letter of Henry Head to Ruth Head, 28 December 1912, *ibid*.
31. Letter of Henry Head to Ruth Head, 16 July 1910, *ibid*.
32. *Ibid*.
33. Letter of Henry Head to Ruth Head, 29 July 1910, *ibid*.
34. Letter of Henry Head to Ruth Head, 27 July 1910, *ibid*.
35. Letter of Henry Head to Ruth Head, 1 August 1910, *ibid*.
36. Letter of Henry Head to Ruth Head, 27 July 1910, *ibid*.
37. Letter of Henry Head to Ruth Head, 28 July 1910, *ibid*.
38. *Ibid*.
39. Letter of Henry Head to Ruth Head, 31 July 1910, *ibid*.
40. *Ibid*.
41. *Ibid*.
42. *Ibid*.
43. Letter of Henry Head to Charles Sherrington, 20 December 1924, Sherrington Papers, Department of Physiology, Oxford. The book in question was: K. Stephen, *The Misuse of Mind: A Study of Bergson's Attack on Intellectualism* (London: Kegan Paul, 1922).
44. Letter of Henry Head to Ruth Head, 11 August 1910, CMAC, PP/HEA D4/19.
45. *Ibid*.
46. Letter of Henry Head to Ruth Head, 27 July 1910, *ibid*.
47. See: R. Wohl, *A Passion for Wings: Aviation and the Western Imagination, 1908–1918* (New Haven, CT: Yale University Press, 1996).
48. Kern, *The Culture of Time and Space*, pp. 242–3.
49. Letter of Henry Head to Ruth Mayhew, 18 August 1910, CMAC, PP/HEA/D4/19.
50. *Ibid*.
51. *Ibid*.
52. *Ibid*.
53. Letter of Ruth Head to Janet Ashbee, 1 June 1915, Ashbee Papers, KCC.
54. Letter of Ruth Head to Janet Ashbee, 15 June 1915, *ibid*.
55. Letter of Ruth Head to Janet Ashbee, 19/21 June 1915, *ibid*.
56. Letter of Ruth Head to Janet Ashbee, 6 December 1915, *ibid*.
57. Letter of Ruth Head to Henry Head, 23 March 1917, CMAC, PP/HEA/D4/19.
58. Letter of Ruth Head to Janet Ashbee, 30 December 1916, Ashbee Papers, KCC.
59. Letter of Henry Head to Ruth Head, 11 November 1914, CMAC, PP/HEA/D4/19.
60. Letter of Ruth Head to Janet Ashbee, 20 February 1916, Ashbee Papers, KCC.
61. Letter of Ruth Head to Janet Ashbee, 27 August 1915, *ibid*. Another of Head's brothers, Christopher, was among those lost with the sinking of the *Titanic*. His mother attended the memorial service held at St Paul's Cathedral in 1912: *Times*, 16 April 1912, p. 9; 20 April 1912.
62. Letter of Henry Head to Ruth Head, 12 March 1915, CMAC, PP/HEA/D4/19.
63. R. Hart-Davies (ed.), *Siegfried Sassoon Diaries 1920–1922* (London: Faber and Faber, 1981), p. 167.
64. *Ibid*., pp. 167–8.
65. Letter of Henry Head to Ruth Head, 16 March 1917, CMAC, PP/HEA/D4/19.
66. Letter of Henry Head to Ruth Head, 17 March 1917, *ibid*.
67. Quoted in: A. Charlton and W. Charlton, *Putting Poetry First*, p. 52.
68. Letter of Henry Head to Ruth Head, 18 March 1917, CMAC, PP/HEA/D4/19.

69. Head also exchanged poems with his fellow scientist, Charles Sherrington. In January 1923, he urged Sherrington to publish, or at least print for private circulation, his verses. Letter of Henry Head to Charles Sherrington, 21 January 1923, Sherrington Papers, Department of Physiology, University of Oxford.
70. Letter of Henry Head to Ruth Head, 26 March 1917, CMAC, PP/HEA/D4/19.
71. Hart-Davies (ed.), *Siegfried Sassoon 1920–1922*, p. 34.
72. *Ibid.*, p. 172.
73. Letter of Henry Head to Siegfried Sassoon, 16 June 1923, Siegfried Lorraine Sassoon Collection, Harry Ransom Center, University of Texas, Austin.
74. Letter of Ruth Head to Siegfried Sassoon, 1 April [1925], *ibid.*
75. Letter of Henry Head to Siegfried Sassoon, 8 March 1926, Sassoon Papers, CUL, Add MS, 9375/364.
76. Hart-Davies (ed.), *Siegfried Sassoon 1920–1922*, p. 34.
77. R. Hart-Davies (ed.), *Siegfried Sassoon Diaries 1923–1925* (London: Faber and Faber, 1985), pp. 34–5.
78. *Ibid.*, p. 82.
79. *Ibid.*, p. 88.
80. Undated postcard, probably 1925, Siegfried Lorraine Sassoon Collection, Harry Ransom Center, University of Texas, Austin.
81. Hart-Davies (ed.), *Siegfried Sassoon 1923–1925*, pp. 194–5.
82. Hart-Davies (ed.), *Siegfried Sassoon 1920–1922*, p. 69.
83. *Ibid.*, p. 132.
84. Letter of Henry Head to Robert Nichols, 19 December 1924, Nichols Papers, BL, uncatalogued.
85. Letter of Ruth Head to Robert Nichols, 11 February 1925, *ibid.*
86. Hart-Davies (ed.), *Siegfried Sassoon 1920–1922*, p. 32.
87. *Ibid.*, p. 194.
88. Letter of Ruth Head to Janet Ashbee, 3 March [1921], Ashbee Papers, KCC.
89. Letter of Ruth Head to Giles Lytton Strachey, BL MSS Add 60669 f.184
90. Hart-Davies (ed.), *Siegfried Sassoon 1923–1925*, p. 44.
91. *Ibid.*, p. 141.
92. *Ibid.*, p. 222.
93. *Ibid.*, p. 221.
94. *Ibid.*, pp. 223–5.
95. Letter of Henry Head to Siegfried Sassoon, Siegfried Lorraine Sassoon Collection, 2 April 1925, Harry Ransom Center, University of Texas, Austin.
96. *The Times*, 14.11.16, p. 9; 16 April 1918, p. 3.
97. Letter of Henry Head to Robert Nichols, 18 December 1923; Letter of Henry Head to Robert Nichols, 29 February 1924, Nichols Papers, BL, uncatalogued.
98. Letter of Ronald Waterhouse to Henry Head, 18 December 1926, CMAC, PP/HEA/D4/19.
99. *The Times*, 1 January 1927, p. 10.
100. Letter of Henry Head to Ruth Mayhew, 19 May 1921, CMAC, PP/HEA/D4/19.
101. Letter of Henry Head to Ruth Mayhew, 20 May 1921, *ibid.*
102. Letter of Henry Head to Robert Nichols, 29 February 1924, Nichols Papers, BL, uncatalogued. It is unclear to which of the numerous youth movements in interwar Germany Head was referring. The National Socialist 'Jugendbund' had been established around

1922; the ideals of this organization, however, hardly coincide with those that Head outlined.
103. Letter of Ruth Head to Janet Ashbee, 6 April [1921], Ashbee Papers, KCC.
104. Letter of Robert Nichols to Henry and Ruth Head, 6 April 1923, Nichols Papers, BL, uncatalogued.
105. Letter of Henry Head to Ruth Mayhew, 15 August 1910, CMAC, PP/HEA/D4/19.
106. Letter of Ruth Head to Janet Ashbee, 3 January [1920], Ashbee Papers, KCC.
107. Letter of Henry Head to Ruth Head, 31 August 1921, CMAC, PP/HEA/D4/19.
108. Letter of Ruth Head to Janet Ashbee, 13 January [1924], Ashbee Papers, KCC.
109. Letter of Ruth Head to Hester Marsden-Smedley, 11 March c. 1926, CMAC, PP/HEA/D10.
110. Hart-Davies (ed.), *Siegfried Sassoon 1923–1925*, p. 193.
111. *Ibid.*, p. 277.
112. Letter of Ruth Head to Siegfried Sassoon, 19 February c. 1926, Sassoon Papers, CUL, Add 9375/365.
113. Letter of Ruth Head to Hester Marsden-Smedley, 13 February c. 1924, CMAC, PP/HEA/D10.
114. Letter of Ruth Head to Robert Nichols, 12 May 1926, Nichols Papers, BL, uncatalogued.
115. *Ibid.*
116. Letter of Ruth Head to Janet Ashbee, 23 December 1925, Ashbee Papers, KCC.
117. Letter of Ruth Head to Robert Nichols, 5 September 1925, Nichols Papers, BL, uncatalogued.
118. Letter of Henry Head to Robert Nichols, 28 July 1925, Nichols Papers, BL, uncatalogued.
119. Letter of Ruth Head to Janet Ashbee, 18 August 1925. Ashbee Papers, KCC. Ashbee annotated this letter with the words: 'Very Pathetic'.
120. Letter of Ruth Head to Janet Ashbee, 23 December 1925, *ibid*.
121. Letter of Ruth Head to Hester Marsden-Smedley, 5 January [1926], CMAC, PP/HEA/D10.
122. Letter of Ruth Head to Hester Marsden-Smedley, 20 April [1926], *ibid*.
123. Letter of Ruth Head to Harvey Cushing, 28 December 1926, Harvey Cushing Correspondence, YUL, folder 670.
124. Letter of Ruth Head to Hester Marsden-Smedley, 16 August 1926, CMAC, PP/HEA/D10.
125. Letter of Janet Ashbee to [Charles Ashbee], 10 March 1927, Ashbee Papers, KCC.
126. *Ibid.*
127. *Ibid.*
128. Letter of Ruth Head to Harvey Cushing, 28 December 1926, Harvey Cushing Correspondence, YUL, folder 670.
129. *Ibid.*
130. Letter of Ruth Head to Siegfried Sassoon, 29 April c. 1927, Sassoon Papers, CUL, Add MS 9375/367.
131. Letter of Ruth Head to Robert Nichols, 29 February 1928, Nichols Papers, BL, uncatalogued.
132. Letter of Ruth Head to Robert Nichols, 13 January 1927, *ibid*.
133. Letter of Ruth Head to Robert Nichols, 18 January 1927, *ibid*.

134. i.e., F.rankie Schuster, with whom Sassoon and Wylde often stayed. See: Jean Moorcroft Wilson, *Siegfried Sassoon*, p. 138.
135. Letter of Ruth Head to Robert Nichols, 17 March 1927, Nichols Papers, BL, uncatalogued.
136. Letter of Ruth Head to Siegfried Sassoon, 22 March *c.* 1927, Sassoon Papers, CUL, Add MS 9375/367.
137. Letter of Ruth Head to Siegfried Sassoon, 24 May *c.* 1927, *ibid.*
138. Letter of Ruth Head to Janet Ashbee, *c.* 1928, Ashbee Papers, KCC.
139. Letter of Ruth Head to Robert Nichols, 28 February 1927, Nichols Papers, BL, uncatalogued.
140. Letter of Ruth Head to John Jay Chapman, 23 May 1930, Chapman Papers, Houghton Library, Harvard, bMS Am 1854.1 (330). Chapman had gone to visit the ailing Head at Robert Nichols's urging. Nichols had opined: 'Poor Henry Head – now 'Sir' – is still alive but I fear he can't last very much longer...'. Letter of Robert Nichols to John Jay Chapman, 9 September 1929, *ibid.*, (556). A similar incident occurred in 1933, when Everard Feilding, 'an old friend but an officious one, actually making an appointment for H. to go up to London! to Baker Street to consult a quack doctor from California who claims to cure Paralysis Agitans by tickling the nerves of the nose!!! I wrote very firmly No and Everard sat down and wrote me the very nastiest letter you can imagine, pouring scorn on all English doctors in general and on H. in particular for his obscurantism. Wasn't that horrid of him'. Letter of Ruth Head to Hester Marsden-Smedley, 11 April [1933], CMAC, PP/HEA/D10.
141. Letter of Hester Marsden-Smedley to Henry and Ruth Head, 17 July [1937], CMAC PP/HEA/D6/3.
142. Letter of Janet Ashbee to Charles Ashbee, 30 January 1929, Ashbee Papers, KCC.
143. Letter of Ruth Head to Siegfried Sassoon, 26 March 1929, Sassoon Papers, CUL, Add MS 9375/369.
144. Letter of Ruth Head to Hester Marsden-Smedley, 23 November [1932], CMAC, PP/HEA/D10.
145. Letter of Ruth Head to Robert Nichols, 13 July 1926, Nichols Papers, BL, uncatalogued.
146. Letter of Ruth Head to Robert Nichols, 12 November 1926, *ibid.*
147. Letter of Ruth Head to Robert Nichols, 29 October 1926, *ibid.*
148. Letter of Ruth Head to Robert Nichols, 19 March 1927, *ibid.*
149. Letter of Ruth Head to Robert Nichols, 7 November 1926, *ibid.*
150. Letter of Ruth Head to Robert Nichols, 28 November 1926, *ibid.*
151. Letter of Ruth Head to Robert Nichols, 24 June 1934, *ibid.*
152. Letter of Ruth Head to Robert Nichols, 26 January 1927, *ibid.*
153. Letter of Ruth Head to Robert Nichols, 14 February 1927, *ibid.*
154. Letter of Ruth Head to Robert Nichols, 7 February 1927, *ibid.*
155. Letter of Ruth Head to Robert Nichols, 22 February 1927, *ibid.*
156. Letter of Ruth Head to Robert Nichols, 28 March 1935, *ibid.*
157. Letter of Ruth Head to Robert Nichols, 23 March 1935, *ibid.*
158. Letter of Janet Ashbee to Ruth Head, 12 June 1935, Ashbee Papers, KCC.
159. Letter of Ruth Head to Janet Ashbee, 21 June 1933, *ibid.*
160. Letter of Ruth Head to Janet Ashbee, 7 November 1933, *ibid.*
161. Letter of Ruth Head to Siegfried Sassoon, 19 September 1930, Sassoon Papers, CUL, Add MS 9375/380.

162. Letter of Ruth Head to Siegfried Sassoon, 25 May 1929, *ibid*.
163. Letter of Ruth Head to Siegfried Sassoon, 4 May 1929, Sassoon Papers, CUL, Add MS 9375/373. The passage in question was from John Bunyan's, *Pilgrim's Progress*.
164. See: P. Hoare, *Serious Pleasures: The Life of Stephen Tennant* (London: Hamish Hamilton, 1990), p. 131.
165. Letter of Ruth Head to Siegfried Sassoon, 28 June 1929, Sassoon Papers, CUL, Add MS 9375/376.
166. Letter of Stephen Tennant to Ruth Head, 6 February 1929, CMAC, PP/HEA/D5.
167. Letter of Stephen Tennant to Ruth Head, 12 [January 1930], *ibid*.
168. Letter of Stephen Tennant to Henry Head, January 1930, *ibid*.
169. Letter of Ruth Head to Siegfried Sassoon, 26 March 1929, Sassoon Papers, CUL, Add MS 9375/369.
170. Letter of Ruth Head to Siegfried Sassoon, 2 September 1936, Sassoon Papers, CUL, Add MS 9375/382..
171. Letter of Ruth Head to Siegfried Sassoon, 7 April 1929, Sassoon Papers, CUL, Add MS 9375/370..
172. *Ibid*.
173. Sassoon had served in Pinney's division in France although he only became personally acquainted with him after the war. When the two met in September 1925, Sassoon noted: 'we got on very well. Fox-hunting, cricket, and infantry warfare were our points of contact and I exploited all three fully'. Head, who was also present, 'looked quite astonished, and must have thought me a bit of a chameleon'. Hart-Davies (ed.), *Siegfried Sassoon 1923–1925*, p. 222.
174. Letter of Ruth Head to Charles Ashbee, 8 January 1932, Ashbee Papers, KCC.
175. Letter of Ruth Head to Charles Ashbee, 30 May 1936, *ibid*.
176. Letter of Ruth Head to Janet Ashbee, 19 [September 1936], *ibid*.
177. Letter of Ruth Head to Janet Ashbee, 10 November 1925, *ibid*.
178. Letter of Ruth Head to Siegfried Sassoon, 24 May 1927, Sassoon Papers, CUL, Add MS 9375/368.
179. Letter of Robert Nichols to Henry Head, 15 August 1925, Nichols Papers, BL, uncatalogued.
180. Letter of Ruth Head to Robert Nichols, 5 September 1925, *ibid*.
181. Letter of Ruth Head to Janet Ashbee, 10 December [1931], Ashbee Papers, KCC.
182. Letter of Ruth Head to Janet Ashbee, [1936], *ibid*.
183. Letter of Ruth Head to Robert Nichols, 6 November 1929, Nichols Papers, BL, uncatalogued.
184. Letter of Ruth Head to Robert Nichols, 2 June 1931, *ibid*.
185. Letter of Janet Ashbee to Charles Ashbee, 30 January 1929, Ashbee Papers, KCC.
186. Letter of Ruth Head to Janet Ashbee, 10 November 1925, *ibid*.
187. Letter of Henry Head to Charles Sherrington, 7 January 1927, Sherrington Papers, Department of Physiology, Oxford.
188. Letter of Henry Head to Charles Sherrington, October 1932, *ibid*.
189. Letter of Ruth Head to Julian Huxley, 14 November *c*. 1928, Nichols Papers, BL, uncatalogued.
190. Letter of Ruth Head to Siegfried Sassoon, 26 March 1929, Sassoon Papers, CUL, Add MS 9375/369.

191. Letter of Ruth Head to Siegfried Sassoon, 22 April 1929, Sassoon Papers, CUL, Add MS 9375/372. Ruth was making a joking allusion to the title of Sassoon's *Memoirs of a Fox-Hunting Man*.
192. Letter of Ruth Head to Siegfried Sassoon, 30 June 1929, Sassoon Papers, CUL, Add MS 9375/377.
193. 'A visit to Henry and Ruth Head. July 16, 1930', Harvey Cushing Papers, YUL, Series I Correspondence, folder 670. An abridged account of this visit is found in: Fulton, *Harvey Cushing*, p. 596.
194. *Ibid.*
195. *Ibid.*
196. Letter of Hester Marsden-Smedley to Henry and Ruth Head, 11 July 1938, CMAC, PP/HEA/D6/3.
197. Letter of Ruth Head to Robert Nichols, 29 October 1926, Nichols Papers, BL, uncatalogued.
198. Letter of Ruth Head to Robert Nichols, 28 November 1926, *ibid*.
199. Letter of Ruth Head to Hester Marsden-Smedley, 9 March [1933], CMAC, PP/HEA/D10.
200. Letter of Ruth Head to Hester Marsden-Smedley, 20 April 1926, *ibid*.
201. Letter of Ruth Head to Robert Nichols, 26 January 1927, Nichols Papers, BL, uncatalogued.
202. Letter of Ruth Head to Janet Ashbee, 20 November 1933, Ashbee Papers, KCC.
203. Letter of Ruth Head to Janet Ashbee, 30 October 1927, *ibid*.
204. Letter of Ruth Head to Janet Ashbee, [c. 1928], *ibid*.
205. Letter of Ruth Head to Robert Nichols, 28 November 1926, Nichols Papers, BL, uncatalogued.
206. Letter of Ruth Head to Hester Pinney junior, 3 July no year given, CMAC, PP/HEA/D10.
207. Letter of Ruth Head to Robert Nichols, 10 January 1934, Nichols Papers, BL, uncatalogued.
208. Letter of Ruth Head to Janet Ashbee, 30 October 1927, Ashbee Papers, KCC.
209. Letter of Ruth Head to Janet Ashbee, [1936], *ibid*.
210. Letter of Ruth Head to Hester Marsden-Smedley, 23 September no year given, CMAC, PP/HEA/D10.
211. Letter of Ruth Head to Robert Nichols, 9 December[1926], Nichols Papers, BL, uncatalogued.
212. See: Hoare, *Serious Pleasures*, p. 161.
213. Letter of Ruth Head to Siegfried Sassoon, 1 August 1930, Sassoon Papers, CUL, Add MS 9375/378.
214. *Ibid.*
215. Letter of Hester Marsden-Smedley to Henry and Ruth Head, 21 January 1938, CMAC, PP/HEA/D6/3.
216. Letter of Ruth Head to Hester Marsden-Smedley, undated, CMAC, PP/HEA/D10.
217. Letter of Stephen Tennant to Ruth Head, 22 January 1936, CMAC, PP/HEA/ D5.
218. Letter of Stephen Tennant to Ruth Head, 3 March 1935, *ibid*.
219. Hoare, *Serious Pleasures*, pp. 170–1.
220. Letter of Stephen Tennant to Ruth Head, 17 May 1935, *ibid*.
221. Letter of Ruth Head to Hester Marsden-Smedley, 14 December 1925, CMAC, PP/HEA/D10.

222. Letter of Ruth Head to Janet Ashbee, 13 January 1918, Ashbee Papers, KCC.
223. Letter of Ruth Head to Janet Ashbee, 18 August 1925, *ibid*.
224. Letter of Ruth Head to Robert Nichols, 28. November 1926, Nichols Papers, BL, uncatalogued.
225. Letter of Ruth Head to Janet Ashbee, [September 1933], Ashbee Papers, KCC.
226. Letter of Ruth Head to Janet Ashbee, 20 November 1933, *ibid*.
227. Letter of Ruth Head to Hester Marsden-Smedley, undated, CMAC, PP/HEA/D10.
228. Letter of Ruth Head to Hester Marsden-Smedley, 18 September no year given, *ibid*.
229. Letter of Ruth Head to Siegfried Sassoon, 24 May 1927, Sassoon Papers, CUL, Add MS 9375/368.
230. Letter of Ruth Head to Janet Ashbee, 20 November 1933, Ashbee Papers, KCC.
231. Letter of Ruth Head to Hester Marsden-Smedley, undated, CMAC, PP/HEA/D10.
232. Letter of Ruth Head to Janet Ashbee, 23 December 1925, Ashbee Papers, KCC.
233. Letter of Ruth Head to Janet Ashbee, 18 August 1925, *ibid*.
234. *Ibid*.
235. Letter of Ruth Head to Hester Pinney, 20 November 1925, CMAC, PP/HEA/D10.
236. Letter of Ruth Head to Robert Nichols, 7 July 1926, Nichols Papers, BL, uncatalogued.
237. Letter of Ruth Head to Janet Ashbee, 30 October 1927, Ashbee Papers, KCC.
238. Letter of Ruth Head to Siegfried Sassoon, 29 April [1927], Sassoon Papers, CUL, Add MS 9375/368.
239. Letter of Ruth Head to Norah Nichols, 26 December 1926, Nichols Papers, BL, uncatalogued.
240. Letter of Ruth Head to Hester Pinney, 28 December 1926, CMAC, PP/HEA/D10.
241. Letter of Ruth Head to Janet Ashbee, *c*. 1928, Ashbee Papers, KCC.
242. Letter of Ruth Head to Siegfried Sassoon, 22 April 1929, Sassoon Papers, CUL, Add MS 9375/372.
243. Letter of Ruth Head to Janet Ashbee, [1933], Ashbee Papers, KCC.
244. Letter of Ruth Head to Janet Ashbee, 16 April 1934, *ibid*.
245. Letter of Ruth Head to Janet Ashbee, 19 September 1936, *ibid*.
246. Letter of Ruth Head to Janet Ashbee, 11 December 1935, *ibid*.
247. Letter of Ruth Head to Janet Ashbee, [September 1933], *ibid*.
248. Letter of Ruth Mayhew to Henry Head, 23 April 1901, CMAC, PP/HEA, D4/8.
249. *The Times*, 15 May 1939, p. 16.
250. Holmes, 'Henry Head', p. 686.
251. Letter of Hester Marsden-Smedley to Basil Marsden-Smedley, 28 September 1939, CMAC, PP/HEA/D6/3.
252. Letter of Hester Marsden-Smedley to Henry Head, 15 April 1940, *ibid*.
253. Letter of Janet Ashbee to Charles Ashbee, 4 April 1940, Ashbee Papers, KCC.
254. Russell Brain, 'Henry Head: the man and his ideas', *Brain*, 1961, 84: 561–6, p. 566.
255. G. Riddoch, 'Sir Henry Head: a further tribute', *The Times*, 18 October 1940, p. 7.
256. Letter of Hester Marsden-Smedley to Henry and Ruth Head, 22 May 1938, CMAC, PP/HEA/D6/3.
257. Letter of Janet Ashbee to Charles Ashbee, [July 1941], Ashbee Papers, KCC.
258. *Ibid*.
259. Letter of Ruth Head to Robert Nichols, 11 February 1925, Nichols Papers, BL, uncatalogued.
260. Letter of Robert Nichols to Hester Pinney, 10 October 1940, *ibid*.
261. Robert Nichols, 'Sir Henry Head: a tribute', *The Times*, 15 October 1940, p. 7.

262. Letter of Redcliffe Nathan Salaman to Hester Pinney, [1940], CMAC, PP/HEA/E3/3. Several words are missing from the surviving text of this letter.
263. Riddoch, 'Sir Henry Head', p. 7.
264. *Ibid.*
265. Letter of Arnold Mayhew to Robert Nichols, 30 March 1941, Nichols Papers, BL, uncatalogued.
266. There is a perhaps unconscious allusion here to a line in Head's poem 'To Courage, Seated':

Like senators of old, with folded hands,
In silence, seated, for the stroke of Fate.
One boon alone an ardent soul demands,
To die before its passion waxes cold,
Enthusiasm fails, or Love grows old.

Henry Head, *Destroyers*, p. 14.
267. Letter of Arnold Mayhew to Robert Nichols, 30 March 1941, Nichols Papers, BL, uncatalogued.
268. Letter of Redcliffe Nathan Salaman to Hester Pinney, [1940], CMAC, PP/HEA/E3/3.

WORKS CITED

Archival Sources

Ashbee Papers, King's College, Cambridge.

John Jay Chapman Papers, Houghton Library, Harvard University, Cambridge, MA.

Harvey Cushing Papers, Yale University Library, New Haven, CT.

Flexner Papers, American Philosophical Society, Philadelphia, PA.

Francis Galton Papers, University College London.

Robert Graves Papers, Morris Library, Southern Illinois University, IL.

Thomas Hardy Papers, Dorchester County Museum, Dorchester.

Henry Head Papers, Contemporary Medical Archives Centre, Wellcome Trust Library for the History of Medicine, London.

Henry Head Papers, Royal College of Physicians of London.

Henry James Papers, Houghton Library, Harvard University, Cambridge, MA.

London Hospital Archives, London.

Adolf Meyer Papers, Alan Mason Chesney Medical Archives, Johns Hopkins University, Baltimore, MA.

Constantin von Monakow Papers, University of Zurich.

Robert Nichols Papers, British Library, London.

C.K. Ogden Papers, McMaster University, Hamilton, ON.

Karl Pearson Papers, University College London.

Royal Society Archives, London.

Bertrand Russell Papers, McMaster University, Hamilton ON.

Redcliffe Nathan Salaman Papers, Cambridge University Library, Cambridge.

Siegfried Lorraine Sassoon Collection, Harry Ransom Center, University of Texas, Austin, TX.

Siegfried Sassoon Papers, Cambridge University Library, Cambridge.

Charles Sherrington Papers, University Library of Physiology, Oxford.

Grafton Elliot Smith Papers, John Rylands Library, Manchester.

Primary Sources

Anon., 'British Institute of Philosophical Studies', *British Medical Journal*, 1 (1927) 699.

Anon., 'Commission on vivisection. Evidence of Dr. Henry Head, F.R.S.,' *British Medical Journal*, 1909, 1: 227–29.

Anon., 'Disorders of symbolic thinking', *Nature*, 106 (1920), pp. 197–8.

Anon., 'Famous doctors get Fordham degrees: Dr. Henry Head of London discovers a unique case in the Montefiore Home', *New York Times*, 12 September 1912, p. 2.

Anon., 'New conceptions of psychology', *Nature*, 105 (1920), pp. 363–364.

Anon., 'Philosophers at Oxford. Henri Bergson. Semantic aphasia', *The Times*, 6 October 1920, p. 8.

Anon., 'Present age decadent but happy, says Dr Henry Head,' *New York Times*, 15 September 1912, p. 5.

Anon., 'The progress of science. Sensation controlled by the brain. Dr. Henry Head's Researches,' *The Times*, 25 October 1921, p. 8.

Barbellion, W. N. P. [i.e., Bruce Frederick Cummings], *The Journal of a Disappointed Man* (London: Chatto & Windus, 1919).

—, with a Preface by A. J. Cummings, *A Last Diary* (London: Chatto & Windus, 1920).

Bartlett, F. C., 'Critical review: *Aphasia and Kindred Disorders of Speech*. By Henry Head,' *Brain*, 49 (1926), pp. 581–7.

Bashford, H. H., *The Corner of Harley Street: Being some Familiar Correspondence of Peter Harding. M.D.* (London: Constable & Co., 1911).

Bazett, H. C. and W. G. Penfield 'A study of the Sherrington decerebrate animal in the chronic as well as the acute condition,' *Brain*, 45 (1922), pp. 185–265.

Campion, G. G., *Elements in Thought & Emotion: An Essay on Education, Epistemology, & the Psycho-Neural Problem* (London: University of London Press, 1923).

Campion, G. G. and G. Elliot Smith, *The Neural Basis of Thought* (London: Kegan Paul, 1934).

Flexner, A., *I Remember: The Autobiography of Abraham Flexner*, (New York, NY: Simon and Schuster, 1940).

Fulton, J. F., *Harvey Cushing: A Biography* (Springfield, IL: Charles C. Thomas, 1946).

Foster, M., *A Text Book of Physiology*, 3rd edn, (London: Macmillan, 1879).

Gaskell, W. H., 'On the relation between the structure, function, distribution and origin of the cranial nerves; together with a theory of the origin of the nervous system of vertebrata,' *Journal of Physiology*, 10 (1889), pp. 153–211.

Graves, R., *Goodbye to All That* (London: Jonathan Cape, 1929).

Hart-Davies, R., (ed.), *Siegfried Sassoon Diaries 1920–1922* (London: Faber and Faber, 1981).

— (ed.), *Siegfried Sassoon Diaries 1923-1925* (London: Faber and Faber, 1985).

Head, H., 'Aphasia: an historical review', *Brain*, 43 (1921), pp. 390–411.

—, *Aphasia and Kindred Disorders*, 2 vols, (Cambridge: Cambridge University Press, 1926).

—, 'Certain mental changes that accompany visceral disease', *Brain*, 24 (1901), pp. 345–403.

—, 'The conception of nervous and mental energy', *British Journal of Psychology*, 1923-4, 14: 126–47.

—, 'The consequences of injury to the peripheral nerves in man', *Brain*, 24 (1905), pp. 116–299.

—, *Destroyers and Other Verses*, (London: Humphrey Milford, Oxford University Press, 1919).

—, 'The diagnosis of hysteria', *British Medical Journal*, 1 (1922), pp. 827–9.

—, 'Discussion of aphasia', *Brain*, 43 (1921), pp. 412–50.

—, 'Disease and diagnosis', *British Medical Journal*, 1 (1919), pp. 365–7.

—, 'Disorders of symbolic thinking and expression', *British Journal of Psychology*, 11, (1920-1), pp. 179–93.

—, 'On disturbances of sensation with especial reference to the pain of visceral disease', *Brain*, 16 (1893), pp. 1–133.

—, 'On disturbances of sensation with especial reference to the pain of visceral disease: Part II—Head and neck', *Brain*, 17 (1894), pp. 339–480.

—, 'The elements of the psycho-neuroses', *British Medical Journal*, 1 (1920), pp. 389–92.

—, 'The evolution of neurology and its bearing on medical education', *British Medical Journal*, 2 (1924), p. 718.

—, 'The grouping of afferent impulses within the spinal cord', *Brain*, 29 (1907), pp. 537–741.

—, 'The physiological basis of the spatial and temporal aspects of sensation', in 'Symposium: Time, space and material: are they, and if so in what sense, the ultimate data of science?', *Aristotelian Society. Supplementary Volume II* (London: Williams and Norgate, 1919), pp. 44–108.

—, 'On the regulation of respiration', *Journal of Physiology*, 10 (1889), pp. 1–70; 279–90.

—, 'Sensation and the cerebral cortex', *Brain*, 41 (1918), pp. 57–253.

—, 'Some principles of neurology', *Lancet* (1918), pp. 657–660.

—, *Studies in Neurology*, 2 vols, (London: Henry Frowde, 1920).

—, 'Über die negativen und positiven Schwankungen des Nervenstromes', *Pflügers Archiv*, 40 (1887), pp. 1–70.

—, 'William Halse Rivers Rivers, 1864-1922', *Proceedings of the Royal Society of London Series B* (1924), pp. xliii–xlvii.

Head, H. and A.W. Campbell 'The pathology of Herpes Zoster and its bearing on sensory localisation', *Brain*, 32 (1900), pp. 353–523.

Head, H. and G. Holmes 'Sensory disturbances from cerebral lesions,' *Brain*, 34 (1911), pp. 102–254.

Head, H. and J. Sherren, 'The consequences of injury to the peripheral nerves in man,' *Brain*, 28 (1905), pp. 116–299.

Head, R. (ed.), *Pictures and Other Passages from Henry James* (New York, NY: Frederick A. Stokes, 1916).

—, *A History of Departed Things* (London: Kegan Paul, 1918).

—, (ed.), *Pages from the Works of Thomas Hardy* (London: Chatto and Windus, 1922).

Hering, K. E. K. *Die Selbststeuerung der Athmung durch den Nervus vagus*. Sitzungsberichte der kaiserlichen Akademie der Wissenschaften. Mathematisch–naturwissenschaftliche Classe, Wien, 1868, 57: 2, pp. 672–677.

Jackson, J. H., 'Reprint of some of Dr. Hughlings Jackson's papers on affections of speech,' *Brain*, 1915, 28–42.

Lee, V., *Music and its Lovers: an Empirical Study of Emotion and Imaginative Responses to Music* (London: George Allen & Unwin, 1932).

Nichols, R. 'Sir Henry Head: a tribute', *The Times*, 15 October 1940, p. 7.

Ogden, C. K. and I. A. Richards, *The Meaning of Meaning: A Study of the Influence of Language upon Thought and of the Science of Symbolism* (London: Kegan Paul, 1923).

Pater, W., *The Renaissance: Studies in Art and Poetry* (1873; Oxford: Oxford University Press, 1998),

Riddoch, G., 'Sir Henry Head: a further tribute,' *The Times*, 18 October 1940, p. 7.

Rivers, W. H. R. *Instinct and the Unconscious: A Contribution to a Biological Theory of the Psycho-Neuroses* (Cambridge: Cambridge University Press, 1922).

Stephen, K., *The Misuse of Mind: A Study of Bergson's Attack on Intellectualism* (London: Kegan Paul, 1922).

Whitehead, A. N., *Adventures of Ideas* (Cambridge: Cambridge University Press, 1933).

Secondary Sources

Ash, M. G., *Gestalt Psychology in German Culture, 1890–1967: Holism and the Quest for Objectivity* (Cambridge: Cambridge University Press, 1995).

Banfield, A., *The Phantom Table: Woolf, Fry, Russell and the Epistemology of Modernism* (Cambridge: Cambridge University Press, 2000).

Baumann, C., *Die Physiologe Ewald Hering (1834–1918)* (Frankfurt: Dr. Hänsel-Hohenhausen, 2002).

—, 'Henry Head in Ewald Hering's laboratory in Prague,' *Journal of the History of the Neurosciences*, 14 (2005), 322–333.

Bell, A. O., *The Diary of Virginia Woolf. Volume III: 1925-30* (Harmondsworth: Penguin, 1982).

Bonner, T. N. *Becoming a Physician: Medical Education in Great Britain, France, Germany, and the United States 1750–1945* (New York, NY: Oxford University Press, 1995).

Brain, R., *Henry Head: the Man and his Ideas* (London: Macmillan, 1961).

Cantor, G., *Quakers, Jews, and Science: Religious Responses to Modernity and the Sciences in Britain, 1650–1900* (Oxford: Oxford University Press, 2005).

Casper, S. T., 'The idioms of practice: British neurology, 1880–1960', PhD Dissertation, University College London, 2006.

Charlton, A. and W. Charlton, *Putting Poetry First: A Life of Robert Nichols* (Norwich: Michael Russell, 2003).

Clark-Kennedy, A. E., *The London: A Study in the Voluntary Hospital System* (London: Pitman Medical Publishing Co. Ltd., 1963).

Cohen, G. B., 'Jews in German liberal politics: Prague, 1880–1914', *Jewish History*, 1 (1986), pp. 55-74.

Cohen, M. N. (ed.), *The Selected Letters of Lewis Carroll* (London: Macmillan, 1982).

—, *Lewis Carroll: A Biography* (London: Macmillan, 1995).

Colebrook L., 'Almroth Edward Wright 1861–1947', *Obituary Notices of the Royal Society*, 6 (1948), pp. 297–314.

Cooter, R., M. Harrison and S. Sturdy, *Medicine and Modern Warfare* (Amsterdam: Rodopi, 1999).

Cranefield, P. F., 'The philosophical and cultural interests of the Biophysics Movement of 1847', *Journal of the History of Medicine*, 21 (1966), pp. 1–7.

Crary, J., *Techniques of the Observer: On Vision and Modernity in the Nineteenth Century* (Cambridge, MA: MIT Press, 1992).

Crary, J., *Suspensions of Perception: Attention, Spectacle, and Modern Culture* (Cambridge, MA: MIT Press, 2000).

Crawford, A., *C. R. Ashbee: Architect, Designer and Romantic Socialist* (New Haven, CT: Yale University Press, 2005).

Critchley, M., *The Black Hole and Other Essays* (London: Pitman, 1964).

Cunningham, A. and N. Jardine (eds), *Romanticism and the Sciences* (Cambridge: Cambridge University Press, 1990).

de Chadarevian, S., 'Graphical method and discipline: Self-recording instruments in nineteenth-century physiology', *Studies in History and Philosophy of Science*, 24 (1993), pp. 267–91.

Dierig, S., J. Lachmund and J. A. Mendelsohn (eds) *Science and the City*, *Osiris*, 18 (2003).

Foucault, M., *The History of Sexuality*, 3 vols, (London: Penguin Books, 1981–4).

Fyfe, A., 'The reception of William Paley's *Natural Theology* in the University of Cambridge', *British Journal of the History of Science*, 30 (1997), pp. 321–35.

Geison, G. L., *Michael Foster and the Cambridge School of Physiology: The Scientific Enterprise in Late Victorian Society* (Princeton, NJ: Princeton University Press, 1978).

Ginzburg, L., *On Psychological Prose*, translated by J. Rosengrant, (Princeton, NJ: Princeton University Press, 1991).

Greenblatt, S., *Renaissance Self-Fashioning: from More to Shakespeare* (Chicago, IL: Chicago University Press, 1984).

Greenslade, W., *Degeneration, Culture, and the Novel, 1880–1940* (Cambridge: Cambridge University Press, 1994).

Harrington, A., (ed.) *So Human a Brain: Knowledge and Values in the Neurosciences* (Boston: Birkhäuser, 1992).

—, *Reenchanted Science: Holism in German Culture from Wilhelm II to Hitler* (Princeton: Princeton University Press, 1996),

Head, H. and K. W. Cross, *Henry Head Centenary: Essays and Bibliography* (London: Macmillan & Co., 1961).

Hillebrand, F., *Ewald Hering: ein Gedenkwort der Psychophysik* (Berlin: Julius Springer, 1918).

Hoare, P., *Serious Pleasures: The Life of Stephen Tennant* (London: Hamish Hamilton, 1990).

Holland, S., *In Black and White* (London: Edward Arnold, 1928).

Holmes, G., 'Henry Head, 1861–1940,' *Obituary Notices of the Fellows of the Royal Society*, 3 (1939–41), 665–689.

Hughes, H. S., *Consciousness and Society: the Re-Orientation of European Social Thought, 1890–1930* (St. Albans: Paladin, 1974).

Jacyna, L. S., *Lost Words: Narratives of Language and the Brain, 1825-1926* (Princeton, NJ: Princeton University Press, 2000).

—, 'A host of experienced microscopists: the establishment of histology in nineteenth-century Edinburgh,' *Bulletin of the History of Medicine*, 75 (2001), pp. 225–253.

—, 'Starting anew: Henry Head's contribution to aphasia studies,' *Journal of Neurolinguistics*, 18 (2005), pp. 327–36.

Jewson, N. D., 'Medical knowledge and the patronage system in 18th century England,' *Sociology*, 8 (1974) 369–385.

—, 'The disappearance of the sick-man from medical cosmology, 1770–1870,' *Sociology*, 10 (1976), pp. 225-244.

Kavka, F., *The Caroline University of Prague* (Prague: Universita Karlova, 1962).

Kern, S., *The Culture of Time and Space 1880-1918* (London: Weidenfeld & Nicolson, 1983).

Klonk, C., *Science and the Perception of Nature: British Landscape Art in the Late Eighteenth and Early Nineteenth Century* (New Haven, CT: Yale University Press, 1996).

Kuklick, H., *The Savage Within: the Social History of British Anthropology, 1885–1945* (Cambridge University Press, 1991).

Lanbourne, L., *The Aesthetic Movement* (London: Phaidon Press, 1996).

Lawrence, C., 'Incommunicable knowledge: science, technology and the clinical art in Britain, 1850–1914', *Journal of Contemporary History*, 20 (1985), pp. 503–20.

—, (ed.), *Medical Theory, Surgical Practice: Studies in the History of Surgery* (London: Routledge, 1992).

—, *Rockefeller Money, the Laboratory and Medicine in Edinburgh 1919-1930: New Science in an Old Country* (Rochester, NY: University of Rochester Press, 2005).

Lawrence, C. and G. Weisz (eds), *Greater than the Parts: Holism in Biomedicine, 1920-1950* (New York, NY: Oxford University Press, 1998).

Lawrence, C. and S. Shapin, *Science Incarnate: Historical Embodiments of Natural Knowledge* (Chicago University Press, 1998).

Lears, T. J. J., *No Place of Grace: Antimodernism and the Transformation of American Culture, 1880–1920* (Chicago, IL: University of Chicago Press, 1994).

Lee, H., *Virginia Woolf* (London: Chatto & Windus, 1996).

Liddell, E. G. T. 'Charles Scott Sherrington 1857–1952', *Obituary Notices of Fellows of the Royal Society*, 8 (1952), pp. 241–70.

Lyons, J. B. 'Correspondence between Sir William Gowers and Sir Victor Horsley', *Medical History*, 9 (1965), pp. 260–7.

Mair, A., *Sir James Mackenzie MD 1853–1925: General Practitioner* (London: Royal College of General Practitioners, 1973).

Mangan, J. A., *Athleticism in the Victorian and Edwardian Public School: The Emergence and Consolidation of an Educational Ideology* (Cambridge: Cambridge University Press, 1981).

Matz, J., *Literary Impressionism and Modernist Aesthetics* (Cambridge: Cambridge University Press, 2001).

Micale, M. S. and P. Lerner (eds), *Traumatic Pasts: History, Psychiatry, and Trauma in the Modern Age, 1870–1930* (Cambridge: Cambridge University Press, 2001).

Morris, E.W., *A History of the London Hospital* (London: Edward Arnold & Co., 1926).

Moscucci, O., *The Science of Woman: Gynaecology and Gender in England 1800–1929* (Cambridge: Cambridge University Press, 1990).

Mussellman, E. G., *Nervous Conditions: Science and the Body Politic in Early Industrial Britain* (Albany, NY: State University of New York, 2006).

Oppenheim, J., *Shattered Nerves: Doctors, Patients, and Depression in Victorian England* (Oxford: Oxford University Press, 1991).

Peterson, M. J., *The Medical Profession in Mid-Victorian London* (Berkeley, CA: University of California Press, 1978).

Pick, D., *Faces of Degeneration: a European Disorder, c.1848–c.1918* (Cambridge: Cambridge University Press, 1989).

Pippin, R. B. *Henry James and Modern Moral Life* (Cambridge: Cambridge University Press, 2000).

Porter, R. (ed.), *Rewriting the Self: Histories from the Renaissance to the Present* (London: Routledge, 1997).

Porter, T. M. *Karl Pearson: the Scientific Life in a Statistical Age* (Princeton, NJ: Princeton University Press, 2004).

Ross, D. (ed.) *Modernist Impulses in the Human Sciences 1870-1930* (Baltimore, MA: Johns Hopkins Press, 1994).

Rothfield, L., *Vital Signs: Medical Realism in Nineteenth-Century Fiction* (Princeton, NJ: Princeton University Press, 1992).

Ryan, J., *The Vanishing Subject: Early Psychology and Literary Modernism* (Chicago, IL: University of Chicago Press, 1991).

Sachs, M., *Historisches Chirurgenlexikon: ein biographisch-bibliographisches Handbuch bedeutender Chirurgen und Wundärtzte*, 5 vols, (Heidelberg: Kaden Verlag, 2002).

Schilpp, P. A., *The Philosophy of Alfred North Whitehead* (Evanston, IL: Northwestern University, 1941).

Schorske, C., *Fin de siècle Vienna: Politics and Culture* (Cambridge: Cambridge University Press, 1979).

Sengoopta, C., *The Most Secret Quintessence of Life: Sex, Glands, and Hormones, 1850-1950* (Springfield, IL: University of Chicago Press, 2006).

Searle, G. R., *Eugenics and Politics in Britain, 1900–1914* (Leyden: Noordhoff, 1976).

Seyfarth, E. A., 'Julius Bernstein (1839-1917): pioneer neurobiologist and biophysicist,' *Biological Cybernetics*, 94 (2006), pp. 2–8.

Shapin, S., *A Social History of Truth: Civility and Science in Seventeenth-Century England* (Chicago, IL: Chicago University Press, 1994).

Shapin, S. and S. Schaffer, *Leviathan and the Air-Pump: Hobbes, Boyle, and the Experimental Life* (Princeton, NJ: Princeton University Press, 1985).

Shortland, M. and R. Yeo (eds), *Telling Lives in Science: Essays in Scientific Biography* (Cambridge: Cambridge University Press, 1996).

Showalter, E., *The Female Malady: Women, Madness and English Culture* (London: Virago Press, 1987).

Shephard, B., *A War of Nerves* (London: Jonathan Cape, 2000).

Siegel, J., *The Idea of the Self: Thought and Experience in Western Europe since the Seventeenth Century* (Cambridge: Cambridge University Press, 2005).

Smith, R., *Inhibition: History and Meaning in the Sciences of Mind and Brain* (London: Free Association Books, 1992).

Söderquist, T. (ed.), *The History and Poetics of Scientific Biography* (Aldershot: Ashgate, 2007).

Stedman Jones, G., *Outcast London: a Study in the Relationship between the Classes in Victorian Society* (Harmondsworth: Penguin, 1984).

Stern, F., *The Politics of Cultural Despair* (Berkeley, CA: University of California Press, 1974).

Stone, D., *Breeding Superman: Nietzsche, Race and Eugenics in Edwardian and Interwar Britain* (Liverpool: Liverpool University Press, 2002).

Taylor, C., *Sources of the Self: The Making of the Modern Identity* (Cambridge: Cambridge University Press, 1989).

Thomson, M., *Mental Hygiene as an International Movement* (Cambridge: Cambridge University Press, 1995).

Todes, D. P. *Pavlov's Physiology Factory: Experiment, Interpretation, Laboratory Enterprise* (Baltimore, MA: Johns Hopkins University Press, 2002).

Trembley, S., *'All that Summer she was Mad': Virginia Woolf and her Doctors* (London: Junction Books, 1981).

Tuchman, A. M., *Science, Medicine, and the State in Germany: the Case of Baden, 1815-1871* (New York, NY: Oxford University Press, 1993).

Turner, R. S., 'Vision studies in Germany: Helmholtz versus Hering,' *Osiris*, 8 (1993), pp. 80–103.

—, *In the Eye's Mind: Vision and the Helmholtz-Hering Controversy* (Princeton University Press 1994).

Walkowitz, J. R. *City of Dreadful Delight: Narratives of Sexual Danger in Late-Victorian London* (London: Virago Press, 1992).

Wangensteen, O. H. and S. D. Wangensteen, *The Rise of Surgery: from Empiric Craft to Scientific Discipline* (Folkestone: Dawson, 1978).

Warwick, A., *Masters of Theory: Cambridge and the Rise of Mathematical Physics* (Chicago, IL: Chicago University Press, 2003).

White, B. V., *Stanley Cobb: A Builder of Modern Neurosciences* (Boston: Francis A. Countway Library, 1984).

Wilson, J. M., *Siegfried Sassoon: The Journey from the Trenches* (London: Duckworth, 2003).

Wohl, R., *A Passion for Wings: Aviation and the Western Imagination, 1908-1918* (New Haven, CT: Yale University Press, 1996).

Wolfe, E. L. A., C. Barger and Saul Benison, *Walter B. Cannon, Science and Society* (Cambridge, MA: Harvard University Press, 2000).

Woolf, V., *Roger Fry: A Biography* (London: Hogarth Press, 1940).

Young, A., *The Harmony of Illusions: Inventing Post-Traumatic Stress Disorder* (Princeton, NJ: Princeton University Press, 1995).

Young, R. M. 'Malthus and the evolutionists: the common context of biological and social theory', *Past and Present*, 43 (1969), pp.109-45.

INDEX

Index note: subsections dealing with life events are indexed in chronological order.

acidosis, 280
Acland, Henry Wentworth, 77
Aesthetic Movement, 206
'Afferent Nervous System, The', 56
antiseptic surgery, 12
aphasia, 1–2
    HH develops interest in, 137–8
    HH's work on: criticism of classic aphasiology, 139–41; impact on contemporaneous thinking, 150–1; language as biological trait, 142–3; monograph, its ramifications, 143–7; officer patient subjects, 138–9; the psychological element, 145; tests to assess nature of, 141–2; verbal, nominal, syntactical and semantic, 142
    John Hughlings Jackson's papers on, 137–8
Aphasia and Kindred Disorders, 1–2, 43, 143–5
'Aphasia: an Historical Review', 138
Armstrong, Henry Edward, 104
Arnim, Elizabeth von, 215
Arters, Frau von, 228
Ashbee, Charles, 58, 87, 207, 273–4, 279
Ashbee, Janet, 62, 87, 167, 271
    cares for HH during illness, 275
    on HH and RH, 266–7
    RH's letters to, 70, 143, 189, 254
    visits Hartley Court after HH's death, 286–7
Ashbee, Mary Elizabeth, 64
auditory delusions, 119

Aue, Carl Eduard, 23
Austria, 231
'Autobiography', 12, 18, 35, 38, 53
autopsies, 72, 114

Babinsky, Joseph Françoise Félix, 246
Baden Powell, Robert, 20
Balfour, Francis Maitland, 40–1
Barker, Pat, *Regeneration,* 1
Barrett–Browning letters, 214
Barrie, J. M., *Quality Street,* 167
Bashford, H. H.
    Corner of Harley Street: Being some Familiar Correspondence of Peter Harding, 60
    on HH, 71
Bastian, H. Charlton, 140
Bateson, William, 21, 104
Bayreuth, 51
Beck, Lucas, 191–2
Beck, Marcus, 12, 22
Beck, Theodore, 37–8, 176
Bennett, Arnold, 94, 259
Berenson, Bernard, 182, 187, 203
Bergson, Henri, 145, 147, 252
Berlin, 157, 159, 160
Bernstein, Julius, 28–30
Bloomsbury Group, 62
Bohemia, 45–7
Bradford, John Rose, 248
Brain, Russell, 71, 286
Brain, 138, 243–4
Bramwell, Byron, 184
Brighton, 159, 162, 173, 241

British Institute of Philosophical Studies, 145
British Psychological Society, 207
Budapest, 246
Bulloch, William, 72, 80
  HH's letter to, 95–6
Buzzard, Thomas, 55

Cambridge
  HH visits, 79, 125–6, 161, 247
  wetlands around, 35, 48
  see also Cambridge University
Cambridge University, 20–1, 100
Cambridge University Press, 143–4
  collegial system, 35–6; physiology teaching, 38–41; Platonic symposia, 36
  Girton and Newham, 34
  Trinity College, 34–5; HH's social life at, 35–6
  tuition fees, 38
  see also Cambridge
Campbell, Alfred Walter, 114
Campbell, Mrs Patrick, 211
Campion, George C., 146
Cannon, Walter Bradford, 144
Carroll, Lewis, 154–5
Cartesian perspectives, 135–6
Chandler, Frederick George, 280
Chapman, John Jay, 268
Charles-Ferdinand University, Prague see German University, Prague
Charterhouse, 16–21, 275
Church of St Ursula, Cologne, 23
Clarence Terrace, costs, 68–9
claustrophobia, 149
Claybury laboratory, 76, 105, 224
Cobb, Stanley, 145
'Conception of Nervous and Mental Energy', 148
Congress of Philosophy, Oxford (1920), 145
Conrad, Joseph, *Lord Jim*, 216
consciousness, 4, 7, 120, 135–7, 150
  man's place in universe, 109–10, 145–6
Copeman, Sydney A. Monckton, 104
Coward, Noel, 259
Craig, Maurice, 274
Critchely, Macdonald, 59
Crookes, William, 104

Cummings, Arthur J., 63–4
Cummings, Bruce Frederick, *Journal of a Disappointed Man*, 63
Curel, François de, *La Nouvelle Idole*, 210–11
Cushing, Harvey, 66, 70, 96, 246, 266, 267, 268
  on Hartley Court and HH, 277–8
Cyrano de Bergerac, 208–9

Daldy, F. F., 19
Darwinism see evolutionary perspectives
Davies, Gerald S., 19, 20
De La Mare, Walter, 218, 268
de Watteville, Armand, 111
Dejerine, Joseph Jules, 246
Derwent, Peter, 268
diagnosis, HH on, 97
'Diagram Makers', 140–1, 151
Dickinson, Goldsworthy Lowes, 275
dissociation of nervous function, 121–2, 124–5, 131, 150
Dodgson, Charles Lutwidge, 154–5
duel, HH witnesses, 24–5
Dürer, Albrecht, 'Christ among the Doctors,' 202
Duse, Eleanora, 204, 211–12

'Elements of the Psycho-Neuroses, The', 85
Empire Hospital, Vincent Square, 138
ethnology and sociology, 46–7, 75, 119, 249, 251
evolutionary perspectives, 40, 110–12, 115–21, 130–7, 146, 150

Feilding, Everard, 148, 209–10
Ferrier, David, 101, 103
finances, summary of HH's (1904), 67–9
Flaubert, Gustave, *Madame Bovary*, 216–17
Fletcher, Walter Morley, 83, 84
Forster, E. M., 62, 274
Forston, Dorset, 266–9, 275–6, 283–4
Foster, Michael, 34, 38–9, 42, 100
France, 250–3
Frazer, James George, *The Golden Bough*, 157, 213
Freudian perspectives, 86, 257, 258
Fry, Helen, 62–3

Fry, Roger, 207
functional (neurotic) patients, 58–60, 65–6, 85–6

Galton, Francis, 12
Gaskell, Walter, 39–40, 102, 103
  work on vertebrate cerebrospinal axis, 111–12, 115
German University, Prague, 42
  Physiological Institute in the *Wenzelsbad*, 48–9
Germany
  Berlin, 157, 159, 160
  nationalist movement, 51
  Neubeurern, Bavaria, 227–30
  'organic physics' movement, 28
  see also Halle; Halle university
Gibbon, Edward, *Decline and Fall of the Roman Empire*, 213–14
Ginzburg, Lydia, 8
Goethe, Johann Wolfgang von, 33, 91
Gosse, Edmund, 260
Goulstonian Lectures (1901), 118–21
Gowers, W. R.
  on HH, 56
Graves, Robert, 265
  'How Many Miles To Babylon', 219
  Poetic Unreason, 218, 219
Grove House School, Tottenham, 16
Gulland, George Lovell, 25

Haldane Commission, 83
Halle
  character of, 24–5
  HH's journey to, 22–3
  Moravian colony, 25
  see also Germany; Halle university
Halle university
  character of faculties, 27; histology teaching, 30–2; physiology teaching, 27–30
  'Klinik', 32
  student fraternities *(Burschenschafteni)*, 24
  see also Germany; Halle
Hardy, Thomas, 156, 189–90, 265, 266
Hartley Court, Reading, 276–7, 278, 282–3, 286–7

Head, Alban (HH's brother), 203–4, 228
Head, Bernard (HH's brother), 255–6
Head, Charles Howard (HH's brother), 14, 16
Head, Henry (HH's father), 12, 16, 21, 29, 55, 166, 252
Head, Hester (HH's sister) *see* Pinney, Hester
Head, Hester (née Beck, HH's mother), 14, 58
  relationship with HH, 12, 86–7, 127–8, 162–4, 166, 169
Head, Jeremiah (HH's paternal grandfather), 12
Head, Sir Henry
  Appearance, *viii*, 2, 71, 89, *126, 291*
  Character: acuity in discerning social distinctions, 5, 122, 131, 137, 138, 226; affection for animals, 29–30; appreciation of natural beauty, 33–4, 74, 105, 224–6, 248; attitude to fellow practitioners, 6, 52–3, 104, 139–40, 246–7; authorial dimension to, 227–8, 231–2; compassion, 64–5, 92–3, 175; contradictions in, 3, 98; cultural elitism, 196–8; idealism, 41, 43, 65; individualism, 13, 71–2, 108, 151, 158, 289–90; intuitiveness, 97, 106–10; lack of avarice, 66, 70; lassitude, 34; modernist elements, 4; modesty, 51; precociousness, 19; sense of alienation and despair, 157–8, 176–7, 199; stoicism, 256
  Courtship and married life: first meeting with Ruth Mayhew, 155–6, 157; relationship deepens, 157–8; early difficulties, 158–9; holiday in Germany, 159, 160–1; parental disapproval, 161–5; love declared, 165; anger at mother's deviousness, 166; correspondence, 169–84, 199–200; Rag Book entries, 185–88; engagement announced, 167–8; marriage, 67, 188–9, 202; social life, 248–54; illness intervenes, 262–4; holiday at Lyme Regis, 264–5; move to Forston, 265–6; Parkinsonism takes hold, 264–7; support of friends during ill-

ness, 267–74; move to Hartley Court, Reading, 276–7
Dedication to science, 99–110
Finances, summary of (1904), 67–9
Interests: aviation, 95, 241, 253–4; botany, 18–19, 35; ecclesiastical architecture, 22–3; German culture, 21–2, 95–6, 201–2, 208, 255; literature, 17–18, 33, 77, 89, 90–1, 155–6, 186, 212–18, 274; music, 201–2, 203–4, 275; pictorial and plastic arts, 202–5; politics, 37, 48; sociology and ethnology, 46–7, 75, 119, 249, 251; sports and cycling, 7, 34, 224–5; theatre, 35, 159, 207–12, 250, 259; utilitarian pursuits, 79–80
Life events: birth, 11; early influences, 12–13; childhood, 13–14; visits grandmother at Stamford Hill, 15; shows interest in medicine, 15–16, 17, 21; formal education begins, 16; attends Charterhouse, 16–17; enthusiasm for science, 16, 19–20; prepares for Cambridge, 21; travels to Germany, 21–3; arrives at Halle university, 23–4; witnesses a duel, 24–5; bicycles at Halle, 26; meets Halle professors, 26; witnesses first vivisection, 29–30; experiences whirlwind, 31–2; introduction to clinical medicine, 32–3; enters Cambridge, 34–5; embarks on Natural Science Tripos, 38–9, 41; appointed as teaching assistant, 41; develops interest in physiology of sensation, 43, 55; internship at Hering's laboratory, Prague, 44–5, 48–53, 112; visits Wilhelm Weiss, 45–7; completes first original scientific research and experiences academic rivalry, 49–51; commits to clinical medicine, 55–7; investigates effectiveness of religious shrines, 57–8; practices as a consulting physician, 58–63, 243; gains reputation as neurologist, 59; works at London Hospital, 61; conducts clandestine autopsy, 72; visits Bavaria, 202–4; disillusionment with working life, 76–9; elected Fellow of Royal Society, 103; involvement in running of Pathological Institute, London, 80–3; conducts research at Royal Army Medical Corps Hospital, Netley, 106–7, 122; visits Neubeurern, Bavaria, 227–30; experiments on own arm, 125–31; travels to North Africa, 128–9; visits Budapest, 246; visits Würzberg, 245–6; assumes editorship of *Brain*, 243–4; develops interest in language disorders, 137–40; visits Paris and the Loire, 250–3; visits America, 145; visits the Bradfords' Merionethshire home, 248–50; war work, 70, 79, 88, 94–5, 131, 261; friendship with Robert Nichols, 88–90; visits France, 246–7; visits Nichols in Devon, 256–7; attends aviation negotiations at Rome, 95; resigns from the London Hospital, is offered Directorship of academic medical unit, 83–5; shows early signs of Parkinsonism, 85; consults Riddoch on illness, 262–3; joins National Council for Mental Hygiene, 97; appears as expert witness, 61–2; Parkinsonism forces retirement, 98; completes monograph on aphasia, 144–7; Parkinsonism worsens, 98, 264–7, 277–8; continues to offer medical advice, 279–81; death, 286, 287–8
Lectures: 'Afferent Nervous System, The' (1912), 56; Aphasia: 'an Historical Review' (1920), 138; in the Netherlands (1921), 261; at New York (1912), 244–5; on effects of WWI on medical practice, 96–7; 'On the Release of function in the Nervous System', 276; at Würzberg (1904), 245
Philosophy: beauty, nature of, 225–6; life as divided reality, 195–200; of art, 205–7, 215; Phenomenalism, 223; professional ethic, 65–6, 99–101; religious, 13, 16, 34, 36–7, 108–10; secular ethic, 256; social, 5, 36; views on women, 45–6, 87, 187, 215, 245

Poems, 232–42; 'Ballade of the Sister,' 236; 'Destroyers,' 239; 'Died of his Wounds,' 239–40; 'Love the Intruder,' 237–8; 'Pegasus,' 240; 'Scent Memory,' 237; 'Spring Tragedy,' 236; 'Sun and Showers' series, 241; 'The Man. The Healer,' 236; 'The Price,' 240; 'The Seeker after Truth,' 235–6
Relationship with mother, 12, 86–7, 127–8, 162–4, 166, 169
Relationship with sister, 195–6
Research: at spinal and encephalic levels, 131–3; in aphasiology, 138–51; in psychoneuroses, 85–95; in respiratory reflex, 49–51; in sensation, 4, 115–25, 133–7; using own arm, 125–31
Works: *Aphasia and Kindred Disorders*, 1–2, 43, 143–5; 'Autobiography', 12, 18, 35, 38, 53; 'Conception of Nervous and Mental Energy', 148; 'Elements of the Psycho–Neuroses, The', 85; paper to 'Time, Space and Material' symposium, 145–6; 'Sensation and the Cerebral Cortex', 131, 133; *Studies in Neurology,* 114–15, 133

Head, Ruth (née Mayhew, HH's wife)
Appearance, *291*
Character and views: exuberance, 260; optimism, 283–4; religious and social views, 156–7, 262; resilience, 155, 164–5, 285; self-deprecation, 170, 172–4
Interests: in literature, 154, 155–6, 185–6, 217, 274; in writing, 190
Letters: 'Dorothea Ladislaw' letter, 153; from Alfred Whitehead, 167; from HH, 14, 18, 51, 58, 64, 81, 118, 125–7, 132, 160–1, 162–3, 168, 169–84, 199–200; on Hester Pinney, 269; on HH's illness, 263–4, 278–9; to Hester Marsden-Smedley, 168, 263–4, 278–9; to HH, 127, 164–5, 169–84; to Siegfried Sassoon, 273
Life events: childhood, 154–5; education, 155; works as schoolteacher, 155; meets HH, 155; relationship with HH deepens, 158–9, 168; realizes depth of HH's dedication to science, 103; meets HH in Berlin, 159, 160–1; appointed headmistress of High School for Girls, Brighton, 159–60, 162, 173; family disapproval of relationship with HH, 161–3; declares her love for HH, 165–6; engagement and marriage, 67, 167–8, 202; move to Montagu Square, 188–9; working life after marriage, 63, 69–70, 189–90; financial pressures, 70; publishes novels, 190–1; reliance upon friends during HH's illness, 267–74; role as carer to HH, 281–5; illness and death, 284–5
on her sensations, 124
Works: *Compensations*, 190–1; *History of Departed Things, A*, 190, 191–4

Heine, Heinrich, 75, 156
Helmhotz, Hermann, 43
Hering, Heinrich Ewald, 42–4, 48–50, 52–3
'Hering-Bruer' reflex, 49
Herpes Zoster (shingles), 40, 113–14
High School for Girls, Brighton, 159
Hillebrand, Franz, 52
His, Wilhelm, 26
Holland, Sydney, 2nd Viscount Knutsford, 81, 83–4, 122, 138
Holmes, Gordon, 13, 55, 57, 243, 245
on RH, 285
Horsley, Victor, 103
Hospital for Nervous Diseases, 56
Housman, Laurence, *Bethlehem*, 209
Hunter, Donald, 71
on HH, 72
Huxley, Aldous, 259, 270
Huxley, Julian, 276

Ibsen, Henrik, 159
Hedda Gabler, 212
John Gabriel Borkman, 159
When We Dead Awaken, 209
Wild Duck, The, 213
International Medical Congress (1904), 246
Irving, Dolly, 173

Jackson, John Hughlings, 1, 101–2, 103, 111
John Hughlings Jackson Lecture, 138

papers on aphasia, 137–8, 140–1
James, Henry, 271
  Aspern Papers, The, 218
  Wings of the Dove, The, 217
James, Monague Rhodes, 35, 119
Jebb, Richard Claverhouse, 35
Jewish inhabitants of Whitechapel, 72, 74–6
Jewson, Nicholas, 60
John Hughlings Jackson Lecture, 138
Jonson, Ben, 'The New Inn,' 207

Kantian categories, 136
Keats, John, 220
Kidd, Percy, 83
Knoll, Philip, 50–1
Knutsford, Lord, 81, 83–4, 122, 138
Koch, Robert, 12
Krivorotov, Woldemar, 25–6

Lambert, Brooke, 37
Langley, J. N., 39, 41, 103
language disorders *see* aphasia
Lawrence, D. H., 93
Lea, Sheridan, 39, 41
Lenbach, Franz von, 204
Lichnowsky, Prince Karl Max, 60
Linke, Frau, 45, 48, 52
List of Diseases (Army medical manual), 96–7
Lister, Joseph, 12, 183
Loathes, Stanley, 35–6
London Hospital, 60–1, 64, 122
  Assistant Director of Pathological Institute, 80–3
  establishment of medical and surgical units, 83–5
  HH's colleagues at, 71–3
  HH's frustration with working conditions, 122–3
  magazine, 72
  Medical Society, 73
  students, 73
Lunatic Asylums, 114
Lyme Regis, 264–5, 282
Lyon, Henry, 86, 256–7

Mackenzie, James, 102, 108, 112, 115
Malvern, 259, 263

Marie, Pierre, 137–8
Marsden-Smedley, Henrietta, 278
Marsden-Smedley, Hester (née Pinney, HH's niece), 266, 269, 280, 285–6
Maude, Cyril, 269
Maugham, William Somerset, *A Man of Honour,* 209
Mayhew, Anthony Lawson (RH's father), 154, 167
  HH's letter to, 168
Mayhew, Arnold (RH's brother), 156, 164
  on HH, 65
Mayhew, Arthur (RH's brother), 163, 233, 273
  on HH, 289–90
Mayhew, Jane Innes (RH's mother), 154, 256
  presses daughter to give up correspondence with HH, 161–2
Mayhew, Ruth *see* Head, Ruth
metaphorical descriptions of sensory nervous function, 121, 134–5, 234–5
Meyer, Adolf, 144, 245, 268
migraine, 117
Milton, Ernest, 259
'modernism', 4
Monakow, Constantin von, 137–8
Montague Square, 71, 89, 188–9, 262, 265–6, 289
Morris, William, 14–15
Mr Brett (Head family's doctor), 15–16

Nantes, 251–3
National Council for Mental Hygiene, 97
natural science
  HH on, 104, 109–10
  HH's dedication to, 99–110
Netherlands, 261
Neubeurern, Bavaria, 227–30
Neurological Society, 56, 102, 103, 110–11
neurology, HH on, 56, 96
Nichols, Robert
  dialogues on 'psychological' conversations with HH, 92–3, 216, 219–20
  HH visits, (1917), 256–7
  HH's letters to, 91, 121, 219
  letters to HH and RH: during HH's illness, 269–71; on Goethe, 91

meets and consults HH, 88–90
on HH, 94, 287–8; and HH's neurological writings, 148–9
on RH, 188, 262
Works: *Guilty Souls,* 159
Niemeyer family, Halle, 23–4, 25, 33

obstetrics, 64
Ogden, Charles Kay, *The Meaning of Meaning,* 146
olfactory delusions, 119
'On the Release of function in the Nervous System', 276
Oppé, Adolph, 157
Oppé, Paul, 58
Osborne, Dorothy, letters, 214

Page-May Lecture, 'The Afferent Nervous System' (1912), 56
Paget, Violet, 147–8
pain, 145
referred pain, 112–16, 120, 130, 133
Palace Green Hospital, 89
Pater, Walter, 110
patients
aphasic, 137, 138–9
diet, 79–80
functional (neurotic), 58–60, 65–6, 85–6
Herpes Zoster subjects, 114
HH's attitude towards, 121–5, 131–2
Jewish, 72, 75–6
nerve injuries from Boer War, 122–3
'referred pain' subjects, 112–13, 118–21
socially prominent, 62–4
spinal cord and brain wounds, 131–2
traumatised WWI soldiers, 88–90, 93, 97
Pearson, Karl, 21
peasants of Bohemia, 47
Pharos Club, 209
Pinney, Hester (née Head, HH's sister), 173, 269
marriage, 195–6
Pinney, Hester, (HH's niece) *see* Marsden-Smedley, Hester, 263
Pinney, Reginald, 196, 273
Plesch, Egon, 280

poetry
HH's receptiveness to impressions and, 224–5
psychological perspectives, 91, 148, 218–25, 257
written by HH, 232–42
Poole, W. H. W., 18–19, 20–1
Poor Law Infirmaries, 114
positivism, 150
Power, D'Arcy, 21
Prague, racial segregation, 44–5
see also German University, Prague
'Pratt', family Butler, 70, 281–2
psychological perspectives, 145
dialogues between HH and Robert Nichols, 92–3, 216, 219–20
functional (neurotic) patients, 58–60, 65–6, 85–6
of novels, 215–18
of poems, 91, 148, 218–25, 257
psychic changes accompanying visceral disease, 117–21
psychoneuroses, 85–95
of sexuality, 87, 92–3

Quakerism, 13, 16, 99–100, 279

Rag Book entries, 73, 75, 77, 101–2, 107, 109, 124, 185–94, 204–5, 206, 207–8, 210, 211–12, 216, 217–18, 221, 223
Rainhill, County Asylum, 57, 114
Ransom, William Bramwell, 25, 35
referred pain, 112–17, 120, 130, 133
respiratory reflexes, 50–1
Reverat, Jacques, 60
Richards, I. A., 146
Riddoch, George, 62, 90, 139, 268
on HH's illness, 288–9
Riemenschneider, Tilman, 204–5
Rivers, William Halse Rivers, 39, 57, 79, 102, *126,* 256
collaborates in experiment on HH's arm, 125–30, 144–5
Rogge family, Berlin, 160
Ross, Dorothy, 4
Ross, George W., 82
Rostand, Edmond, *Cyrano de Bergerac,* 208–9

Royal Army Medical School Hospital, Netley, 106–7, 122
Royal College of Physicians, 55
Royal College of Surgeons, 55
Royal Society of Medicine, 15, 103–4, 183
  Neurological Section, 96
Ruskin, John, 77
Russell, Bertrand, 78, 145, 203

Salaman, Redcliffe Nathan, 74, 81–2, 113, 127, 288, 290
Sassoon, Siegfried, 256, 257–60, 280
  on HH and RH, 258, 260; and HH's illness, 263
  joins HH and RH at Lyme Regis, 265
  visits Forston, 267–8
  visits Garmisch, Bavaria, 271–2
  works: 'Concert Impressions,' 259258; *Grandeur of Ghosts,* 258; *Lingual Exercises,* 260; *Memoirs,* 273; *Memoirs of a Fox-Hunting Man,* 196
'schemata', 136, 146
science, HH's dedication to, 99–110
self *see* consciousness
Seligman, Brenda, 107
sensation, 55–6
  autonomic part of nervous system, 120–1
  dissociation, 121–2, 124–5, 131, 150
  Herpes Zoster (shingles) as model, 113–14
  HH's experiment on own arm, 125–30; effect on his creative activities, 132; Lateral view of forearm and hand, *129*
  localization as psychical phenomenon, 117–21
  Location of eruptions on pre-printed schematic body outline, *116*
  metaphorical formulation, 134–5
  referred pain, 112–17, 120, 130, 133, 145
  sensory elitism, 116–17; protopathic and epicritic modes, 129–31, 132–3, 135, 144–5
'Sensation and the Cerebral Cortex', 131, 133
sensory elitism, 116–17

protopathic and epicritic modes, 129–31, 132–3, 135, 144–5
sexuality, psychology of, 87, 92–3
Shakespeare's plays, 159, 207
Sherren, James, 125
Sherrington, Charles, 39, 109, 111, 113, 143, 276
shingles, 40, 113–14
Slade, Geoffrey, 72–3
Smith, Grafton Elliot, 146
Society of Friends *see* Quakerism
sociology and ethnology, 46–7, 75, 119, 249, 251
Sorley, William Ritchie, 35
Special Hospitals for Officers, 138
Sprigge, Samuel Squire, 108
Stage Society, The, 209
Stamford Hill, 14–15
Stephen, Jim, 36
Strachey, Lytton, 259
Streeter, Burnett Hillman, 275
Studies in Neurology, 114–15, 133
Swift, Jonathan, 215

Taggart, John Ellis, 79
Tennant, Stephen, 272–3, 280, 281
Thackeray, William Makepeace, 215
Thompson, D'Arcy Wentworth, 35, 36
'Time, Space and Material' symposium (1919), 145–6
Times, The,
  'Sensation Controlled by the Brain' (Oct 1921), 146–7
  (July 1901), 81
  tribute to HH, 287–8
  tribute to RH, 285
Trinity College, Cambridge, 34–6
Turnbull, Hubert Maitland, 71, 82–3

Unites States of America, 244–5
University College of London, 41, 55, 140, 146
  Page-May Lecture (1912), 56

Vandervelde, Emile, 93
Venice, 223
Verhaeren, Emile, 260

Victoria Park Hospital for Diseases of the Chest, 57, 112, 118
visualization, 43, 119, 124, 128, 136, 141–2
Volkmann, Richard, 32

Waggett, Philip, 21
Wagner, Richard
   Ring cycle, 201–2
   Tristan and Isolde, 202
Waldegrave, John, 214
Weiss, Fraulein, 45–6
Weiss, Wilhelm, 45
Wellcome Library, 2
Wendelstadt, Baron Jan, and Baroness, 227, 228–9
Whistler, James McNeill, 206
Whitehead, Alfred North, 35, 36, 41–2, 78, 145–6, 161, 167

Whitehead, Evelyn, 197
Whitman, Walt, 'Song of Myself,' 225
Woolfe, Virginia, 62
   Mrs Dalloway, 274
World War I
   aphasic patients, 138–9
   early fatalities, 255–6
   HH's poems, written during, 239–41
   outbreak of, 79, 189
   Special Hospitals for Officers, 138
   spinal cord and brain wounds, 131–2
   traumatised soldiers, 88–90, 93, 97
   zeppelin raids, 254–5
World War II, 285–6
Wright, Almroth, 122, 147
Wylde, Leslie, 267

Young, Thomas, 12